High-Density Solid-State Memory Devices and Technologies

High-Density Solid-State Memory Devices and Technologies

Editors

Christian Monzio Compagnoni
Riichiro Shirota

MDPI • Basel • Beijing • Wuhan • Barcelona • Belgrade • Manchester • Tokyo • Cluj • Tianjin

Editors

Christian Monzio Compagnoni
Politecnico di Milano
Italy

Riichiro Shirota
National Yang Ming Chiao Tung University
Taiwan

Editorial Office
MDPI
St. Alban-Anlage 66
4052 Basel, Switzerland

This is a reprint of articles from the Special Issue published online in the open access journal *Electronics* (ISSN 2079-9292) (available at: https://www.mdpi.com/journal/electronics/special_issues/HDSMDT_electronics).

For citation purposes, cite each article independently as indicated on the article page online and as indicated below:

LastName, A.A.; LastName, B.B.; LastName, C.C. Article Title. *Journal Name* **Year**, *Volume Number*, Page Range.

ISBN 978-3-0365-3359-9 (Hbk)
ISBN 978-3-0365-3360-5 (PDF)

Contents

About the Editors

Christian Monzio Compagnoni (Professor) received his Laurea (cum laude) degree in Electronic Engineering and Ph.D. (cum laude) degree in Information Technology from the Politecnico di Milano, Milan, Italy, in 2001 and 2005, respectively. Since 2006, he has been with the Dipartimento di Elettronica, Informazione e Bioingegneria, Politecnico di Milano, Milan, Italy, first in the capacity of Assistant Professor and then of Professor of Electronic Engineering. His research interests are in micro/nanoelectronic devices and technologies. In particular, he has been performing research activities on the basic physics, operation and reliability of solid-state devices and technologies for data storage for nearly 20 years, with emphasis on NAND Flash technology. Activities were conducted in collaboration with some of the most important semiconductor companies in the world, which has led to more than 130 papers published in international journals and conference proceedings and to 2 US patents.

Riichiro Shirota (Professor) received his Bachelor of Science and Ph.D. degrees following research in the Department of Physics at Nagoya University in Japan, in 1977 and 1982, respectively. In 1982, he joined Toshiba Corporation and worked on the development of DRAM until 1986. From 1986, he was in charge of the development of nonvolatile memory and started research to set up NAND Flash in 1987. In 2006, he resigned from Toshiba Corp. and became Professor at National Yang Ming Chiao Tung University in Taiwan. In 2010, he moved to National Yang Ming Chiao Tung University. His main research fields are nanoscale memory and logic devices and power devices using GaN. He has been particularly focusing on the realization of scaled memory devices and their reliability issues for over 30 years. He is currently collaborating with a memory company who is support him in executing research of memory devices. He has published more than 80 papers in international journals and conference proceedings and has registered more than 170 US patents.

electronics

Editorial

High-Density Solid-State Memory Devices and Technologies

Christian Monzio Compagnoni [1,*] and Riichiro Shirota [2]

1 Dipartimento di Elettronica, Informazione e Bioingegneria, Politecnico di Milano, 20133 Milan, Italy
2 Department of Electrical and Computer Engineering, National Yang Ming Chiao Tung University, Hsinchu 300, Taiwan; rshirota@faculty.nctu.edu.tw
* Correspondence: christian.monzio@polimi.it

1. Introduction

The relevance of solid-state memories in the world of electronics is on the constant rise. On one hand, the continuous increase in the integration density of semiconductor technologies has been making solid-state storage the dominant storage solution of the 21st century, thanks to a successful trade-off against cost, performance and reliability. On the other hand, new memory-centric computing scenarios based on solid-state memories are appearing on the horizon to overcome the limitations of the mainstream von Neumann computing architecture. The 3D NAND Flash memory technology, the NOR Flash memory technology, the phase-change memory (PCM) technology, the resistive random-access memory (ReRAM) technology, the magnetoresistive random-access memory (MRAM) technology, and the ferroelectric memory technology are the most important players at the heart of the ongoing memory revolution, along with the dynamic random-access memory (DRAM) technology and the static random-access memory (SRAM) technology.

In this context, this Special Issue aims to examine high-density solid-state memory devices and technologies from various standpoints, in the attempt to foster their continuous success in the future. Considering that the broadening of the range of applications will likely offer different types of solid-state memories the chance to come to the spotlight, the Special Issue is not focused on a specific storage solution, but it embraces all the most relevant solid-state memory devices and technologies currently on stage. Even the subjects dealt with in the Special Issue are widespread, going from process and design issues/innovations to the experimental and theoretical analysis of the operation, the performance and the reliability of memory devices and arrays and to the exploitation of solid-state memories to pursue new computing paradigms.

Citation: Compagnoni, C.M.; Shirota, R. High-Density Solid-State Memory Devices and Technologies. *Electronics* **2022**, *11*, 538. https://doi.org/10.3390/electronics11040538

Received: 7 February 2022
Accepted: 9 February 2022
Published: 11 February 2022

2. Overview of the Papers in the Special Issue

This Special Issue includes six review papers and three original research papers focused on the most important solid-state memory devices and technologies.

The first review paper in the Special Issue is by Pedretti and Ielmini [1] and summarizes the current status of analog in-memory computing with RRAM devices, representing a promising non-von Neumann computing approach. In the paper, the fundamentals of RRAM devices and of the new computing concept are first reviewed, highlighting the importance of achieving a tight control over the analog conductance of the resistive memory elements. The constraints to that control are then discussed, considering programming variations, conductance drifts/time-dependent fluctuations and array-level issues. The options for coding the computational coefficients in the RRAM array are also presented, discussing the trade-off between precision and memory area, and examples of circuit primitives for analog in-memory computing are provided. Finally, the prospects and challenges of the new computing approach are debated, pointing out that RRAM represents the most mature and promising technology to pursue it.

The second review paper in the Special Issue is by Watanabe and Lin [2] and presents an overview of the reliability issues arising from traps in the dielectric layers of solid-state memory cells. Trap-related phenomena are discussed in the time and frequency domain, pointing out the intrinsic discreteness involved in them. The possibility to get some relevant information about traps in dielectrics through ad hoc experimental analyses and theoretical investigations is demonstrated, paving the way to possible technology and process optimizations.

The third review paper in the Special Issue is by Teramoto [3] and discusses low-frequency noise in metal-oxide-semiconductor field-effect transistors (MOSFETs), representing the building blocks of the memory cells of the most important solid-state storage technologies. Due to the relevant role played by the phenomenon on the reliability of deeply scaled devices, emphasis is on random telegraph noise (RTN) and on its statistical experimental characterization and theoretical analysis. In the paper, the amplitude of RTN fluctuations is first addressed, considering its dependence on device parameters and operating conditions and taking into account both two-state and multi-state waveforms. The time constants of the process are then analyzed not only for stationary but also for switched biasing conditions. Finally, RTN in advanced device structures, such as buried-channel and asymmetric source-drain MOSFETs, is debated, pointing out possible solutions to mitigate the phenomenon in future technology nodes.

The fourth review paper in the Special Issue is by Chiu and Shirota [4] and summarizes a method to investigate the endurance of NAND Flash memories starting from cell transconductance. Nowadays, NAND Flash memories represent the dominant nonvolatile storage solution, and techniques allowing us to get some information about the microscopic mechanisms constraining its endurance are of utmost importance for the development of next-generation technology nodes. In the paper, first, a mix of experimental and theoretical analyses is adopted to come to the build-up of charge in the cell tunnel-oxide as a result of the program/erase cycles performed on the memory array. Then, the charge build-up is studied as a function of the cycling conditions of the array, e.g., the cycling temperature and the idle time in-between cycles. Finally, a physical picture for the generation of oxide charge is proposed, allowing us to deepen the understanding of NAND Flash reliability.

The fifth review paper in the Special Issue is by Goda [5] and comprehensively examines the evolution of the 3D NAND Flash technology. In the paper, the historical trend of the storage density of NAND Flash memory chips is first discussed, showing how 3D technologies achieved the extraordinary figures that are revolutionizing the nonvolatile storage scenario. The role played by the growth of the number of memory layers vertically stacked in the array, the miniaturization of cell dimensions and the increase in the stored bits per cell is then discussed, along with innovative integration schemes such as the CMOS-under array solution. Finally, the future challenges and opportunities of the technology are highlighted.

The sixth review paper in the Special Issue is by Ohba et al. [6] and summarizes the state-of-the-art of the Bumpless Build Cube (BBCube) using a Wafer-on-Wafer (WOW) and a Chip-on-Wafer (COW) approach, representing a new process solution for Tera-Scale Three-Dimensional Integration (3DI). In the paper, the BBCube architecture is explicitly considered for memory applications, considering BBCube DRAM and BBCube NAND Flash solutions. The challenges and prospects of the new integration approach are clearly highlighted, unveiling the promise of a next big step in the semiconductor roadmap.

The first research paper in the Special Issue is by Zambelli et al. [7] and investigates the benefits and shortcomings of program suspend operations in 3D NAND Flash Solid-State Drives. Through experimental analyses and system-level simulations, the impact of including program suspend in the command set of the storage device on its speed, reliability and power consumption is clarified.

The second research paper in the Special Issue is by Rao et al. [8] and demonstrates the benefits of alloying conventional free-layer materials such as CoFeB in STT-MRAM cells with nonmagnetic metals (such as W). The solution is presented as an alternative approach to achieve write performance improvements in STT-MRAM technologies.

The third research paper in the Special Issue is by Malavena et al. [9] and proposes the adoption of a pulse-width modulation scheme to implement hardware artificial neural networks based on NOR Flash memory arrays. The new operating scheme is shown to be highly immune to noise and temperature variations, paving the way to the development of highly reliable, noise-resilient neuromorphic systems.

Funding: This research received no external funding.

Acknowledgments: The guest editors of this Special Issue want to thank all the authors who contributed to the papers published in it and all the reviewers and associate editors who supported the review process of the papers.

Conflicts of Interest: The authors declare no conflict of interest.

References

1. Pedretti, G.; Ielmini, D. In-Memory Computing with Resistive Memory Circuits: Status and Outlook. *Electronics* **2021**, *10*, 1063. [CrossRef]
2. Watanabe, H.; Lin, H.-J. Trap-Related Reliability Problems of Dielectrics in Memory Cells. *Electronics* **2021**, *10*, 1287. [CrossRef]
3. Teramoto, A. Evaluation of Low-Frequency Noise in MOSFETs Used as a Key Component in Semiconductor Memory Devices. *Electronics* **2021**, *10*, 1759. [CrossRef]
4. Chiu, Y.-Y.; Shirota, R. Technique for Profiling the Cycling-Induced Oxide Trapped Charge in NAND Flash Memories. *Electronics* **2021**, *10*, 2492. [CrossRef]
5. Goda, A. Recent Progress on 3D NAND Flash Technologies. *Electronics* **2021**, *10*, 3156. [CrossRef]
6. Ohba, T.; Sakui, K.; Sugatani, S.; Ryoson, H.; Chujo, N. Review of Bumpless Build Cube (BBCube) Using Wafer-on-Wafer (WOW) and Chip-on-Wafer (COW) for Tera-Scale Three-Dimensional Integration (3DI). *Electronics* **2022**, *11*, 236. [CrossRef]
7. Zambelli, C.; Zuolo, L.; Aldarese, A.; Scommegna, S.; Micheloni, R.; Olivo, P. Assessing the Role of Program Suspend Operation in 3D NAND Flash Based Solid State Drives. *Electronics* **2021**, *10*, 1394. [CrossRef]
8. Rao, S.; Couet, S.; Van Beek, S.; Kundu, S.; Sharifi, S.H.; Jossart, N.; Kar, G.S. A Systematic Assessment of W-Doped CoFeB Single Free Layers for Low Power STT-MRAM Applications. *Electronics* **2021**, *10*, 2384. [CrossRef]
9. Malavena, G.; Spinelli, A.S.; Monzio Compagnoni, C. A Noise-Resilient Neuromorphic Digit Classifier Based on NOR Flash Memories with Pulse–Width Modulation Scheme. *Electronics* **2021**, *10*, 2784. [CrossRef]

electronics MDPI

Review

In-Memory Computing with Resistive Memory Circuits: Status and Outlook

Giacomo Pedretti and Daniele Ielmini *

Dipartimento di Elettronica, Informazione e Bioingegneria, Politecnico di Milano and IU.NET, 20133 Milano, Italy; giacomo.pedretti@polimi.it
* Correspondence: daniele.ielmini@polimi.it

Abstract: In-memory computing (IMC) refers to non-von Neumann architectures where data are processed *in situ* within the memory by taking advantage of physical laws. Among the memory devices that have been considered for IMC, the resistive switching memory (RRAM), also known as memristor, is one of the most promising technologies due to its relatively easy integration and scaling. RRAM devices have been explored for both memory and IMC applications, such as neural network accelerators and neuromorphic processors. This work presents the status and outlook on the RRAM for analog computing, where the precision of the encoded coefficients, such as the synaptic weights of a neural network, is one of the key requirements. We show the experimental study of the cycle-to-cycle variation of set and reset processes for HfO_2-based RRAM, which indicate that gate-controlled pulses present the least variation in conductance. Assuming a constant variation of conductance σ_G, we then evaluate and compare various mapping schemes, including multilevel, binary, unary, redundant and slicing techniques. We present analytical formulas for the standard deviation of the conductance and the maximum number of bits that still satisfies a given maximum error. Finally, we discuss RRAM performance for various analog computing tasks compared to other computational memory devices. RRAM appears as one of the most promising devices in terms of scaling, accuracy and low-current operation.

Keywords: resistive switching memory; in-memory computing; crosspoint array; artificial intelligence; deep learning

Citation: Pedretti, G.; Ielmini, D. In-Memory Computing with Resistive Memory Circuits: Status and Outlook. *Electronics* **2021**, *10*, 1063. https://doi.org/10.3390/electronics10091063

Academic Editor: Gaetano Palumbo

Received: 1 April 2021
Accepted: 26 April 2021
Published: 30 April 2021

Publisher's Note: MDPI stays neutral with regard to jurisdictional claims in published maps and institutional affiliations.

1. Introduction

According to More's law, novel computing concepts are being researched to mitigate the memory bottleneck typical of von Neumann architectures. Among these new concepts, in-memory computing (IMC) has attracted an increasing interest due to the ability to execute computing tasks directly within the memory array [1–3]. Various IMC circuits have been proposed so far, including digital gates [4–7], physical unclonable functions (PUF) [8–11] and neuromorphic neurons and synapses [12–16]. In these circuits, the computational function stems from a physical property of the memory device and circuit, such as set/reset dynamics of resistive switching memory (RRAM) for synaptic potentiation/depression and neuron integration and fire. As a result of the physics-based computation, most IMC circuits operate in the analog domain and in a continuous time scale. A typical example of analog IMC primitive is the matrix vector multiplication (MVM), which can be executed in one step in a crosspoint memory array [17–19]. The parallel and analog IMC operation allows the MVM operation to be significantly sped up with respect to the conventional multiply-accumulate (MAC) algorithm of digital computers [20,21]. Crosspoint-based MVM has been adopted in a number of computing applications, ranging from deep neural networks (DNNs) [22–25] to linear algebra accelerators [26–29].

Figure 1 illustrates the hierarchical design approach for IMC accelerators. Computation is enabled at the device level by transport phenomena, e.g., the Ohm's law enabling multiplication of voltage and conductance or the threshold switching enabling comparison between voltages [30]. Devices are connected within circuits, allowing parallel flow of input and output signals, Kirchhoff's current summation and feedback connections, usually in the analog domain. Circuit primitives are organized within novel architectures to harness the full potential of the computing cores and accelerate data-intensive workloads such as neural networks. Based on such a hierarchical structure, it is clear that the proper optimization of the analog accelerators requires a co-design approach from materials to applications to take into account device characteristics, circuit/device non-idealities and possible architecture limitations. Algorithms such as training of neural networks can also be used to optimize precision in view of device and circuit non-idealities. In this regard, a key point is the precision of the physical representation or mapping of the computational coefficients and the overall computation, which might be affected by noise, instability and parasitic elements in the crosspoint array circuit [31].

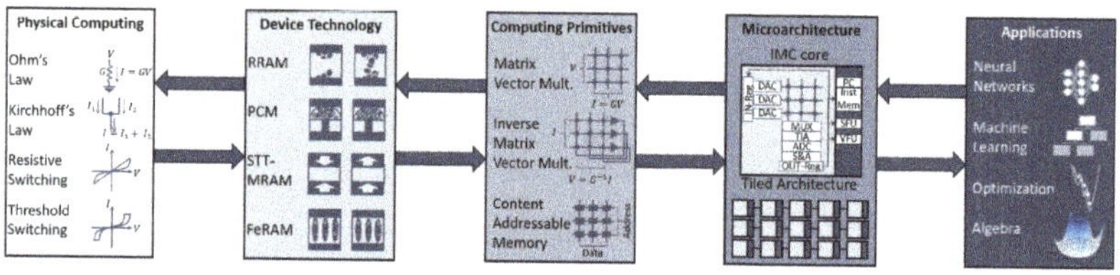

Figure 1. A conceptual illustration of the different scales of in-memory computing. Computing relies on fundamental physical laws that are implemented in various types of memory devices and circuit designs. To perform large-scale computation, new architectures have to be developed for accelerating real world applications. New applications can also arise given the possibility of performing highly-parallel computation. The design and optimization of each different level should be performed by considering all the hierarchical stack.

This work presents an overview of analog IMC with RRAM devices. We first address the programming characteristics of a typical HfO_2 RRAM devices to highlight the intrinsic conductance variations for various programming algorithms. Then we focus on the various options for mapping computational parameters, such as synaptic weights in a DNN, discussing the trade-off between precision and memory area. Finally, we extend our study to various memory parameters, including low-current operation, programming speed and cycling endurance, by discussing their importance for various computational applications and the memory devices that maximize these performance metrics.

2. RRAM Device Structure

Various nanoscale devices have been proposed as a new class of non-volatile memory, where information is stored as the physical configuration of active material, resulting in different conductance [1]. For example, information can be stored and retrieved by the device spin magnetization in a magnetic random access memory (MRAM) [32], as the phase structure of the materials in phase change memories [33,34] or as the atomic configuration of conductive defects in resistive switching memories, or RRAM [35]. The latter has attracted particular interest thanks to the simple structure, compatibility with CMOS process, fast operation and high density [36,37]. Figure 2a shows the typical structure and operation of a RRAM device, consisting of an insulating metal-oxide layer interposed between a metallic top electrode (TE) and a metallic bottom electrode (BE). The insulating layer results in a typical high resistance following fabrication. A forming process consisting of the application of a relatively high positive voltage pulse on the TE induces a local modification of the material composition with the growth of a metallic filament resulting in the low resistance state (LRS). A high resistance state (HRS) can be recovered by means of a negative voltage applied to the TE, which results in the creation of a depletion region across the conductive filament, thus resulting in a decrease of conductance. In addition to the LRS and the HRS intermediate states can be programmed. Thus it is possible, e.g., by controlling the filament size via the maximum current flowing across the device during the set operation, i.e., the compliance current I_C. This is relatively straightforward in the 1-transistor-1-resistor (1T1R) structure as shown in the inset Figure 2b, or by controlling the maximum voltage applied during the reset operation which results in a different thickness of the depletion gap. For instance, Figure 2b shows typical current-voltage (I-V) curves of a RRAM device in 1T1R configuration for increasing I_C, controlled with the transistor gate voltage V_G [38], demonstrating the possibility of analog programming.

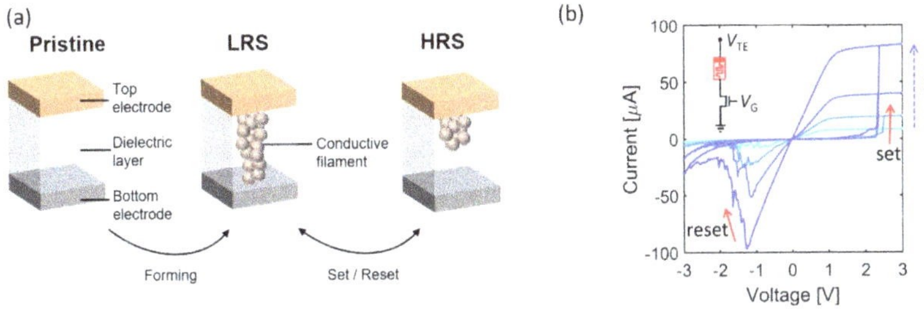

Figure 2. RRAM structure and operation. (**a**) RRAM device made of a metallic TE and BE, with an interposed dielectric layer. After a forming procedure it is possible to set/reset the device and switch from LRS to HRS and vice versa. (**b**) Typical I-V curve of a RRAM device with 1T1R configuration for increasing gate voltage V_G during set operation demonstrating analog programmability. Adapted from [31,38].

The multilevel cell (MLC) capability of Figure 2b is interesting not only for increasing the bit density of the memory, but also for enabling computing applications [39]. In fact, by arranging RRAM in crosspoint configuration as shown in Figure 3, it is possible to directly encode arbitrary matrix entry A_{ij} into the conductance value G_{ij} [1,31]. Then, by applying a voltage vector V as input on the columns and collecting the current flowing in each row I_j, it is possible to compute the matrix vector multiplication (MVM) according to:

$$I_j = \sum_{i=1}^{N} G_{ij} V_i \tag{1}$$

where N is the dimension of the input vector, thus G is a $N \times N$ matrix. MVM is at the backbone of a variety of data-intensive applications that can be accelerated by IMC acceleration, such as neural network training and inference [40–43], neuromorphic computing [15,16,44,45], image processing [18,46], optimization problems [47–50] and the iterative solution of linear equations [26,27]. Circuits able of solving matrix equations without iteration thanks to analog feedback have also been demonstrated [28,29,38]. All these applications have in common the need for reliable analog memory, which still appears a challenge due to the inherent stochasticity of switching behavior in RRAM devices.

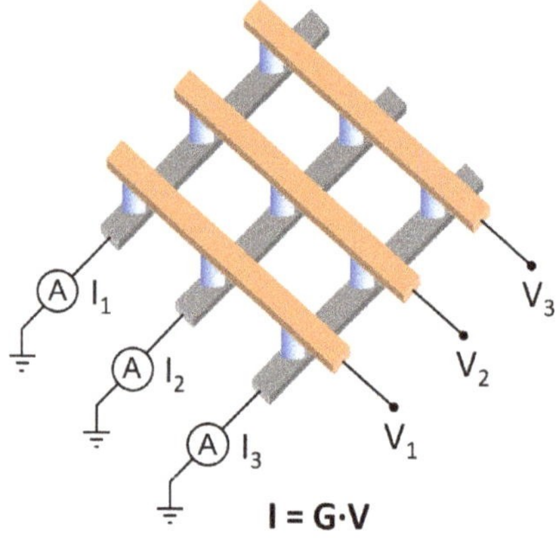

Figure 3. Crosspoint memory structure to perform analog MVM. At the intersection of each TE lines (orange) with each BE line (grey), a RRAM is placed (blue). By programming the RRAM conductance to the matrix entries of A and applying a voltage vector V on the columns, the resulting current flowing in each row j tied to ground according to Kirchoff's law is $I_j = \sum_{i=1}^{N} G_{ij} V_i$. Adapted from [29].

3. Analog Memory Programming Techniques and Variations

Programming the same device several times, in the same condition (or initial state) of programming pulse-width and amplitude, results in various conductance due to the stochastic process of ionic migration during set/reset operation, which is usually referred to as cycle-to-cycle variability [35]. Figure 4a shows an example of a programming algorithm during set operation, namely incremental gate pulse programming (IGPP), where the TE voltage V_{TE} is kept constant while the gate voltage V_G is increased at each time step. This process was repeated 1200 times and the conductance was read with a relatively low voltage after each cycle [51]. Figure 4b shows that the single traces (grey) suffer from relatively large variations, while the median value (blue) seems to grow linearly with pulse number (i.e., V_G). The cumulative distribution of the conductance for each cycle is reported in Figure 4c for increasing V_G. The standard deviation of the conductance σ_G as a function of the median conductance is reported in Figure 4d, indicating a fairly constant value $\sigma_G = 3.8$ μS. This indicates a linear increase in relative resistance variation σ_R/R with resistance R, since $R = 1/G$ and, on a first approximation, variations obey the same relationship between differential quantities, hence $\sigma_R/R = \sigma_G/G$ [35]. The linear increase in σ_R with R is generally observed in variability measurements of RRAM [52,53] and has been attributed to variation in the shape of the conductive filament. Other variability data have been reported indicating an increase in σ_R/R as $R^{0.5}$, which can be interpreted in terms of Poissonian variation of the number of defects in the conductive filament [54–56].

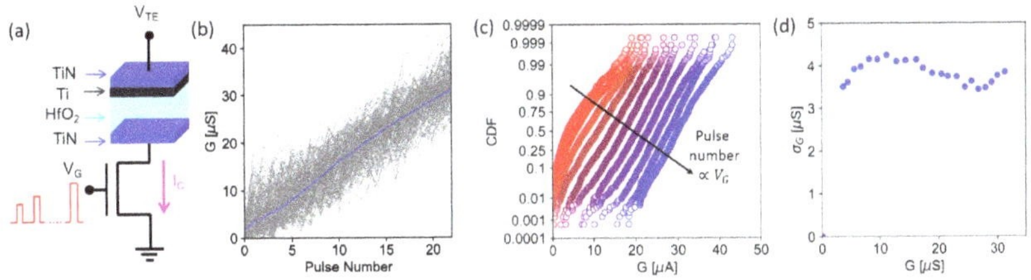

Figure 4. Analog programming with set pulses at increasing I_C according to the IGPP algorithm. (**a**) Conceptual schematic of the IGPP algorithm. (**b**) Conductance as a function of pulse number for multiple iterations (grey lines) and the average behavior (blue). (**c**) Cumulative distribution function (CDF) of the conductance for increasing gate voltage V_G. (**d**) Standard deviation of the conductance σ_G as a function of the average conductance G. Adapted from [51].

Analog programming of RRAM is also possible in the opposite polarity, namely by increasing the negative reset voltage V_{TE} applied on the TE for a fixed V_G, an algorithm referred to as incremental reset pulse programming (IRPP), as shown in Figure 5a. In this case, by first initializing the conductance into the LRS, it is possible to characterize the variability of analog programming with reset voltage by applying IRPP several times (i.e., 1200 as in Figure 5) [51]. Figure 5b again evidences that single traces (grey) corresponding to conductance read after each pulse of a single reset ramp, show high variability and fluctuation, while the median value (blue) shows a gradual decrease with the pulse number. The cumulative distributions of such conductance are reported in Figure 5c for increasing applied TE voltage $|V_{STOP}|$. The resulting standard deviation of conductance σ_G as a function of median conductance G is shown in Figure 5d (red): from the comparison with IGPP results of Figure 5 (blue), it is possible to see that σ_G is generally larger in the reset process. We can conclude that gradual set programming is more convenient than gradual reset for accurate tuning of resistance; however, accurate program/verify (PV) algorithms are needed for reducing the error to acceptable levels (i.e., for having $\sigma_G < 1$ μS).

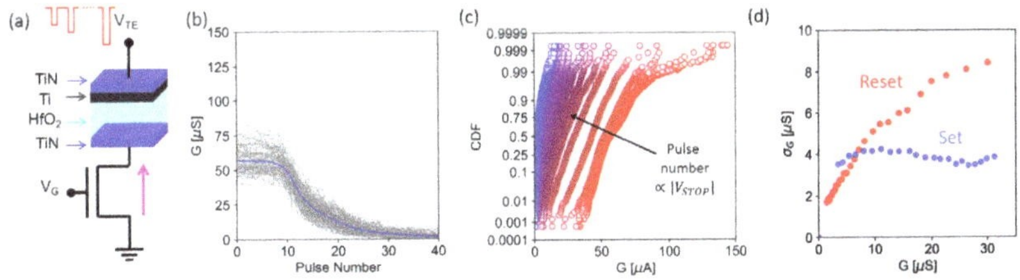

Figure 5. Analog programming with reset pulses. (**a**) Conceptual schematic of the IRPP algorithm. (**b**) Conductance as a function of pulse number for multiple iterations (grey lines) and the average behavior (blue). (**c**) Cumulative distribution function (CDF) of the conductance for increasing stop voltage V_{STOP}. (**d**) Standard deviation of the conductance σ_G as a function of the average conductance G for IRPP and IGPP algorithms. Adapted from [51].

3.1. Program-Verify Algorithms and Device-to-Device Variations

To date, only cycle-to-cyle variation on a single device was considered. To address device-to-device variation, conductance is typically characterized in a relatively large array (e.g., >1 kB), which allows one to study the main variability features with a relatively simple circuit and short experimental time. Figure 6a illustrates a conceptual schematic of a PV algorithm for 1T-1R RRAM, namely incremental-step program-verify algorithm (ISPVA) [43] applied on a 4 kB array. For a given V_G (or I_C), the TE is incremented until the read value of G after programming reaches the target value $G = L_i$. Figure 6b shows conductance traces as a function of V_{TE} for increasing V_G obtained with ISPVA for target levels L_{2-5}. The experiment was repeated on all the devices in the array, and the read current probability distributions are shown in Figure 6d [43]. The final average standard deviation is $\sigma_G = 7.5$ µS, which is slightly larger than the case of Figure 4 despite the PV algorithm where the number of pulses is adapted to reach a certain conductance. The larger σ_G can be understood by the superposition of cycle-to-cycle variation and device-to-device variation, the latter having the larger contribution to variability within the array.

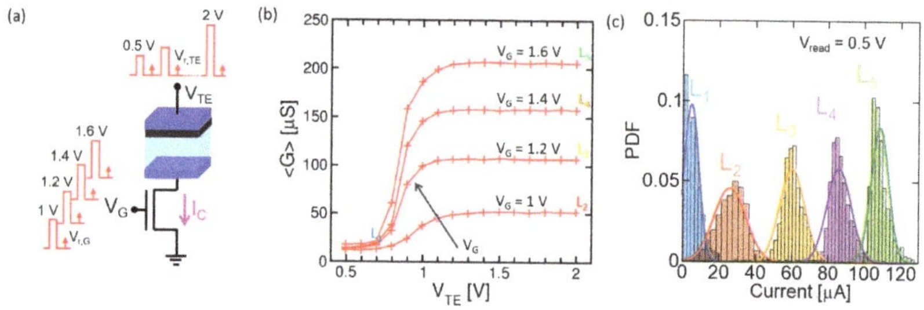

Figure 6. Program and verify algorithm. (**a**) Conceptual schematic of ISPVA program and verify algorithm. (**b**) Mean conductance as a function of set voltage V_{TE} for multiple values of the gate voltage V_G. (**c**) Probability density function (PDF) of programmed conductance levels. Reprinted from [43,57].

3.2. Conductance Drift and Fluctuations

Unfortunately, once the device is programmed with a given precision, the conductance might still change due to time-dependent drift and fluctuation that affects the reliability of IMC. Figure 7a shows measured traces of conductance programmed with 4 levels on a 1Mb RRAM array as a function of time during annealing at 150 °C [58]. The change of conductance can be explained by the thermally-activated atomic diffusion in the conductive filament. Note that, in addition to conductance drift with time, random fluctuations across the median value are present, as shown in Figure 7a. This is better illustrated in Figure 7b [59], where the resistance of a RRAM device in HRS is plotted as a function of time. The results show that the resistance can experience abrupt variations, called random walk, in addition to the typical random telegraph noise (RTN). For instance, the accuracy of neural network can decrease with time due to the introduction of a bias in the conductance as a result of the drift [43,58,60]. Figure 7c shows the cumulative distribution of the initial and drifted conductance by annealing at T = 125 °C of Figure 6c [43], which confirms the time-dependent drift of weights of a two-layer neural network. The network schematic is represented in Figure 7d, and it is composed of a 14 × 14 input layer (corresponding to a downsized version of the MNIST dataset), a hidden layer of 20 neurons and 10 output neurons corresponding to 10 handwritten digits and resulting in a total of 14 × 14 × 20 + 20 × 10 weights, which can be mapped in a 4 kB array. Figure 7e shows the confusion map for testing the MNIST dataset before annealing, showing a relatively good average accuracy of 83%. However, this accuracy drops to 72% after annealing as shown in Figure 7f, demonstrating the need of stable states for reliable neural network inference.

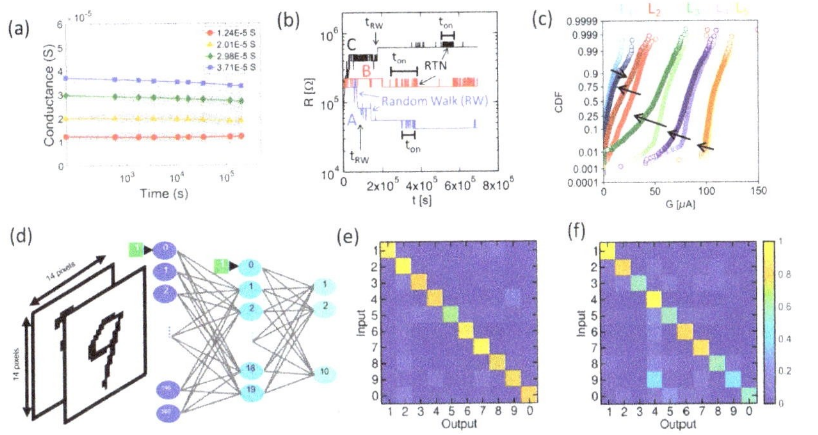

Figure 7. Drift and fluctuations in RRAM devices. (**a**) Conductance as function of time for 4 different analog levels after heating at 150 °C. (**b**) Different fluctuations' effect as a function of time. (**c**) Cumulative distributions of 5 programmed levels before and after annealing at T = 125 °C. (**d**) Conceptual schematic of the neural network used to evaluate the effect of drift and its accuracy in classifying the MNIST dataset before (**e**) and after (**f**) annealing. Adapted from [43,58,59].

4. RRAM Conductance Mapping Techniques

While Figures 2–5 focus on multilevel conductance mapping, aiming at an increase in the number of programmable levels and a reduction of the programming error, other techniques can be adopted to map a given computing coefficient, such as a synaptic weight, into one or multiple memory devices. Figure 8 summarizes the main programming methodologies for IMC [51], including multilevel (a) [43,61], binary (b) [62], unary (c) [63], multilevel with redundancy (d) [64] and slicing (e) [23]. In the following, we compare the various techniques in terms of mapping accuracy, maximum number of bits and number of devices. The mapping accuracy is evaluated by the standard deviation of the error σ_ϵ in programming a certain coefficient with a given number of bits N. The accuracy is evaluated by analytical formulas assuming a constant σ_G in programming an individual memory device with a maximum conductance G_{max} [51].

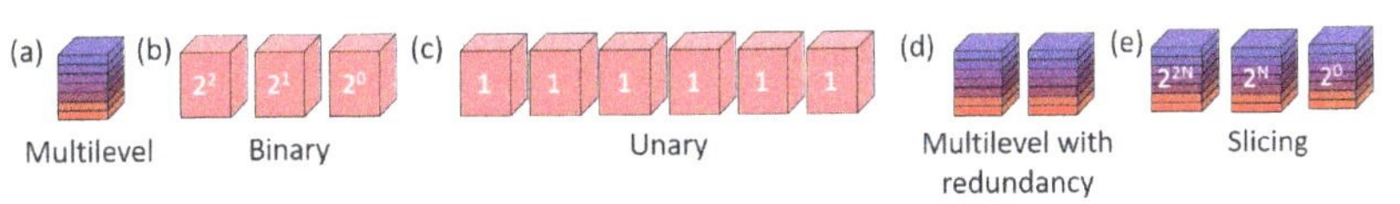

Figure 8. Programming mapping techniques. Conceptual representation of multilevel (**a**), binary (**b**), unary (**c**), multilevel with redundancy (**d**) and multilevel with slicing (**e**) programming. Adapted from [51].

4.1. Multilevel

Analog memories can naturally map discretized levels, which is referred to as multilevel mapping. To store N bits, 2^N equally spaced conductance levels between 0 and G_{max} are needed. As a result, each level is separated from the adjacent ones by a conductance step $\Delta G = G_{max}/(2^N - 1)$. Given a standard deviation of the programming error σ_G, such as the value σ_G = 3.8 µS in Figure 4d, the resulting standard deviation σ_ϵ of the error in programming N bits can be obtained as:

$$\sigma_\epsilon = \frac{\sigma_G}{\Delta G} = \frac{(2^N - 1)\sigma_G}{G_{max}}. \tag{2}$$

Equation (2) allows one to estimate the maximum number of bits N_{max}, which can be mapped in a RRAM device with a given acceptable error $\sigma_\epsilon << 1$, which yields $N_{max} = log_2(1 + \sigma_\epsilon G_{max}/\sigma_G)$. For example, by considering $\sigma_G = 2.2$ µS, $G_{max} = 225$ µS [57] and targeting a maximum error of $\sigma_\epsilon = 10\%$, the resulting maximum number of bits is $N_{max} = 3.35$ corresponding to 10 levels that can be written in the memory to satisfy the precision requirement.

4.2. Binary

Binary storage is the typical mapping of conventional digital memories, where a value x is converted in its binary representation with N bits written in N memory cells, each containing two states for 0 and 1. For instance, $x = 14$ can be written in binary representation as $x_{bin} = 1110$ with 4 RRAM cells programmed to $[G_{max}, G_{max}, G_{max}, 0]$, respectively. A weighted summation of the conductance values is possible by multiplying the current flowing in each cell by the corresponding power of 2, namely $2^3, 2^2, 2^1, 2^0$, respectively, thus allowing one to reconstruct the correct number as $[2^3 + 2^2 + 2^1]V_{read}G_{max} = 14V_{read}G_{max}$. To estimate σ_ϵ, we consider the average imprecision divided by the least significant bit (LSB), namely:

$$\sigma_\epsilon = \frac{1}{G_{max}}\sqrt{\sum_{i=0}^{N-1} 2^{2i}\sigma_G^2} = \frac{\sigma_G}{G_{max}}\sqrt{\frac{2^{2N}-1}{3}}, \tag{3}$$

where the square-root term combines the independent variation of each memory cell multiplied by its weight. The maximum number of bits thus can be obtained as:

$$N = \frac{1}{2}log_2\left(1 + 3\left(G_{max}\frac{\sigma_\epsilon}{\sigma_G}\right)^2\right). \tag{4}$$

Assuming the same σ_G, G_{max} and σ_ϵ of the estimation for multilevel mapping, we obtain $N_{max} = 4.15$ with binary RRAM.

4.3. Unary

To increase the precision of binary mapping, unary (or thermometric) coding uses $2^N - 1$ devices to represent the information, each one having equal weight, which requires no bit-specific gain in the current summation. In unary coding, the error is given by:

$$\sigma_\epsilon = \frac{\sigma_G}{G_{max}}\sqrt{2^N - 1} \tag{5}$$

which leads to a $N_{max} = 7.7$ with the same σ_G, G_{max} and σ_ϵ used in the previous estimation. However, note that the higher precision has been achieved at the cost of a larger number of RRAMs, namely $2^{N_{max}} - 1 = 207$ memory devices.

4.4. Multilevel with Redundancy

To reduce the impact of σ_G on multilevel coding, M memory devices having the same nominal conductance can be programmed and operated in parallel. As a result, the error is reduced by a factor $\sqrt{M}$ thanks to the averaging among the M redundant cells (see Table 1). The maximum number of bits is equivalently enhanced. Assuming $M = 4$ and the same σ_G, G_{max} and σ_ϵ used in the previous estimation, we obtain $N_{max} = 4.36$ bits, i.e., one additional bit compared to the pure multilevel case with no redundancy.

4.5. Slicing

By encoding a given number in base l, with $l = 2^N$ number of levels stored in a single memory element with multilevel mapping, and using M different memories to represent the data, it is possible to significantly increase the precision in a compact implementation. For example, $x = 14$ encoded in base $l = 4$ yields to $x_4 = 32$ such that by using two memory elements with weights 4^1 and 4^0, the current summation yields $x = 4^1 \times 3 + 4^0 \times 2$. Slicing can thus increase the number of addressable levels l by the number of used cells. The error can be evaluated in the same way as the binary scheme, by summing the weighted error contribution of each cell, which yields:

$$\sigma_\epsilon = \frac{\sigma_G}{G_{max}}\sqrt{1 + 2^{2\left(\frac{N}{M}-1\right)}}. \tag{6}$$

Assuming $M = 4$ and the same values of σ_G, G_{max} and σ_ϵ as before, we obtain $N_{max} = 13.41$ bits for slicing.

4.6. Simulation Results

Table 1 summarizes the formulas for calculating the error σ_ϵ and the maximum number of bits N_{max} for different programming techniques. To validate the analytical formulas, we performed Monte Carlo simulations of the various programming conditions and compared the results to the analytical calculations [51].

Table 1. Comparison of various mapping schemes, in terms of conductance range, variability-induced error and resulting maximum number of programmable bits.

Technique	G_{range}	σ_ϵ	N_{max}
Multilevel	G_{max}	$2^N \frac{\sigma_G}{G_{max}}$	$log_2(\frac{\sigma_\epsilon G_{max}}{\sigma_G})$
Binary	$(2^N-1)G_{max}$	$\sqrt{\frac{2^{2N}-1}{3}\frac{\sigma_G}{G_{max}}}$	$log_2\left[1+3\left(\frac{\sigma_\epsilon G_{max}}{\sigma_G}\right)^2\right]/2$
Unary	$(2^N-1)G_{max}$	$\sqrt{2N-1}\frac{\sigma_G}{G_{max}}$	$2log_2(\frac{\sigma_\epsilon G_{max}}{\sigma_G})+1$
Multilevel with redundancy	MG_{max}	$2^N\frac{\sigma_G}{G_{max}\sqrt{M}}$	$log_2(\frac{\sigma_\epsilon G_{max}\sqrt{M}}{\sigma_G})$
Slicing	$(2^M N-1)G_{max}$	$\frac{\sqrt{1+2^{2\left(\frac{N}{M}-1\right)}}\sigma_G}{G_{max}}$	$\frac{M}{2}\left[1+log_2\left(\frac{\sigma_\epsilon^2 G_{max}^2}{\sigma_G^2}-1\right)\right]$

Figure 9 shows the results of the analytical formulas (top) compared with the results of MC simulations (bottom) for the standard deviation of the error σ_ϵ as a function of the standard deviation of conductance σ_G and the number of bits to encode N, for multilevel (a,f), binary (b,g), unary (c,h), multilevel with redundancy $M = 4$ (d,i) and slicing with $M = 2$ (e,j) [51]. The analytical formulas and the MC simulations show a good agreement, thus confirming the correctness of the models in Table 1.

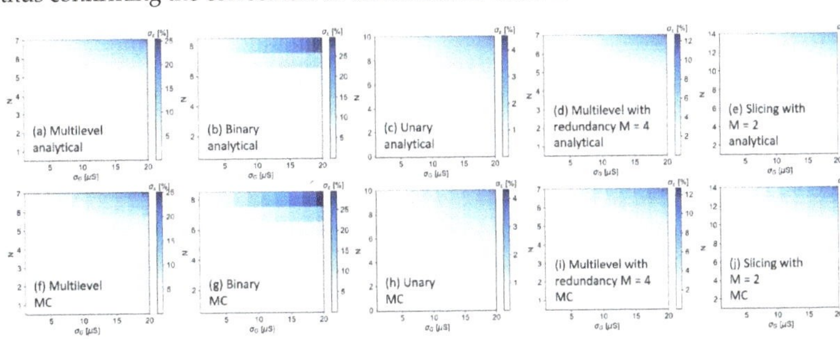

Figure 9. Comparison between analytical formula (top) and MC simulations (bottom) of standard deviation of the programming error σ_ϵ as a function of the number of bits N and the standard deviation of the conductance σ_G for multilevel (**a**,**f**), binary (**b**,**g**), unary (**c**,**h**), multilevel with redundancy factor $M = 4$ (**d**,**i**) and slicing in $M = 2$ units (**e**,**j**). Adapted from [51].

Figure 10a shows the calculated σ_ϵ as a function of σ_G for a target number $N = 7$ of bits. The results indicate that slicing and unary programming have a significant advantage over the other techniques by approximately one order of magnitude. This result is confirmed in

Figure 10b showing the maximum number of bits N_{max} as a function of σ_G for $\sigma_\epsilon = 1\%$. However, unary and slicing techniques require several memory elements to encode the coefficients, which thus increase the energy consumption and chip area per bit. To address the precision/area trade-off, Figure 10c shows the bit density, evaluated as N_{max} divided by the number of memory cells as a function of σ_G. The results demonstrate the good trade-off for the case of the slicing technique.

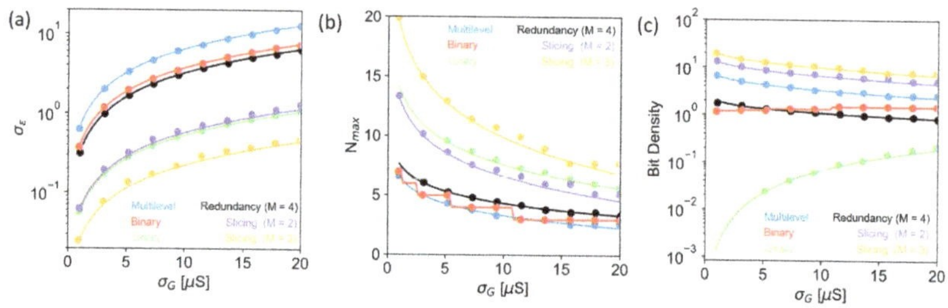

Figure 10. Figures of merit of various programming strategies. (**a**) Standard deviation of the overall programming error σ_G as a function of the conductance error σ_G assuming $N = 7$ bits. (**b**) Maximum number of bits N_{max} as a function of the conductance error σ_G considering a programming error $\sigma_\epsilon = 1\%$. (**c**) Bit density as a function of the conductance error σ_G. Adapted from [51].

5. Array-Level Reliability Issues

Device variations and errors are not the only origin for accuracy degradation in IMC circuits. In large memory arrays, the interconnect lines are generally affected by non-ideal behavior due to the parasitic resistance and capacitance that can deteriorate the analog signal integrity. In particular, parasitic wire resistance introduces a significant error, due to the current-resistance (IR) drop on the array rows/columns, which results in a difference between the applied voltage and the one across each memory cell. Figure 11a shows a sketch of a 4×4 memory array highlighting the wire resistance, namely the input resistance R_{in}, the output resistance R_{out} and the row/column wire resistance r between each memory cell [65]. Assuming for instance an inference operation with a typical read voltage $V_{read} = 0.1$ V, an average RRAM resistance $R = 100$ kΩ, corresponding to a current $I = 100$ μA for each cell within a a 100×100 crosspoint array with wire resistance $r = 1$ Ω, then the overall IR drop can be computed as $rI + 2rI + 3rI + \cdots + NrI = \frac{rIN^2}{2} = 5$ mV corresponding to a 5% error in the summation current [66]. Note that this error does not include any contribution due to variations in RRAM conductance.

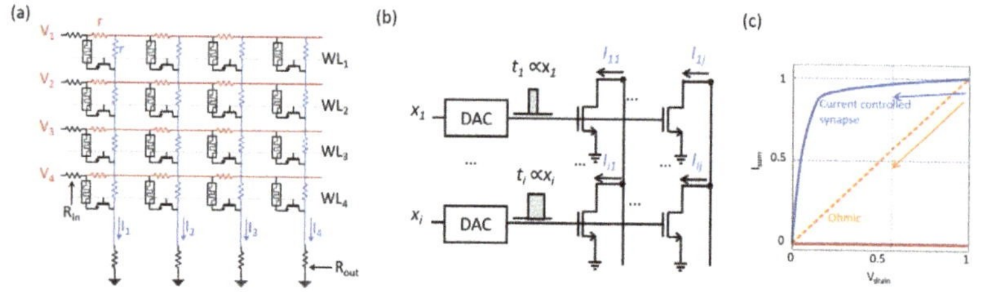

Figure 11. IR drop in crosspoint arrays. (**a**) Explicit representation of the parasitic resistance in a crosspoint array, namely the input resistance R_{in}, output resistance R_{out} and wire resistance r. (**b**) Memory array with current controlled memory element programmed to various saturation currents. (**c**) Comparison of the impact of IR drop for ohmic and saturated characteristics. Reprinted from [67].

To mitigate this effect, it is possible to increase the device resistance to decrease the current at the array wires. Unfortunately, large device resistances are usually more prone to variations, drift, fluctuations and noise [52,55]. Furthermore, a high cell resistance requires a longer readout time because of CMOS noise in the sensing circuit, thus resulting in longer program/verify algorithms. Finally, the higher cell resistance could increase the *RC* delay time for charging the BL. IR drop can be mitigated by changing the memory device topology, for example, by inserting a current-controlled synaptic element [67], such as a Flash memory, ionic transistor or FeFET, as shown in Figure 11b. A three-terminal memory transistor can be programmed to various saturation currents, each representing a synaptic weight. By encoding the input in the gate pulsewidth and integrating the current flowing in each column, it is possible to perform a current-based computation where the resulting charge is given by:

$$Q_i = \sum_{j=1}^{N} i_{ij} t_j \tag{7}$$

which corresponds to a MVM of a current matrix times a pulsewidth vector. Figure 11c shows a a comparison of the impact of IR drop for ohmic and saturated characteristics, demonstrating a much smaller impact of the current controlled devices [67].

At algorithmic level, IR drop and other non-idealities can be taken into account during training/inference or programming of computational memory devices. For instance, parasitic-aware program-and-verify algorithms have been presented to minimize the impact of IR drop [65,68]. By iteratively programming the target conductance matrix G resulting in G', evaluating the current $i' = VG'$ and comparing it to the ideal current $i = vG$, a new target matrix can be computed and programmed. The algorithm is repeated until the error is reduced below a certain tolerance, e.g., $\epsilon = |i - i'| < 1\%$.

At circuit architecture level, various techniques have been proposed to mitigate the impact of IR drop. Typically, the synaptic weights in a neural network have a differential structure, thus two separate 1T1R memory devices are used for representing a single weight. This is shown in Figure 12a where two contiguous columns are used for representing positive and negative weights that are summed up in the digital domain [69]. This configuration results in a relatively large impact of the IR drop since two array locations are used for each matrix entry. To mitigate this issues, Figure 12b shows a signed-weighted 2-transistor/2-resistor (SW-2T2R) structure to represent the positive and negative part of each weight, where the current summation is performed directly within the memory cell, thus effectively reducing the impact of IR drop by roughly a factor 2 [69]. Another approach is to use small-size crosspoint arrays and/or organize the IMC architecture in computing tiles [70,71], where the overall problem is divided in smaller operations with smaller summation currents, hence a lower IR drop.

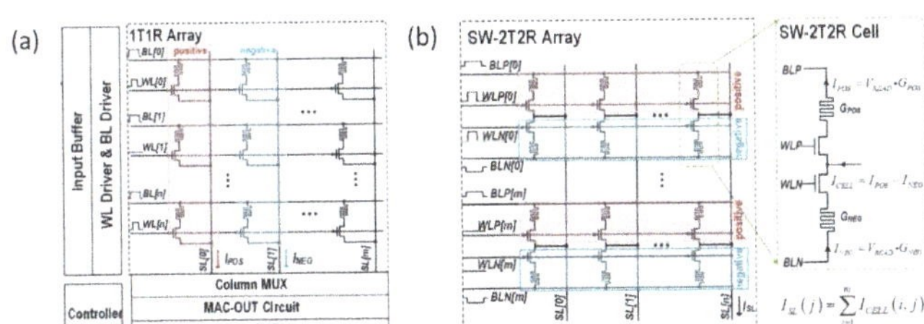

Figure 12. Bipolar conductance mapping and IR drop. (**a**) Typical bipolar conductance mapping in two adjacent 1T1R columns with the positive one (red) encoding the positive part of the weights and the negative one (blue) encoding the negative part of the weights. Currents are then subtracted in the digital domain after conversion. (**b**) To reduce the impact of IR drop, conductance representing the positive and negative weights can be summed up in the analog domain with a dedicated ST-2T2R circuit structure. Reprinted from [69].

6. Circuit Primitives for Analog Computing

Various computing primitives based on similar array organization of memory devices have been proposed. Figure 13 summarizes the main circuit primitives, including the MVM accelerator [1,18,22,26,27,42,46–49,65], the closed-loop circuit for inverse algebra problems [28,29,38] and the analog content addressable memory (CAM) [72,73].

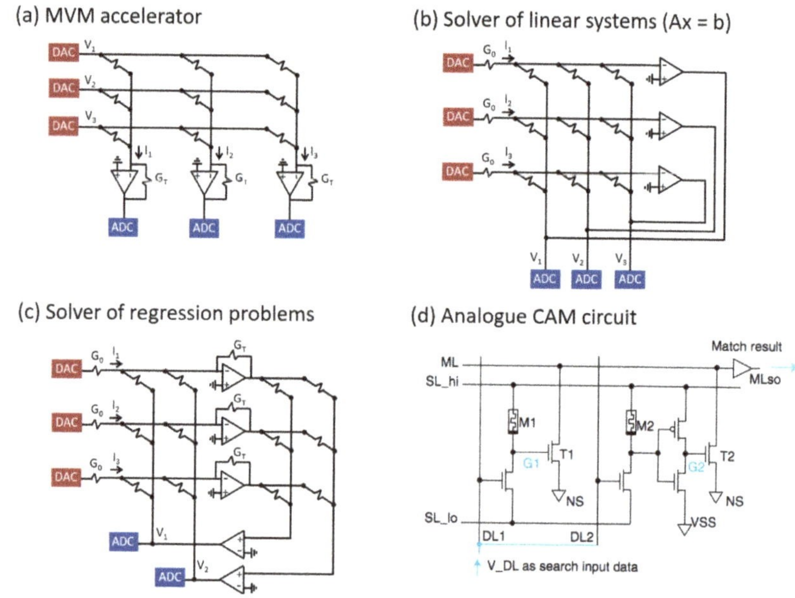

Figure 13. Schematic of IMC circuits for various applications. (**a**) MVM accelerator performing $I = AV$, where V is the input voltage vector, A is the conductance stored in the crosspoint array and I is the vector of the current in each row. (**b**) Linear system solver performing $V = A^{-1}I$ where I is the vector of the row input currents and V is the output vector of column voltages. (**c**) Regression problem solver performing $v = -(X^TX)^{-1}X^Ti$ where i is the vector of the input row currents at the left crosspoint array, X is the matrix of conductances in the two crosspoint arrays and v is the output voltage of the second stage of amplifiers. (**d**) Analog CAM cell, a range is stored in the memory conductance $M1$ and $M2$, the ML is pre-charged and an analog input is applied to the DL line. If the analog input is in the range of acceptance given by $M1$ and $M2$ the ML remains charged otherwise it discharges. Reprinted from [31,73].

6.1. MVM Accelerator

Figure 13a shows a circuit schematic for implementing the MVM accelerator of Equation (1). The conductance matrix G is stored in the RRAM device of the crosspoint array and the input voltage vector V is applied with a digital-to-analog converter (DAC) connected to the rows of the array. Columns are connected to the sensing circuit, consisting of a trans-impedance amplifier (TIA) that converts currents into voltages, and an analog-to-digital converter (ADC), which encodes the analog signals into digital words. MVM is performed in a single step, irrespective of the matrix size, although typically the sensing operation is multiplexed between various columns to reduce the peripheral overhead [60]. Since the forward propagation in a neural networks basically consists in extensive MVM of activation signal multiplied by synaptic weights [74], MVM crosspoint circuits have been heavily used for accelerating both the inference [22,43] and the training [41,42] stage of neural networks. Since the neuron activation function is typically performed in the digital domain, various training algorithms have been developed, including supervised training [42], unsupervised learning [75] and reinforcement learning [76]. MVM can serve as the core operation of various neural networks, including fully-connected neural networks [43], convolutional neural networks (CNNs) [77] and recurrent neural networks, such as long short term memory (LSTM) networks [78]. Integrated circuit comprising the crosspoint memory array for MVM and the routing/sensing units have been reported [24,79,80], demonstrating strong improvement in throughput and energy efficiency compared to conventional digital accelerators.

Among recurrent neural networks, the Hopfield neural network (HNN) [81] provides a brain-inspired structure that is capable of storing and recalling attractors, thus allowing an associative memory in hardware to be realized [50,82,83]. Interestingly, HNN can also naturally perform gradient descent algorithms, thus accelerating the solution of optimization problems such as constraint satisfaction problems (CSPs) [84]. By performing inference of an appropriate Hamiltonian representing the problem [85], the HNN converges to a stable state representing a minimum of the energy function of the problem, given by:

$$E = -\frac{1}{2}\sum_{i,j}^{N} G_{ij} v_i v_j \tag{8}$$

where G is the encoded matrix and v_i, v_j are the states of neurons i and j, respectively. However, the most difficult problems are usually not expressed by a convex energy landscape; thus even HNN cannot solve them, as the steady state remains trapped in a local minimum of the energy function. To prevent locking of the HNN in a local minimum, noise can be added to the system to perform simulated annealing [86], thus allowing the state to escape from the local minimum and converge to the global minimum corresponding to the optimization problem solution. Since memristive devices are inherently stochastic, various implementations combining MVM acceleration with noise generation have been proposed for accelerating the solution of CSPs [47–50,87,88]. Within this type of IMC accelerator, a major challenge consists of the extension of the circuit size to the scale of real-world intractable problems.

In addition to neural networks, MVM have been used to accelerate image processing [18], sparse coding [46], compressed sensing [89], linear programming [90] and the solution of linear systems of equations [26,27].

The MVM circuit can also be used for accelerating the training of a neural network according to the concept of outer product [91]. According to the back-propagation technique of the stochastic gradient descent (SGD) training, the error ϵ between the true output and the actual output at a certain stage during training is used to train each synaptic weight according to the formula [74]:

$$\Delta w_{ij} = \eta x_i \epsilon_j \tag{9}$$

where Δw_{ij} is the synaptic weight increase/decrease, η is a learning rate, x_i is the input or activation value and ϵ_j is the back-propagated error. Equation (9), which represents the outer product between the input vector x and the error vector ϵ, can be executed in hardware, e.g., by applying fixed pulse-width pulses with amplitude proportional to x at the array rows and fixed-amplitude pulses with pulse-width proportional to ϵ at the array columns [91]. This concept, which is the main idea for the hardware acceleration of time- and energy-consuming DNN training, relies on the linearity of the weight update on both the time and the voltage amplitude, which is rarely demonstrated in practical memory devices [3,92].

6.2. Analog Closed-Loop Accelerators

Analog in-memory circuits can solve algebraic problems without the need for digital iterations [28,29,38]. Figure 13b shows a circuit for the non-iterative solution of a system of linear equations [28]. Given a matrix problem $Ax = b$, it is possible to encode the coefficients A in the conductance matrix G, while the input vector b is applied as currents i at the crosspoint rows, which are connected to the negative input of the operational amplifiers. Since the output of the operational amplifiers are connected to the array columns, the Kirchhoff's law for the crosspoint array reads $i + Gv = 0$, where v is the output voltage vector on the columns and where we have neglected the input currents entering the high-impedance input nodes of the operational amplifiers. This leads to

$$v = -G^{-1}i, \tag{10}$$

which corresponds to the solution of the system of linear equations $Ax = b$. Note that the solution is obtained in one step, without iterations. It has been shown, on a first approximation, that the time complexity for solving linear systems in this circuit does not depend explicitly on the matrix size [93]; i.e., it displays an $O(1)$ complexity for solving linear systems, since the speed of solution solely depends on the configuration of poles which are limited by the smallest eigenvalue of the conductance matrix G. The circuit can implement both positive and negative coefficients of matrix A by adding an inverting buffer along each array column and connecting a second crosspoint array with the negative coefficients [28]. In addition to linear systems $Ax = b$, closed-loop crosspoint arrays allow one to solve eigenvector problems in the form $(A - \lambda I)x = 0$, where λ is the principal eigenvalue and I is the identity matrix [38]. For instance, the Pagerank algorithm [94], which is at the backbone of many computing tasks for searching, ranking and recommendation, relies on the calculation of the principal eigenvector of a given link matrix. Similar to the linear system solution, the IMC-accelerated computation of eigenvectors has a time complexity that does not depend explicitly on the matrix size, thus displaying $O(1)$ time complexity [95].

Figure 13c illustrates the IMC circuit for analog closed-loop linear-regression problems [29]. Assuming a set of N data points where the values of M independent variables are stored in matrix X of dimension $N \times M$ and y contains the values of the dependent variable, a regression consists of the calculation of the M coefficients α, which minimize the square error of the matrix equation $X\alpha = y$. This problem is non-iteratively solved by the circuit in Figure 13c where the input current vector i is applied to the rows of a rectangular crosspoint array that maps the matrix X, which are in turn connected to the negative input terminals of operational amplifiers A_1 with TIA configuration and gain G_T. The TIA's output terminals are connected to the rows of a second rectangular crosspoint array X. The columns of the second array are connected to the positive input terminals of operational amplifiers A_2, whose output terminals are connected to the columns of the first array. From applying Kirchhoff's laws at the first array, the output voltage of the set of TIAs is given by $v_\epsilon = G_T(i + Xv)$, where v is the output voltage of the operational amplifiers A_2. Since no current can flow in the high-impedance input terminals of operational amplifiers A_2, we can assume $X^T v_\epsilon = 0$, which can be rearranged as follows:

$$v = -(X^T X)^{-1} X^T i. \tag{11}$$

Here, the output voltage v is given by the product of the Moore–Penrose pseudo-inverse of X times the input vector y mapped by i, which provides the least-square solution α by minimizing the norm $||Xw - y||$ [29]. Linear and logistic regression accelerators with the circuit configuration of Figure 13c have been demonstrated [29], where the application can range from predicting the price of a house based on a set of descriptions to the training of the output layer of an extreme learning machine, a particular neural network with a wide and random input layer and an output layer that can be trained with logistic regression [29].

6.3. Analog CAM

In a conventional random access memory (RAM), an address is given as input and the stored word is returned as output. CAMs work on the opposite direction; i.e., the data content is provided as input word, while its location in the memory (or address) is returned as output, thus serving as data search and data matching circuits [96]. CAM is generally implemented by SRAM-based circuits, which can be relatively bulky and power-hungry;

thus RRAM-based CAMs have been proposed [97–99]. A distinctive advantage of RRAM with respect to SRAM is the possibility for multilevel CAM [72,100] and analog CAM [73], thus allowing to increase significantly the memory density.

Figure 13d shows the schematic of an analog CAM cell [73] based on RRAM devices. By storing in the conductance of RRAMs M_1 and M_2 two different values, pre-charging the match line (ML) and applying an analog search input data to the data line (DL), ML will remain charged only if the voltage on DL is such that $f(M_1) < V_{DL} < f(M_2)$. This property can be used for implementing a multilevel CAM; e.g., if the stored number to be searched is $x = 15$, M_1 could be set to the level corresponding to 14.5 while M_2 to the level corresponding to 15.5 where the 0.5 range represents the acceptance tolerance within an error of $LSB/2$. Interestingly, analog CAMs have been used for accelerating machine learning tasks such as one-shot learning in memory augmented neural networks [72], or tree-based models [101]. In the latter case, each threshold of a root-to-leaf path in a decision tree is mapped to an analog CAM row, thus allowing the inference in parallel within a large amount of trees to be accelerated.

7. Outlook on Memory Technologies and Computing Applications

For each specific application of analog IMC, such as the training of a neural network or the accelerated solution of algebra problems, different properties are required from a device and circuit perspective. To illustrate the device- and application-dependent requirements, Figure 14a shows a radar/spider chart summarizing the device properties in terms of cycling endurance, low-current operation, scaling and possibility of 3D integration, ohmic/linear conduction behavior, programming speed, analog precision and linear conductance upgrade. Each property is shown on a relative scale for various IMC tasks, including algebra accelerators, DNN training/inference accelerators and spiking neural networks (SNNs). Among these computing applications, DNN training accelerators is one with the highest demand for high-performance memory devices. This is because training acceleration relies on the synaptic weights to be updated online, typically in parallel via the outer product operation, which requires a linearity in both time and voltage for accurate and fast convergence [41,102]. This property is extremely difficult to achieve with resistive memory technologies due to the tendency for abrupt increase/decrease of conductivity, followed by a saturating change of conductance [3,92]. A typical figure of merit for linearity is the exponent n_{pot} in the formula:

$$G = G_{min} + G_0(1 - e^{-n_{pot}p}) \tag{12}$$

which describes the increase in conductance G as a function of the number p of potentiation pulses, where G_{min} and G_0 are fitting parameters. A similar exponent n_{dep} describes the linearity under depression. Parameters n_{pot} and n_{dep} should generally have similar values to allow for symmetric potentiation/depression, which is another key requirement for online training. Previous simulation results have shown that the recognition accuracy can range between about 82% for PCMO-based RRAM and a highest possible accuracy of about 94% for perfectly linear and symmetric characteristics in the case of a 3-layer neural network for MNIST recognition [103]. On-line training accelerators also generally require a high programming speed and cycling endurance, due to the programming-intensive update of the synaptic elements. On the one hand, gradual set/reset operation reduces the stress on the memory device compared with full set/reset in the binary or memory application. However, the ability for analog-type programming generally degrades with cycling [104,105].

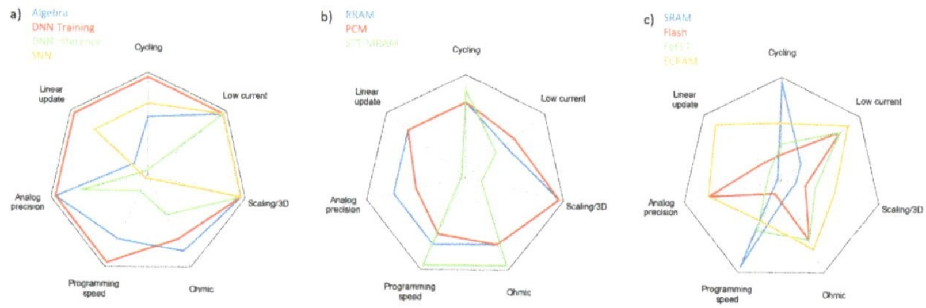

Figure 14. Application requirements (**a**) and figures of merit for various memory technologies with 2-terminal (**b**) and -terminal structure (**c**).

Analog closed-loop algebra accelerators also show a high demand of device properties, including highly-linear conduction characteristics to prevent unstable and oscillatory behavior of the system. DNN inference accelerators show less stringent requirements, thanks to the mostly-read operation of the memory array device for accelerating the MVM, while a relatively small number of program/verify operations are needed to reconfigure the system for a new AI task, which considerably reduces the requirements in terms of endurance, programming speed and linear weight update. In the case of non-ohmic conduction, the array can be operated in shift-and-add fashion such that the input is applied as digital word and the output is reconstructed with post-processing [23]. Non-linear characteristics are generally observed in RRAM devices with low conductance, which is essential for all computing schemes. When a device is programmed in the low conductance range, close to the HRS, transport typically takes place via Poole–Frenkel phenomena, which have a non-linear dependence on voltage [106]. The desired conductance range for parallel MVM is generally below 10 μS, which would allow for an overall error due to IR drop around 5% for a 100 × 100 array (see Section 5). Achieving a lower conductance in the sub-μS range would enable the scaling-up of the computational array, with advantages of throughput, energy efficiency and area efficiency due to the smaller peripheral circuits. Another key general requirement is the precision of conductance, i.e., ensuring a low σ_G. For the case of inference accelerators, it has been recently shown that the network accuracy decreases only from 91.6% to 91.2% for a σ_G between 0 and 10 μS for a 2-layer perceptron for MNIST recognition [57]. Studies on deeper networks have indicated that the sensitivity to conductance variation can vary widely depending on the specific neural network [107]. For instance, a ResNet model shows an increasing sensitivity to conductance for increasing number of layers, which can be understood by error accumulation in the forward propagation. On the other hand, AlexNet CNN shows a decreasing sensitivity at increasing size of the convolutional filter, due to error averaging within larger filters.

SNN show the most relaxed requirements thanks to neuro-biological frequencies in the 100 Hz range, which significantly relaxes the demand in terms of programming speed. Furthermore, update/conduction non-linearity and stochasticity are generally well tolerated or even potentially harnessed to perform brain-inspired adaptation and computations [50,108]. All applications generally require high scalability and 3D integration of the memory elements to take advantage of high density of information for data-intensive computing. For instance, a recent neural network model for natural language processing (NLP) called generative pre-trained transformer 3 (GPT-3) includes 175 billion parameters, which approximately corresponds to 175 GB of memory devices [109].

Figure 14b shows the figures of merit for two-terminal devices, namely RRAM, phase change memory (PCM) [33,110,111] and spin-transfer torque (STT) magnetic random access memory (MRAM) [112]. RRAM and PCM show comparable properties, the main difference being the analog precision, which is typically lower in PCM devices because of drift phenomena [113]. STT-MRAM offers high programming speed, good endurance and highly-ohmic conduction [11]; however, the conductance is generally limited to two states, corresponding to the parallel and the antiparallel magnetic polarization in the magnetic tunnel junction. As a result, use of the STT-MRAM device is limited to digital computing, such as binarized neural networks (BNNs) [114,115]. Figure 14c shows the relative performance of three-terminal devices for accelerating analog IMC, including both CMOS-based memory technologies and memristive technologies [116]. Static random access memory (SRAM) is typically limited to digital operation, whereas Flash, ferroelectric field-effect transistor (FEFET) [117] and electrochemical random access memory (ECRAM) [118] show well-tunable analog conductance and low current operation. Compared to 2-terminal devices, transistor-type memory devices can display a lower conductance thanks to the sub-threshold operation regime [119]. The relatively low conductance in ECRAM transistors can be explained by the use of low-mobility channels, usually consisting of metal oxides such as WO_3 [120] and TiO_2 [121]. ECRAM devices have also shown well-tunable conductance levels, which translates in a large number of multilevel states [120]. The precision of conductance can be attributed to the bulk-type conduction process within the switchable metal-oxide channel, as opposed to the filamentary conduction in typical 2-terminal RRAM [121]. The weight of ECRAM can be updated with extremely high linearity, thus offering an ideal device for online training accelerators [118]. For instance, the non-linearity exponents in Equation (12) are $n_{pot} = 0.347$ and $n_{dep} = 0.268$ in Li-based ECRAM, compared to a minimum n_{pot} of about 2 for most RRAM and PCM devices [118]. While CMOS technology has limited capability for 3D integration, the memristive FEFET and ECRAM seems most suitable for high density, back-end integrated memory arrays for analog computation of large-scale problems.

8. Conclusions

This work reviews the status and outlook of in-memory computing with RRAM devices. The RRAM device concept and the programming techniques aimed at high-precision analog conductance are presented. The possible coding of computational coefficient in the RRAM, including binary, unary and various multilevel approaches, are compared in terms of precision and circuit area. The typical challenges for analog precision of conductance are discussed, in terms of both device reliability (programming variations, drift and time-dependent fluctuations) and circuit-level parasitic resistance leading to IR drop errors. The most relevant analog IMC circuit primitives, including MVM, linear algebra accelerators and CAM, are presented. Finally, RRAM is compared to other computational memory devices in terms of reliability, precision, low current operation and scaling. From this overview, RRAM appears as one of the most mature and most promising technologies, despite significant challenges remain in reducing the operational current and controlling time-dependent fluctuations and programming variations.

Author Contributions: G.P. and D.I. performed literature review and wrote the paper with equal contributions. All authors have read and agreed to the published version of the manuscript.

Funding: This research was funded by the European Research Council (ERC) under the European Union's Horizon 2020 Research and Innovation Program under Grant 648635.

Conflicts of Interest: The authors declare no conflict of interest.

References

1. Ielmini, D.; Wong, H.S.P. In-memory computing with resistive switching devices. *Nat. Electron.* **2018**, *1*, 333–343. [CrossRef]
2. Zidan, M.A.; Strachan, J.P.; Lu, W.D. The future of electronics based on memristive systems. *Nat. Electron.* **2018**, *1*, 22–29. [CrossRef]
3. Yu, S. Neuro-Inspired Computing with Emerging Nonvolatile Memorys. *Proc. IEEE* **2018**, *106*, 260–285. [CrossRef]
4. Borghetti, J.; Snider, G.S.; Kuekes, P.J.; Yang, J.J.; Stewart, D.R.; Williams, R.S. 'Memristive' switches enable 'stateful' logic operations via material implication. *Nature* **2010**, *464*, 873–876. [CrossRef] [PubMed]
5. Reuben, J.; Ben-Hur, R.; Wald, N.; Talati, N.; Ali, A.H.; Gaillardon, P.E.; Kvatinsky, S. Memristive logic: A framework for evaluation and comparison. In Proceedings of the 2017 27th International Symposium on Power and Timing Modeling, Optimization and Simulation (PATMOS), Thessaloniki, Greece, 25–27 September 2017; pp. 1–8. [CrossRef]
6. Jeong, D.S.; Kim, K.M.; Kim, S.; Choi, B.J.; Hwang, C.S. Memristors for Energy-Efficient New Computing Paradigms. *Adv. Electron. Mater.* **2016**, *2*, 1600090. [CrossRef]

7. Balatti, S.; Ambrogio, S.; Ielmini, D. Normally-off Logic Based on Resistive Switches—Part I: Logic Gates. *IEEE Trans. Electron Devices* **2015**, *62*, 1831–1838. [CrossRef]
8. Chen, A. Utilizing the Variability of Resistive Random Access Memory to Implement Reconfigurable Physical Unclonable Functions. *IEEE Electron Device Lett.* **2015**, *36*, 138–140. [CrossRef]
9. Gao, L.; Chen, P.Y.; Liu, R.; Yu, S. Physical Unclonable Function Exploiting Sneak Paths in Resistive Cross-point Array. *IEEE Trans. Electron Devices* **2016**, *63*, 3109–3115. [CrossRef]
10. Nili, H.; Adam, G.C.; Hoskins, B.; Prezioso, M.; Kim, J.; Mahmoodi, M.R.; Bayat, F.M.; Kavehei, O.; Strukov, D.B. Hardware-intrinsic security primitives enabled by analogue state and nonlinear conductance variations in integrated memristors. *Nat. Electron.* **2018**, *1*, 197–202. [CrossRef]
11. Carboni, R.; Ambrogio, S.; Chen, W.; Siddik, M.; Harms, J.; Lyle, A.; Kula, W.; Sandhu, G.; Ielmini, D. Modeling of Breakdown-Limited Endurance in Spin-Transfer Torque Magnetic Memory Under Pulsed Cycling Regime. *IEEE Trans. Electron Devices* **2018**, *65*, 2470–2478. [CrossRef]
12. Jo, S.H.; Chang, T.; Ebong, I.; Bhadviya, B.B.; Mazumder, P.; Lu, W. Nanoscale Memristor Device as Synapse in Neuromorphic Systems. *Nano Lett.* **2010**, *10*, 1297–1301. [CrossRef]
13. Yu, S.; Wu, Y.; Jeyasingh, R.; Kuzum, D.; Wong, H.S.P. An Electronic Synapse Device Based on Metal Oxide Resistive Switching Memory for Neuromorphic Computation. *IEEE Trans. Electron Devices* **2011**, *58*, 2729–2737. [CrossRef]
14. Yu, S.; Gao, B.; Fang, Z.; Yu, H.; Kang, J.; Wong, H.S.P. A Low Energy Oxide-Based Electronic Synaptic Device for Neuromorphic Visual Systems with Tolerance to Device Variation. *Adv. Mater.* **2013**, *25*, 1774–1779. [CrossRef]
15. Pedretti, G.; Milo, V.; Ambrogio, S.; Carboni, R.; Bianchi, S.; Calderoni, A.; Ramaswamy, N.; Spinelli, A.S.; Ielmini, D. Memristive neural network for on-line learning and tracking with brain-inspired spike timing dependent plasticity. *Sci. Rep.* **2017**, *7*, 5288. [CrossRef]
16. Wang, Z.; Joshi, S.; Savel'ev, S.; Song, W.; Midya, R.; Li, Y.; Rao, M.; Yan, P.; Asapu, S.; Zhuo, Y.; et al. Fully memristive neural networks for pattern classification with unsupervised learning. *Nat. Electron.* **2018**, *1*, 137–145. [CrossRef]
17. Truong, S.N.; Min, K.S. New Memristor-Based Crossbar Array Architecture with 50-% Area Reduction and 48-% Power Saving for Matrix-Vector Multiplication of Analog Neuromorphic Computing. *JSTS J. Semicond. Technol. Sci.* **2014**, *14*, 356–363. [CrossRef]
18. Li, C.; Hu, M.; Li, Y.; Jiang, H.; Ge, N.; Montgomery, E.; Zhang, J.; Song, W.; Dávila, N.; Graves, C.E.; et al. Analogue signal and image processing with large memristor crossbars. *Nat. Electron.* **2018**, *1*, 52–59. [CrossRef]
19. Hu, M.; Graves, C.E.; Li, C.; Li, Y.; Ge, N.; Montgomery, E.; Davila, N.; Jiang, H.; Williams, R.S.; Yang, J.J.; et al. Memristor-Based Analog Computation and Neural Network Classification with a Dot Product Engine. *Adv. Mater.* **2018**, *30*, 1705914. [CrossRef] [PubMed]
20. Chi, P.; Li, S.; Xu, C.; Zhang, T.; Zhao, J.; Liu, Y.; Wang, Y.; Xie, Y. PRIME: A Novel Processing-in-Memory Architecture for Neural Network Computation in ReRAM-Based Main Memory. In Proceedings of the 2016 ACM/IEEE 43rd Annual International Symposium on Computer Architecture (ISCA), Seoul, Korea, 18–22 June 2016; pp. 27–39. [CrossRef]
21. Gokmen, T.; Vlasov, Y. Acceleration of Deep Neural Network Training with Resistive Cross-Point Devices: Design Considerations. *Front. Neurosci.* **2016**, *10*, 333. [CrossRef]
22. Yao, P.; Wu, H.; Gao, B.; Eryilmaz, S.B.; Huang, X.; Zhang, W.; Zhang, Q.; Deng, N.; Shi, L.; Wong, H.S.P.; et al. Face classification using electronic synapses. *Nat. Commun.* **2017**, *8*, 15199. [CrossRef] [PubMed]
23. Shafiee, A.; Nag, A.; Muralimanohar, N.; Balasubramonian, R.; Strachan, J.P.; Hu, M.; Williams, R.S.; Srikumar, V. ISAAC: A Convolutional Neural Network Accelerator with In-Situ Analog Arithmetic in Crossbars. In Proceedings of the 2016 ACM/IEEE 43rd Annual International Symposium on Computer Architecture (ISCA), Seoul, Korea, 18–22 June 2016; pp. 14–26. [CrossRef]
24. Yao, P.; Wu, H.; Gao, B.; Tang, J.; Zhang, Q.; Zhang, W.; Yang, J.J.; Qian, H. Fully hardware-implemented memristor convolutional neural network. *Nature* **2020**, *577*, 641–646. [CrossRef]
25. Xue, C.X.; Chiu, Y.C.; Liu, T.W.; Huang, T.Y.; Liu, J.S.; Chang, T.W.; Kao, H.Y.; Wang, J.H.; Wei, S.Y.; Lee, C.Y.; et al. A CMOS-integrated compute-in-memory macro based on resistive random-access memory for AI edge devices. *Nat. Electron.* **2021**, *4*, 81–90. [CrossRef]
26. Le Gallo, M.; Sebastian, A.; Mathis, R.; Manica, M.; Giefers, H.; Tuma, T.; Bekas, C.; Curioni, A.; Eleftheriou, E. Mixed-precision in-memory computing. *Nat. Electron.* **2018**, *1*, 246–253. [CrossRef]
27. Zidan, M.A.; Jeong, Y.; Lee, J.; Chen, B.; Huang, S.; Kushner, M.J.; Lu, W.D. A general memristor-based partial differential equation solver. *Nat. Electron.* **2018**, *1*, 411–420. [CrossRef]
28. Sun, Z.; Pedretti, G.; Ambrosi, E.; Bricalli, A.; Wang, W.; Ielmini, D. Solving matrix equations in one step with cross-point resistive arrays. *Proc. Natl. Acad. Sci. USA* **2019**, *116*, 4123–4128. [CrossRef] [PubMed]
29. Sun, Z.; Pedretti, G.; Bricalli, A.; Ielmini, D. One-step regression and classification with cross-point resistive memory arrays. *Sci. Adv.* **2020**, *6*, eaay2378. [CrossRef]
30. Cassinerio, M.; Ciocchini, N.; Ielmini, D. Logic Computation in Phase Change Materials by Threshold and Memory Switching. *Adv. Mater.* **2013**, *25*, 5975–5980. [CrossRef] [PubMed]
31. Ielmini, D.; Pedretti, G. Device and Circuit Architectures for In-Memory Computing. *Adv. Intell. Syst.* **2020**, *2*, 2000040. [CrossRef]
32. Chappert, C.; Fert, A.; Van Dau, F.N. The emergence of spin electronics in data storage. *Nat. Mater.* **2007**, *6*, 813–823. [CrossRef] [PubMed]

33. Raoux, S.; Wełnic, W.; Ielmini, D. Phase Change Materials and Their Application to Nonvolatile Memories. *Chem. Rev.* **2010**, *110*, 240–267. [CrossRef] [PubMed]
34. Burr, G.W.; Breitwisch, M.J.; Franceschini, M.; Garetto, D.; Gopalakrishnan, K.; Jackson, B.; Kurdi, B.; Lam, C.; Lastras, L.A.; Padilla, A.; et al. Phase change memory technology. *J. Vac. Sci. Technol. Nanotechnol. Microelectron. Mater. Process. Meas. Phenom.* **2010**, *28*, 223–262. [CrossRef]
35. Ielmini, D. Resistive switching memories based on metal oxides: mechanisms, reliability and scaling. *Semicond. Sci. Technol.* **2016**, *31*, 063002. [CrossRef]
36. Govoreanu, B.; Kar, G.; Chen, Y.Y.; Paraschiv, V.; Kubicek, S.; Fantini, A.; Radu, I.; Goux, L.; Clima, S.; Degraeve, R.; et al. 10 × 10 nm^2 Hf/HfOx crossbar resistive RAM with excellent performance, reliability and low-energy operation. In *2011 International Electron Devices Meeting*; IEEE: Washington, DC, USA, 2011; pp. 31.6.1–31.6.4. [CrossRef]
37. Pi, S.; Li, C.; Jiang, H.; Xia, W.; Xin, H.; Yang, J.J.; Xia, Q. Memristor crossbar arrays with 6-nm half-pitch and 2-nm critical dimension. *Nat. Nanotechnol.* **2019**, *14*, 35–39. [CrossRef] [PubMed]
38. Sun, Z.; Ambrosi, E.; Pedretti, G.; Bricalli, A.; Ielmini, D. In-Memory PageRank Accelerator With a Cross-Point Array of Resistive Memories. *IEEE Trans. Electron Devices* **2020**, *67*, 1466–1470. [CrossRef]
39. Yang, J.J.; Strukov, D.B.; Stewart, D.R. Memristive devices for computing. *Nat. Nanotechnol.* **2013**, *8*, 13–24. [CrossRef]
40. Prezioso, M.; Merrikh-Bayat, F.; Hoskins, B.D.; Adam, G.C.; Likharev, K.K.; Strukov, D.B. Training and operation of an integrated neuromorphic network based on metal-oxide memristors. *Nature* **2015**, *521*, 61–64. [CrossRef]
41. Ambrogio, S.; Narayanan, P.; Tsai, H.; Shelby, R.M.; Boybat, I.; di Nolfo, C.; Sidler, S.; Giordano, M.; Bodini, M.; Farinha, N.C.P.; et al. Equivalent-accuracy accelerated neural-network training using analogue memory. *Nature* **2018**, *558*, 60–67. [CrossRef]
42. Li, C.; Belkin, D.; Li, Y.; Yan, P.; Hu, M.; Ge, N.; Jiang, H.; Montgomery, E.; Lin, P.; Wang, Z.; et al. Efficient and self-adaptive in situ learning in multilayer memristor neural networks. *Nat. Commun.* **2018**, *9*, 2385. [CrossRef]
43. Milo, V.; Zambelli, C.; Olivo, P.; Pérez, E.; K. Mahadevaiah, M.; G. Ossorio, O.; Wenger, C.; Ielmini, D. Multilevel HfO_2 -based RRAM devices for low-power neuromorphic networks. *APL Mater.* **2019**, *7*, 081120. [CrossRef]
44. Prezioso, M.; Mahmoodi, M.R.; Bayat, F.M.; Nili, H.; Kim, H.; Vincent, A.; Strukov, D.B. Spike-timing-dependent plasticity learning of coincidence detection with passively integrated memristive circuits. *Nat. Commun.* **2018**, *9*, 5311. [CrossRef]
45. Wang, Z.; Zeng, T.; Ren, Y.; Lin, Y.; Xu, H.; Zhao, X.; Liu, Y.; Ielmini, D. Toward a generalized Bienenstock-Cooper-Munro rule for spatiotemporal learning via triplet-STDP in memristive devices. *Nat. Commun.* **2020**, *11*, 1510. [CrossRef]
46. Sheridan, P.M.; Cai, F.; Du, C.; Ma, W.; Zhang, Z.; Lu, W.D. Sparse coding with memristor networks. *Nat. Nanotechnol.* **2017**, *12*, 784–789. [CrossRef]
47. Shin, J.H.; Jeong, Y.J.; Zidan, M.A.; Wang, Q.; Lu, W.D. Hardware Acceleration of Simulated Annealing of Spin Glass by RRAM Crossbar Array. In Proceedings of the 2018 IEEE International Electron Devices Meeting (IEDM), San Francisco, CA, USA, 1–5 December 2018; pp. 3.3.1–3.3.4.
48. Mahmoodi, M.R.; Kim, H.; Fahimi, Z.; Nili, H.; Sedov, L.; Polishchuk, V.; Strukov, D.B. An Analog Neuro-Optimizer with Adaptable Annealing Based on 64x64 0T1R Crossbar Circuit. In Proceedings of the 2019 IEEE International Electron Devices Meeting (IEDM), San Francisco, CA, USA, 7–11 December 2019; pp. 14.7.1–14.7.4. [CrossRef]
49. Cai, F.; Kumar, S.; Van Vaerenbergh, T.; Sheng, X.; Liu, R.; Li, C.; Liu, Z.; Foltin, M.; Yu, S.; Xia, Q.; et al. Power-efficient combinatorial optimization using intrinsic noise in memristor Hopfield neural networks. *Nat. Electron.* **2020**. [CrossRef]
50. Pedretti, G.; Mannocci, P.; Hashemkhani, S.; Milo, V.; Melnic, O.; Chicca, E.; Ielmini, D. A Spiking Recurrent Neural Network With Phase-Change Memory Neurons and Synapses for the Accelerated Solution of Constraint Satisfaction Problems. *IEEE J. Explor. Solid State Comput. Devices Circ.* **2020**, *6*, 89–97. [CrossRef]
51. Pedretti, G.; Ambrosi, E.; Ielmini, D. Conductance variations and their impact on the precision of in-memory computing with resistive switching memory (RRAM). In Proceedings of the 2021 IEEE International Reliability Physics Symposium (IRPS), live virtual conference, 21–24 March 2021; pp. 2C.1-1–2C.1-4.
52. Ambrogio, S.; Balatti, S.; Cubeta, A.; Calderoni, A.; Ramaswamy, N.; Ielmini, D. Statistical Fluctuations in HfO_x Resistive-Switching Memory: Part II—Random Telegraph Noise. *IEEE Trans. Electron Devices* **2014**, *61*, 2920–2927. [CrossRef]
53. Bricalli, A.; Ambrosi, E.; Laudato, M.; Maestro, M.; Rodriguez, R.; Ielmini, D. Resistive Switching Device Technology Based on Silicon Oxide for Improved ON—OFF Ratio—Part I: Memory Devices. *IEEE Trans. Electron Devices* **2018**, *65*, 115–121. [CrossRef]
54. Balatti, S.; Ambrogio, S.; Ielmini, D.; Gilmer, D.C. Variability and failure of set process in HfO2 RRAM. In Proceedings of the 2013 5th IEEE International Memory Workshop, Monterey, CA, USA, 26–29 May 2013; pp. 38–41. [CrossRef]
55. Balatti, S.; Ambrogio, S.; Gilmer, D.C.; Ielmini, D. Set Variability and Failure Induced by Complementary Switching in Bipolar RRAM. *IEEE Electron Device Lett.* **2013**, *34*, 861–863. [CrossRef]
56. Fantini, A.; Goux, L.; Degraeve, R.; Wouters, D.; Raghavan, N.; Kar, G.; Belmonte, A.; Chen, Y.Y.; Govoreanu, B.; Jurczak, M. Intrinsic switching variability in HfO2 RRAM. In Proceedings of the 2013 5th IEEE International Memory Workshop, Monterey, CA, USA, 26–29 May 2013; pp. 30–33. [CrossRef]
57. Milo, V.; Anzalone, F.; Zambelli, C.; Perez, E.; Mahadevaiah, M.; Ossorio, O.; Olivo, P.; Wenger, C.; Ielmini, D. Optimized programming algorithms for multilevel RRAM in hardware neural networks. In Proceedings of the 2021 IEEE International Reliability Physics Symposium (IRPS), live virtual conference, 21–24 March 2021; pp. 2C.4-1–2C.4-4.
58. Lin, Y.H.; Wang, C.H.; Lee, M.H.; Lee, D.Y.; Lin, Y.Y.; Lee, F.M.; Lung, H.L.; Wang, K.C.; Tseng, T.Y.; Lu, C.Y. Performance Impacts of Analog ReRAM Non-ideality on Neuromorphic Computing. *IEEE Trans. Electron Devices* **2019**, *66*, 1289–1295. [CrossRef]

59. Ambrogio, S.; Balatti, S.; McCaffrey, V.; Wang, D.C.; Ielmini, D. Noise-Induced Resistance Broadening in Resistive Switching Memory—Part II: Array Statistics. *IEEE Trans. Electron Devices* **2015**, *62*, 3812–3819. [CrossRef]
60. Peng, X.; Huang, S.; Luo, Y.; Sun, X.; Yu, S. DNN+NeuroSim: An End-to-End Benchmarking Framework for Compute-in-Memory Accelerators with Versatile Device Technologies. In Proceedings of the 2019 IEEE International Electron Devices Meeting (IEDM), San Francisco, CA, USA, 7–11 December 2019; pp. 32.5.1–32.5.4.
61. Alibart, F.; Gao, L.; Hoskins, B.D.; Strukov, D.B. High Precision Tuning of State for Memristive Devices by Adaptable Variation-Tolerant Algorithm. *Nanotechnology* **2012**.. p. 8. [CrossRef]
62. Yu, S.; Li, Z.; Chen, P.Y.; Wu, H.; Gao, B.; Wang, D.; Wu, W.; Qian, H. Binary neural network with 16 Mb RRAM macro chip for classification and online training. In Proceedings of the 2016 IEEE International Electron Devices Meeting (IEDM), San Francisco, CA, USA, 3–7 December 2016; pp. 16.2.1–16.2.4. [CrossRef]
63. Ma, C.; Sun, Y.; Qian, W.; Meng, Z.; Yang, R.; Jiang, L. Go Unary: A Novel Synapse Coding and Mapping Scheme for Reliable ReRAM-based Neuromorphic Computing. In Proceedings of the 2020 Design, Automation & Test in Europe Conference & Exhibition (DATE), Grenoble, France, 9–13 March 2020; pp. 1432–1437. [CrossRef]
64. Boybat, I.; Le Gallo, M.; Nandakumar, S.R.; Moraitis, T.; Parnell, T.; Tuma, T.; Rajendran, B.; Leblebici, Y.; Sebastian, A.; Eleftheriou, E. Neuromorphic computing with multi-memristive synapses. *Nat. Commun.* **2018**, *9*, 2514. [CrossRef]
65. Hu, M.; Williams, R.S.; Strachan, J.P.; Li, Z.; Grafals, E.M.; Davila, N.; Graves, C.; Lam, S.; Ge, N.; Yang, J.J. Dot-product engine for neuromorphic computing: programming 1T1M crossbar to accelerate matrix-vector multiplication. In *Proceedings of the 53rd Annual Design Automation Conference on-DAC '16*; ACM Press: Austin, TX, USA, 2016; pp. 1–6. [CrossRef]
66. Gokmen, T.; Rasch, M.J.; Haensch, W. The marriage of training and inference for scaled deep learning analog hardware. In Proceedings of the 2019 IEEE International Electron Devices Meeting (IEDM), San Francisco, CA, USA, 7–11 December 2019; pp. 22.3.1–22.3.4. [CrossRef]
67. Cosemans, S.; Verhoef, B.; Doevenspeck, J.; Papistas, I.A.; Catthoor, F.; Debacker, P.; Mallik, A.; Verkest, D. Towards 10000TOPS/W DNN Inference with Analog in-Memory Computing—A Circuit Blueprint, Device Options and Requirements. In Proceedings of the 2019 IEEE International Electron Devices Meeting (IEDM), San Francisco, CA, USA, 7–11 December 2019; pp. 22.2.1–22.2.4. [CrossRef]
68. Zhang, F.; Hu, M. Mitigate Parasitic Resistance in Resistive Crossbar-based Convolutional Neural Networks. *ACM J. Emerg. Technol. Comput. Syst.* **2020**, *16*, 1–20. [CrossRef]
69. Liu, Q.; Gao, B.; Yao, P.; Wu, D.; Chen, J.; Pang, Y.; Zhang, W.; Liao, Y.; Xue, C.X.; Chen, W.H.; et al. 33.2 A Fully Integrated Analog ReRAM Based 78.4TOPS/W Compute-In-Memory Chip with Fully Parallel MAC Computing. In Proceedings of the 2020 IEEE International Solid- State Circuits Conference-(ISSCC), San Francisco, CA, USA, 16–20 February 2020; pp. 500–502. [CrossRef]
70. Ankit, A.; Hajj, i.e.,; Chalamalasetti, S.R.; Ndu, G.; Foltin, M.; Williams, R.S.; Faraboschi, P.; Hwu, W.m.W.; Strachan, J.P.; Roy, K.; et al. PUMA: A Programmable Ultra-efficient Memristor-based Accelerator for Machine Learning Inference. In Proceedings of the Twenty-Fourth International Conference on Architectural Support for Programming Languages and Operating Systems, Providence, RI, USA, 13–17 April 2019; pp. 715–731. [CrossRef]
71. Wang, Q.; Wang, X.; Lee, S.H.; Meng, F.H.; Lu, W.D. A Deep Neural Network Accelerator Based on Tiled RRAM Architecture. In Proceedings of the 2019 IEEE International Electron Devices Meeting (IEDM), San Francisco, CA, USA, 7–11 December 2019; pp. 14.4.1–14.4.4. [CrossRef]
72. Ni, K.; Yin, X.; Laguna, A.F.; Joshi, S.; Dünkel, S.; Trentzsch, M.; Müller, J.; Beyer, S.; Niemier, M.; Hu, X.S.; et al. Ferroelectric ternary content-addressable memory for one-shot learning. *Nat. Electron.* **2019**, *2*, 521–529. [CrossRef]
73. Li, C.; Graves, C.E.; Sheng, X.; Miller, D.; Foltin, M.; Pedretti, G.; Strachan, J.P. Analog content-addressable memories with memristors. *Nat. Commun.* **2020**, *11*, 1638. [CrossRef]
74. LeCun, Y.; Bengio, Y.; Hinton, G. Deep learning. *Nature* **2015**, *521*, 436–444. [CrossRef]
75. Oh, S.; Shi, Y.; Liu, X.; Song, J.; Kuzum, D. Drift-Enhanced Unsupervised Learning of Handwritten Digits in Spiking Neural Network With PCM Synapses. *IEEE Electron Device Lett.* **2018**, *39*, 1768–1771. [CrossRef]
76. Wang, Z.; Li, C.; Song, W.; Rao, M.; Belkin, D.; Li, Y.; Yan, P.; Jiang, H.; Lin, P.; Hu, M.; et al. Reinforcement learning with analogue memristor arrays. *Nat. Electron.* **2019**, *2*, 115–124. [CrossRef]
77. Wang, Z.; Li, C.; Lin, P.; Rao, M.; Nie, Y.; Song, W.; Qiu, Q.; Li, Y.; Yan, P.; Strachan, J.P.; et al. In situ training of feed-forward and recurrent convolutional memristor networks. *Nat. Mach. Intell.* **2019**, *1*, 434–442. [CrossRef]
78. Li, C.; Wang, Z.; Rao, M.; Belkin, D.; Song, W.; Jiang, H.; Yan, P.; Li, Y.; Lin, P.; Hu, M.; et al. Long short-term memory networks in memristor crossbar arrays. *Nat. Mach. Intell.* **2019**, *1*, 49–57. [CrossRef]
79. Cai, F.; Correll, J.M.; Lee, S.H.; Lim, Y.; Bothra, V.; Zhang, Z.; Flynn, M.P.; Lu, W.D. A fully integrated reprogrammable memristor–CMOS system for efficient multiply—Accumulate operations. *Nat. Electron.* **2019**, *2*, 290–299. [CrossRef]
80. Li, C.; Ignowski, J.; Sheng, X.; Wessel, R.; Jaffe, B.; Ingemi, J.; Graves, C.; Strachan, J.P. CMOS-integrated nanoscale memristive crossbars for CNN and optimization acceleration. In Proceedings of the 2020 IEEE International Memory Workshop (IMW), Dresden, Germany, 17–20 May 2020; pp. 1–4. [CrossRef]
81. Hopfield, J.; Tank, D. Computing with neural circuits: A model. *Science* **1986**, *233*, 625–633. [CrossRef] [PubMed]
82. Eryilmaz, S.B.; Kuzum, D.; Jeyasingh, R.; Kim, S.; BrightSky, M.; Lam, C.; Wong, H.S.P. Brain-like associative learning using a nanoscale non-volatile phase change synaptic device array. *Front. Neurosci.* **2014**, *8*, 205. [CrossRef]

83. Milo, V.; Ielmini, D.; Chicca, E. Attractor networks and associative memories with STDP learning in RRAM synapses. In Proceedings of the 2017 IEEE International Electron Devices Meeting (IEDM), San Francisco, CA, USA, 2–6 December 2017; pp. 11.2.1–11.2.4. [CrossRef]
84. Tank, D.; Hopfield, J. Simple 'neural' optimization networks: An A/D converter, signal decision circuit, and a linear programming circuit. *IEEE Trans. Circ. Syst.* **1986**, *33*, 533–541. [CrossRef]
85. Lucas, A. Ising formulations of many NP problems. *Front. Phys.* **2014**, *2*, 5. [CrossRef]
86. Kirkpatrick, S.; Gelatt, C.; Vecchi, M. Optimization by Simulated Annealing. *Science* **1983**, *220*, 671–680. [CrossRef] [PubMed]
87. Kumar, S.; Strachan, J.P.; Williams, R.S. Chaotic dynamics in nanoscale NbO2 Mott memristors for analogue computing. *Nature* **2017**, *548*, 318–321. [CrossRef]
88. Mahmoodi, M.R.; Prezioso, M.; Strukov, D.B. Versatile stochastic dot product circuits based on nonvolatile memories for high performance neurocomputing and neurooptimization. *Nat. Commun.* **2019**, *10*, 5113. [CrossRef]
89. Le Gallo, M.; Sebastian, A.; Cherubini, G.; Giefers, H.; Eleftheriou, E. Compressed Sensing with Approximate Message Passing Using In-Memory Computing. *IEEE Trans. Electron Devices* **2018**, *65*, 4304–4312. [CrossRef]
90. Cai, R.; Ren, A.; Soundarajan, S.; Wang, Y. A low-computation-complexity, energy-efficient, and high-performance linear program solver based on primal–dual interior point method using memristor crossbars. *Nano Commun. Netw.* **2018**, *18*, 62–71. [CrossRef]
91. Agarwal, S.; Plimpton, S.J.; Hughart, D.R.; Hsia, A.H.; Richter, I.; Cox, J.A.; James, C.D.; Marinella, M.J. Resistive memory device requirements for a neural algorithm accelerator. In Proceedings of the 2016 International Joint Conference on Neural Networks (IJCNN), Vancouver, BC, Canada, 24–29 July 2016; pp. 929–938. [CrossRef]
92. Ielmini, D.; Ambrogio, S. Emerging neuromorphic devices. *Nanotechnology* **2019**, *31*, 092001. [CrossRef]
93. Sun, Z.; Pedretti, G.; Mannocci, P.; Ambrosi, E.; Bricalli, A.; Ielmini, D. Time Complexity of In-Memory Solution of Linear Systems. *IEEE Trans. Electron Devices* **2020**, *67*, 2945–2951. [CrossRef]
94. Bryan, K.; Leise, T. The $25,000,000,000 Eigenvector: The Linear Algebra behind Google. *SIAM Rev.* **2006**, *48*, 569–581. [CrossRef]
95. Sun, Z.; Pedretti, G.; Ambrosi, E.; Bricalli, A.; Ielmini, D. In-Memory Eigenvector Computation in Time *O* (1). *Adv. Intell. Syst.* **2020**, 2000042. [CrossRef]
96. Pagiamtzis, K.; Sheikholeslami, A. Content-Addressable Memory (CAM) Circuits and Architectures: A Tutorial and Survey. *IEEE J. Solid State Circ.* **2006**, *41*, 712–727. [CrossRef]
97. Guo, Q.; Guo, X.; Bai, Y.; İpek, E. A resistive TCAM accelerator for data-intensive computing. In *Proceedings of the 44th Annual IEEE/ACM International Symposium on Microarchitecture—MICRO-44 '11*; ACM Press: Porto Alegre, Brazil, 2011; p. 339. [CrossRef]
98. Guo, Q.; Guo, X.; Patel, R.; Ipek, E.; Friedman, E.G. AC-DIMM: Associative Computing with STT-MRAM. In *Proceedings of the 40th Annual International Symposium on Computer Architecture*; Association for Computing Machinery: New York, NY, USA, 2013; pp. 189–200. [CrossRef]
99. Graves, C.E.; Li, C.; Sheng, X.; Miller, D.; Ignowski, J.; Kiyama, L.; Strachan, J.P. In-Memory Computing with Memristor Content Addressable Memories for Pattern Matching. *Adv. Mater.* **2020**, *32*, 2003437. [CrossRef] [PubMed]
100. Li, C.; Muller, F.; Ali, T.; Olivo, R.; Imani, M.; Deng, S.; Zhuo, C.; Kampfe, T.; Yin, X.; Ni, K. A Scalable Design of Multi-Bit Ferroelectric Content Addressable Memory for Data-Centric Computing. In Proceedings of the 2020 IEEE International Electron Devices Meeting (IEDM), San Francisco, CA, USA, 12–18 December 2020; pp. 29.3.1–29.3.4. [CrossRef]
101. Pedretti, G.; Graves, C.E.; Li, C.; Serebryakov, S.; Sheng, X.; Foltin, M.; Mao, R.; Strachan, J.P. Tree-based machine learning performed in-memory with memristive analog CAM. *arXiv* **2021**, arXiv:2103.08986.
102. Burr, G.W.; Shelby, R.M.; Sidler, S.; di Nolfo, C.; Jang, J.; Boybat, I.; Shenoy, R.S.; Narayanan, P.; Virwani, K.; Giacometti, E.U.; et al. Experimental Demonstration and Tolerancing of a Large-Scale Neural Network (165,000 Synapses) Using Phase-Change Memory as the Synaptic Weight Element. *IEEE Trans. Electron Devices* **2015**, *62*, 3498–3507. [CrossRef]
103. Jang, J.W.; Park, S.; Burr, G.W.; Hwang, H.; Jeong, Y.H. Optimization of Conductance Change in Pr_{1-x} Ca_x MnO_3 -Based Synaptic Devices for Neuromorphic Systems. *IEEE Electron Device Lett.* **2015**, *36*, 457–459. [CrossRef]
104. Wang, Z.; Ambrogio, S.; Balatti, S.; Sills, S.; Calderoni, A.; Ramaswamy, N.; Ielmini, D. Postcycling Degradation in Metal-Oxide Bipolar Resistive Switching Memory. *IEEE Trans. Electron Devices* **2016**, *63*, 4279–4287. [CrossRef]
105. Chen, P.Y.; Yu, S. Reliability perspective of resistive synaptic devices on the neuromorphic system performance. In Proceedings of the 2018 IEEE International Reliability Physics Symposium (IRPS), Burlingame, CA, 11–15 March 2018; pp. 5C.4–1–5C.4–4. [CrossRef]
106. Nardi, F.; Larentis, S.; Balatti, S.; Gilmer, D.C.; Ielmini, D. Resistive Switching by Voltage-Driven Ion Migration in Bipolar RRAM—Part I: Experimental Study. *IEEE Trans. Electron Devices* **2012**, *59*, 2461–2467. [CrossRef]
107. Yang, T.J.; Sze, V. Design Considerations for Efficient Deep Neural Networks on Processing-in-Memory Accelerators. In Proceedings of the 2019 IEEE International Electron Devices Meeting (IEDM), San Francisco, CA, USA, 7–11 December 2019; pp. 22.1.1–22.1.4. [CrossRef]
108. Pedretti, G.; Milo, V.; Ambrogio, S.; Carboni, R.; Bianchi, S.; Calderoni, A.; Ramaswamy, N.; Spinelli, A.S.; Ielmini, D. Stochastic Learning in Neuromorphic Hardware via Spike Timing Dependent Plasticity With RRAM Synapses. *IEEE J. Emerg. Sel. Top. Circ. Syst.* **2018**, *8*, 77–85. [CrossRef]
109. Brown, T.B.; Mann, B.; Ryder, N.; Subbiah, M.; Kaplan, J.; Dhariwal, P.; Neelakantan, A.; Shyam, P.; Sastry, G.; Askell, A.; et al. Language Models are Few-Shot Learners. *arXiv* **2020**, arXiv:cs.CL/2005.14165.

110. Wong, H.S.P.; Raoux, S.; Kim, S.; Liang, J.; Reifenberg, J.P.; Rajendran, B.; Asheghi, M.; Goodson, K.E. Phase Change Memory. *Proc. IEEE* **2010**, *98*, 2201–2227. [CrossRef]
111. Le Gallo, M.; Sebastian, A. An overview of phase-change memory device physics. *J. Phys. D Appl. Phys.* **2020**, *53*, 213002. [CrossRef]
112. Dieny, B.; Prejbeanu, I.L.; Garello, K.; Gambardella, P.; Freitas, P.; Lehndorff, R.; Raberg, W.; Ebels, U.; Demokritov, S.O.; Akerman, J.; et al. Opportunities and challenges for spintronics in the microelectronics industry. *Nat. Electron.* **2020**, *3*, 446–459. [CrossRef]
113. Ielmini, D.; Sharma, D.; Lavizzari, S.; Lacaita, A.L. Reliability Impact of Chalcogenide-Structure Relaxation in Phase-Change Memory (PCM) Cells—Part I: Experimental Study. *IEEE Trans. Electron Devices* **2009**, *56*, 1070–1077. [CrossRef]
114. Chang, C.; Wu, M.; Lin, J.; Li, C.; Parmar, V.; Lee, H.; Wei, J.; Sheu, S.; Suri, M.; Chang, T.; et al. NV-BNN: An Accurate Deep Convolutional Neural Network Based on Binary STT-MRAM for Adaptive AI Edge. In Proceedings of the 2019 56th ACM/IEEE Design Automation Conference (DAC), Las Vegas, NV, USA, 2–6 June 2019; pp. 1–6.
115. Hirtzlin, T.; Penkovsky, B.; Bocquet, M.; Klein, J.O.; Portal, J.M.; Querlioz, D. Stochastic Computing for Hardware Implementation of Binarized Neural Networks. *IEEE Access* **2019**, *7*, 76394–76403. [CrossRef]
116. Milo, V.; Malavena, G.; Monzio Compagnoni, C.; Ielmini, D. Memristive and CMOS Devices for Neuromorphic Computing. *Materials* **2020**, *13*, 166. [CrossRef]
117. Jerry, M.; Chen, P.; Zhang, J.; Sharma, P.; Ni, K.; Yu, S.; Datta, S. Ferroelectric FET analog synapse for acceleration of deep neural network training. In Proceedings of the 2017 IEEE International Electron Devices Meeting (IEDM), San Francisco, CA, USA, 2–6 December 2017; pp. 6.2.1–6.2.4. [CrossRef]
118. Tang, J.; Bishop, D.; Kim, S.; Copel, M.; Gokmen, T.; Todorov, T.; Shin, S.; Lee, K.T.; Solomon, P.; Chan, K.; et al. ECRAM as Scalable Synaptic Cell for High-Speed, Low-Power Neuromorphic Computing. In Proceedings of the 2018 IEEE International Electron Devices Meeting (IEDM), San Francisco, CA, USA, 1–5 December 2018; pp. 13.1.1–13.1.4. [CrossRef]
119. Guo, X.; Bayat, F.M.; Bavandpour, M.; Klachko, M.; Mahmoodi, M.R.; Prezioso, M.; Likharev, K.K.; Strukov, D.B. Fast, energy-efficient, robust, and reproducible mixed-signal neuromorphic classifier based on embedded NOR flash memory technology. In Proceedings of the 2017 IEEE International Electron Devices Meeting (IEDM), San Francisco, CA, USA, 26 December 2017; pp. 6.5.1–6.5.4. [CrossRef]
120. Kim, S.; Ott, J.A.; Ando, T.; Miyazoe, H.; Narayanan, V.; Rozen, J.; Todorov, T.; Onen, M.; Gokmen, T.; Bishop, D.; et al. Metal-oxide based, CMOS-compatible ECRAM for Deep Learning Accelerator. In Proceedings of the 2019 IEEE International Electron Devices Meeting (IEDM), San Francisco, CA, USA, 7–11 December 2019; pp. 35.7.1–35.7.4. [CrossRef]
121. Li, Y.; Fuller, E.J.; Sugar, J.D.; Yoo, S.; Ashby, D.S.; Bennett, C.H.; Horton, R.D.; Bartsch, M.S.; Marinella, M.J.; Lu, W.D.; et al. Filament-Free Bulk Resistive Memory Enables Deterministic Analogue Switching. *Adv. Mater.* **2020**, *32*, 2003984. [CrossRef]

electronics

MDPI

Review

Trap-Related Reliability Problems of Dielectrics in Memory Cells

Hiroshi Watanabe * and Hsin-Jyun Lin

Department of Electrical and Computer Engineering, National Yang Ming Chiao Tung University, Hsinchu 30010, Taiwan; sinjyunlin.eed03g@nctu.edu.tw
* Correspondence: hwhpnabe@mail.nctu.edu.tw

Abstract: A basic mechanism for storing data in memory cells is to record changes in electronic charges, material phases, resistivities, magnetic properties, and so forth. The change in electronic charge has been widely used in the majority of mass-produced memories, such as dynamic random-access memory (DRAM), static random-access memory (SRAM), NOR Flash, and NAND Flash. Other emerging memories have collected widespread attention for acquiring extra advantages which cannot be achieved using the change in electronic charge. Many years of studies have told us that reliability problems are critically important in the development of both conventional and emerging memories, in order to improve the product yield. However, the topics related to these problems are too wide to cover in these limited pages. In this review chapter, we address several interesting examples of trap-related problems in dielectrics for use in various memory cells. For engineering purposes, it is very important to grasp the relation of the achieved physical intuitions and electronic characteristics of dielectrics.

Keywords: dielectric; RTN; TAT; Wiener–Khinchin; transient analysis; phonon; surface roughness; spectral index; power spectrum

Citation: Watanabe, H.; Lin, H.-J. Trap-Related Reliability Problems of Dielectrics in Memory Cells. *Electronics* **2021**, *10*, 1287. https://doi.org/10.3390/electronics10111287

Academic Editor: Tae-Yeon Seong

Received: 7 April 2021
Accepted: 25 May 2021
Published: 28 May 2021

1. Introduction

There are two major innovations that human being have made since the appearance of mankind. One is the improvement of energy conversion efficiency. The other is the improvement of information communication efficiency. In the past few decades before 2020, semiconductor memory technologies have been regarded as belonging to the latter. However, as the energy that integrated circuit (IC) chips consume in data centers increases rapidly, the reduction of IC power consumption is also becoming meaningful for the global environmental problem. By multiplying the absolute temperature (T) to Shannon's entropy, we can obtain the equality of energy and information quantity, where p_i is the probability for state-i (that is, $0 \leq p_i \leq 1$ for each i) and k_B is the Boltzmann constant,

$$|\Delta E|_{min} = -k_B T \sum_i p_i lnnp_i. \tag{1}$$

During the period that we process information by $\sum_i p_i lnp_i$ at T, we consume energy at least by $|\Delta E|_{min}$. By shortening this period following device scaling [1], we have so far improved the efficiency of information communication. However, this also tells us that the minimum energy that we consume per time unit, named the lower bound of power consumption, increases with the efficiency of information communication. If this continues, then the improvement in communication efficiency may surpass the improvement in energy conversion efficiency. From 2010 to 2020, the worldwide energy use of data centers stayed around 1% [2]. However, in the next decade, there will be a risk of breaking the balance of efficiency improvement and the growth in energy use demands. This may be due to emerging applications in data intensive technologies such as artificial intelligence, automotive, and the internet-of-things. In response to this, memory technologies are being aggressively studied to reduce the power consumption of data centers [3]. From a similar viewpoint of power consumption, as well as big data accessibility [3], storage class memory is being extensively investigated [3–5]. In general, some part of the quantity of information must be noise in the wide range of electronic systems. The information entropy not related to the signal must be involved in the power consumption of memory systems. Therefore, basic research of the fluctuation sources caused by memory cells is critically important. In this chapter, we discuss electronic perturbations regarding dielectric films for use in memory cells.

2. Electronic Perturbation Is Very Small and Discrete

Following the law of large numbers, the average dominates fluctuations as the number of events increases. If the leakage current through a gate dielectric film is composed of tunneling events of electrons (depicted by the red and black arrows in Figure 1), then we can expect that the summed tunneling events turn out a continuous gate leakage current, as follows [6]: where q is the elementary charge , $\hbar$ is Dirac's constant, A is the gate area, $g(E)$ is the density-of-state for a given energy E, $D(E)$ is the tunneling probability, $f(E)$ is the Fermi distribution function, and V_g is the gate voltage applied to the top electrode, while the bottom electrode is grounded,

$$I = \frac{2\pi q}{\hbar} \int dA \int dE g(E) D(E) (f(E + qV_g) - f(E)). \tag{2}$$

In this equation, according to [6], the energy level in the cathode (bottom electrode) is higher by qV_g, because the energy level in the anode (top electrode) is lowered by qV_g. The Fermi distribution function is written as follows, where μ is the chemical potential,

$$f(E) = \frac{1}{1 + exp\left(\frac{E-\mu}{k_B T}\right)}. \tag{3}$$

In this figure, we assumed that there is an electronic perturbation, which is a local trap depicted by the yellow square. The trap-assisted tunneling may substantially enhance the tunneling, as depicted by the black bulk arrow. The red arrows depict the tunneling events without the trap-assistance. While the gate area (A) is large, we can regard the number of tunneling events as very large. If the tunneling events without the trap-assistance prevail, then (2) is valid.

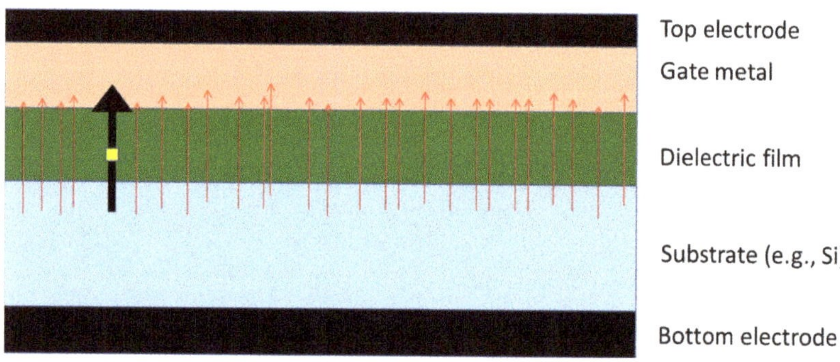

Figure 1. Continuum leakage [7]. The red arrows depict tunneling events with no trap-assistance. The black bulk arrow depicts tunneling event with trap-assistance. The yellow square depicts a local trap. The number of tunneling events with no trap-assistance prevails but the tunneling event is enhanced by the trap. As the gate area increases, the tunneling events with no trap-assistance dominate.

However, in real engineering, we divide cells by etching, as illustrated in Figure 2. If most of cells have no local trap, then the trap-assisting can substantially increase the leakage only in the cell with a local trap. With device scaling, the geometry of the memory cells is shrunk so that the leakage through the cell with a local trap can be described by a discrete formula, as follows: where $M_{i,f}$ is the transition matrix element from the initial state-i to the final state-f, and E_i and E_f are the energies of tunneling electrons at the initial state-i and the final state-f, respectively,

$$I = \sum_{i,f} \frac{2\pi q}{\hbar} \left| M_{i,f} \right|^2 \delta\left(E_i - E_f + \Delta E \right). \tag{4}$$

The E_i and E_f correspond to the energy levels in the cathode and anode electrodes in (2), respectively. If $\Delta E \neq 0$, then it is an inelastic process with the energy loss being ΔE. If $\Delta E = 0$, then it is an elastic process. The ΔE is not the applied voltage.

What is important is that a smaller memory cell is more affected by a local event. Thus, it turns out to be easier to investigate a discrete electronic perturbation by using smaller memory cells than larger ones. For either two-dimensional (2D) or three-dimensional (3D) integration, with or without specific dielectric films, this is the common nature of the electronic perturbation. In order to investigate a discrete electronic perturbation by using a larger memory cell, a higher precision is necessary in the measurement. This, however, is usually difficult.

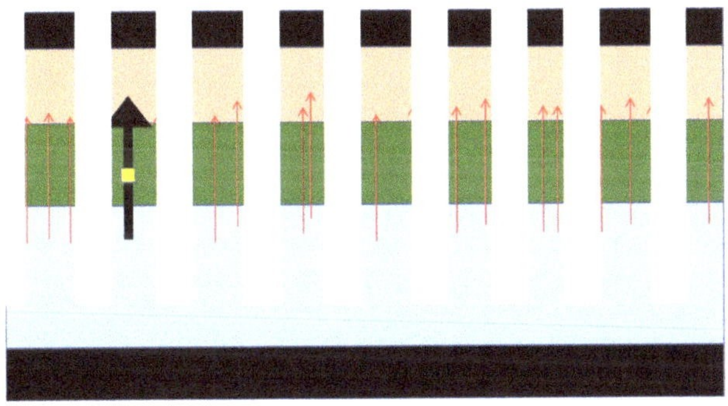

Figure 2. Discrete leakage [7]. After the etching, the capacitor is divided into many cells. The trap-assisted tunneling occurs through a cell having a local trap shows. The other cells are free from trap-assistance.

3. Discrete Electronic Perturbation Is Unstable

To carefully investigate a discrete electronic perturbation, a reliable oxide film is convenient for excluding non-targeted fluctuation factors, as far as possible. In [8], we prepared such a convenient sample using a 130 nm standard complementary metal-oxide-semiconductor (CMOS) process. Figure 3 contains an illustration and transmission electron microscope (TEM) images describing the sample geometry. The left-hand side (a) is a cross-sectional view along the cigar-shaped polysilicon above the substrate. The substrate surface is divided into two parts, labeled "S/D" and "G" by using the shallow trench isolation (STI). C1 is the capacitance between the cigar-shaped polysilicon and the substrate surface G. C2 is the capacitance between the cigar-shaped polysilicon and the substrate surface S/D. If the width of S/D is larger than that of G, then C2 is greater than C1, then the following capacitance coupling ratio (CR) becomes greater than 0.5.

$$CR = \frac{C2}{C1 + C2} > 0.5 \tag{5}$$

The center (b) is a bird's eye view of this sample. Along the perpendicular to (a), we have the source and drain diffusion layers, labeled "S" and "D", respectively, on the substrate surface S/D. These diffusion layers sandwich the channel. We have two diffusion layers labeled "G" on the substrate surface G. The diffusion layers, G, can connect to each other by the back-end of line (BEOL). The gate lengths on both substrate surfaces are the same as the width of the cigar-shaped polysilicon (130 nm). Thus, we can easily design CR by tuning the gate width of both substrate surfaces (S/D and G). In this geometry, the gate oxide on the substrate surface G can serve as a tunnel oxide, while $CR > 0.5$. If a positive voltage is applied to the diffusion layers S/D, then the vertical electric field is concentrated above the substrate surface G. When the voltage is large enough, an inversion layer is generated between the two diffusion layers G. Electrons coming from the diffusion layers G to the inversion layer can be injected into the cigar-shaped polysilicon by Fowler–Nordheim tunneling (FNT).

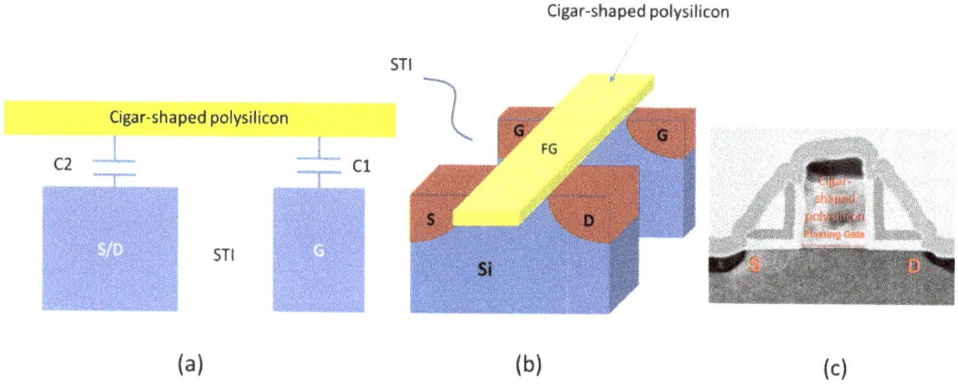

Figure 3. Sample structure. See Figures 3 and 5b of [8]. In (**a**), we show the cross-sectional view of the sample cell, in which a cigar-shaped polysilicon expands. The cigar-shaped polysilicon is a floating gate (FG). In (**b**), we show bird's view of the sample cell. The TEM image in (**c**) is the same as that of metal-oxide-semiconductor field-effect transistor (MOSFET) of the standard CMOS process.

The righthand side image (c) is a TEM image of a cross-section in the plane perpendicular to the cigar-shaped polysilicon. It is the same as the cross-sectional view of the metal-oxide-semiconductor field effect transistor (MOSFET) of the 130 nm standard CMOS process, except for having no contact with the cigar-shaped polysilicon. Since the cigar-shaped polysilicon can in this way serve as the floating gate (FG), this sample is useful for various experiments of embedded type non-volatile memories (e.g., a battery-less timer [8]). Since the standard CMOS process is used to fabricate it, the quality of oxide and the substrate interface is within the scope of mass-production. Furthermore, we can obtain a sufficient number of sample cells in a mass-production line in which the cell-to-cell variation is under control.

We investigated the current flow of electrons from the diffusion layers, G to FG. The measurement scheme is as follows. We chose a sample cell with the gate lengths of 0.52 µm and 0.4 µm in the substrate surface S/D and the substrate surface G, respectively. This means that $CR = 0.52/(0.4 + 0.52) \cong 0.565$. We applied a voltage to the source and drain diffusion layers S/D, while the substrate and the diffusion layers G were grounded. The substrate voltage was actually applied to the well contact, which is far from the channel between the diffusion layers S/D. Thus, we observed that $CR \times (applied\ voltage) - Vt$ approximates the FG potential to the inversion layer being generated between the diffusion layers G. While sweeping the voltage from zero to a voltage at which an abrupt increase appears (i.e., stopped by the ampere limiter), we measured the current flowing to the diffusion layers G by subtracting the substrate current from it. This procedure was repeated multiple times in succession for each cell. However, between any two sequential measurements of each cell, the voltage became zero once. The breakdown voltage of a 130 nm node is about 7.2 V [9]. The repeated applications of nearly 7 V to the tunnel oxide may generate an electronic perturbation. The oxide thickness is 3 nm or less in a 130 nm node [9]. This is a little bit thinner than necessary to generate the FNT at 10 MV/cm, which could cause the FNT in flash memories. The FNT is not caused by damage to the oxides. The tunnel gate area was about 0.052 μm^2 (=400 nm × 130 nm). The number of repetitions of voltage application was 30 for the measurement of each cell.

In Figure 4a, we show the results of the first measurement of a chosen cell (along the vertical axis) over the electric field until 11.3 MV/cm (along the horizontal axis). The signal of the measured current was lower than the measurement limit. This limit was due to the equipment noise, and not of interest in this work. The leakage, if it exists, can be regarded as direct tunneling with no trap-assistance, that is, tunneling not assisted by traps (i.e., direct tunneling), as illustrated in Figure 4b. Since we were interested in a discrete electronic perturbation in the oxide, we further continued the sequential measurement of this cell.

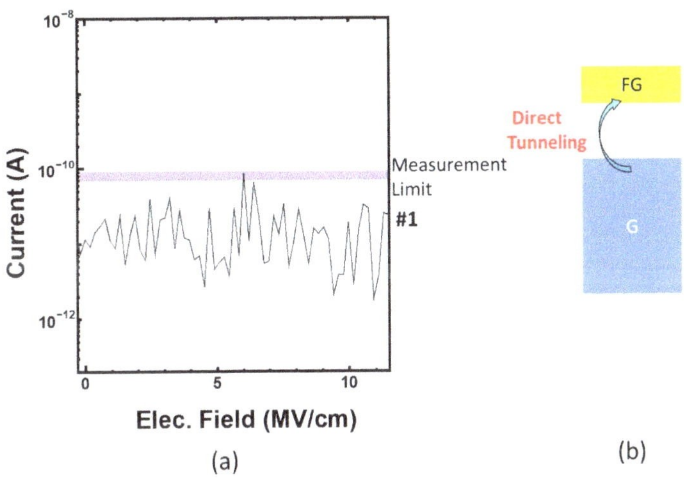

Figure 4. Experiments of stress-induced leakage current. See Figure 1 in [8]. In (**a**), the vertical axis is for the measured current and the horizontal axis is for the electric field applied to measure the current. The current obtained in the first measurement of the cell was smaller than the measurement limit. In (**b**), we illustrate the tunneling, which is tunneling not assisted by traps, that is, the direct tunneling. However, the direct tunneling is too small to be measurable with a small gate area.

In Figure 5a, we show the results of the 2nd to 18th measurements of the same cell as the first measurement. The vertical axis is for current and the horizontal is for electric field. A moderate increase of the current occurs from 5 MV/cm. This moderate increase may be attributable to 1-trap process in a local trap having been generated in the previous measurement (stressing by more than 10 MV/cm), as illustrated in Figure 5b. This is called the stress-induced leakage current (SILC) [10]. The leakage current is sensitive to the location and the energy level of the traps. Therefore, if the location and level of traps vary over the sequential measurement of the same cell, then the leakage current may vary in proportion to the fluctuation of the trap location and level. In (c), we consider the appearance of a second trap. If the location of the second trap is in a different tunnel path from that of the first trap, then it can increase the leakage current by about two-fold. Thus, the current of the 2nd to 18th measurements makes a bundle, as in Figure 5a. We, furthermore, continued the sequential measurement of the same cell.

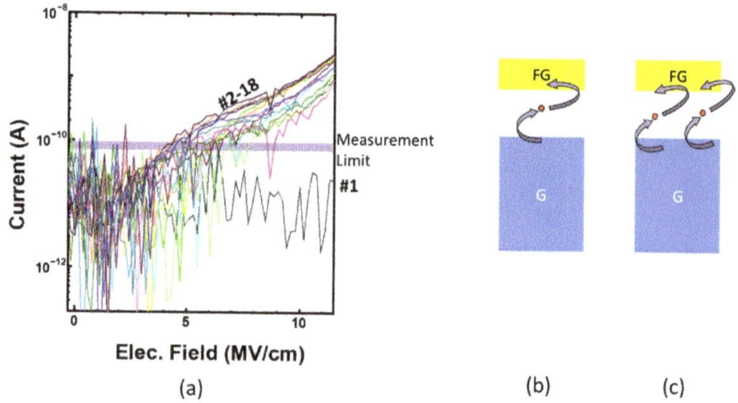

Figure 5. Experiments of stress-induced leakage current. See Figure 1 in [8]. In (**a**), the vertical axis is the measured current and the horizontal axis is the electric field applied to measure the current. The current obtained in the 2nd to 18th measurement of the same cell was larger than the measurement limit while the electric field was larger than 5 MV/cm. In (**b**), trap-assisted tunneling is illustrated using a trap. In (**c**), there are two traps, which are in different tunnel paths from each other.

In Figure 6a, we show the results of the 19th and 20th measurements of the same cell (along the vertical axis) over the electric field (along the horizontal axis). There was a discontinuous increase from the moderate increase by ten-fold in the current level in the 19th and 20th measurements of the same cell from 2 MV/cm. Ielmini et al. found that it is hard to explain this by using two traps which are in different tunnel paths [11]. Following them, we can regard this phenomenon as related to the two-trap process (i.e., two traps existing in a tunnel path [11]), as illustrated in Figure 6b. Accordingly, a second trap may be generated due to the stressing in the 18th measurement of the same cell. This implies that a local trap is a discrete perturbation. In (c), we can consider that there is another trap in a different tunnel path. However, the two-trap process may dominate it. We, furthermore, continued the sequential measurement of the same cell.

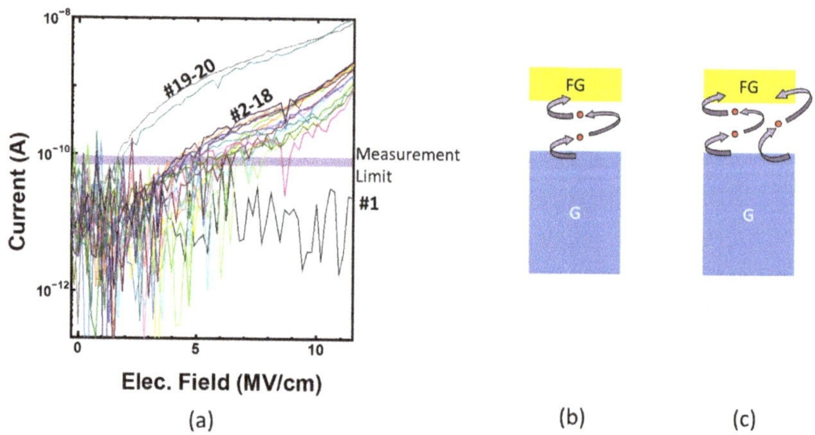

Figure 6. Experiments of stress-induced leakage current. See Figure 1 in [8]. In (**a**), the vertical axis is for the measured current and the horizontal axis is for the electric field applied to measure the current. The current obtained in the 19th and 20th measurement of the same cell was one-order larger than those obtained in the 2nd to 18th measurements of the same cell. In (**b**,**c**), we illustrate two-trap process for the mechanism for 19th to 20th measurement.

In Figure 7a, we show the results of the 19th and 20th measurements of the same cell along the vertical axis over the electric field along the horizontal axis. What is surprising is that the current became lower than the measuring limit in the 21st to 30th measurements of the same cell. It is possible that the traps (i.e., discrete electronic perturbations) became inactivated or vanished. Therefore, we can regard the current in the 21st to 30th measurements as tunneling not assisted by traps (i.e., returning back to #1), as illustrated in Figure 7b.

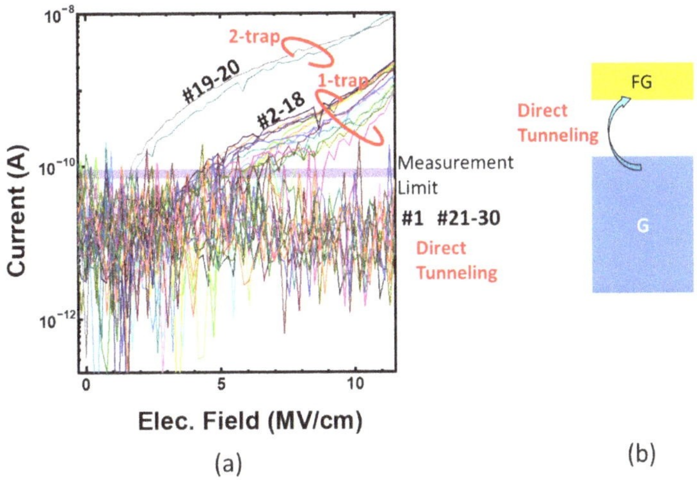

Figure 7. Experiments of stress-induce leakage current. See Figure 1 in [8]. In (**a**), the vertical axis is for the measured current and the horizontal axis is for the electric field applied to measure the current. The current obtained in the 21st to 30th measurements of the same cell was lower than the measurement limit. In (**b**), the mechanics for the 21st to 30th measurement goes back to the direct tunneling.

4. What Is a Local Trap?

This is a fairly difficult question because no one has observed (watched) one by a physical inspection method. As described in the above, the traps are unstable. It sometimes becomes activated, as shown in Figures 5a and 6a. On another occasion, it becomes inactivated, as shown in the 21st to 30th measurements of the same cell in Figure 7a. Suppose that one tries to inspect a sample having shown a discrete increase in the leakage current; how can we acquire a spot to be clipped for obtaining a cross-section wherein a trap exists? Even if we could successfully know in which cross-section a trap exists, can unstable traps still exist after segmenting the cross-section? Or will another trap be generated by the segmentation of the cross-section? If someone succeeds in solving this problem, then he or she could win the Nobel prize.

The difficulty in the above can increase as the size of sample cells decreases, because the average number of traps decreases. If the gate area increases, then the number of traps may increase. Is it helpful to solve this problem? As mentioned above, a higher precision in the measurement is necessary if the gate area increases. In [12], the atto-ampere measurement is demonstrated. Such a method may be helpful for inspecting what is veiled under the measurement limit. It can be investigated if more trapping processes are hidden below the measurement limit [13]. If yes, the leakage process, which has been considered as tunneling not assisted by traps in Figure 7, could be further divided into several trapping processes. In this event, the one-trap and two-trap processes may turn out to be three-trap and four-trap processes, respectively, for example. Nevertheless, the basic research of such unstable and discrete electronic perturbations is critically important for memory device technologies.

5. Trap-Related Phenomena

From experience, we have found that transient phenomena are more sensitive to discrete electronic perturbations. However, the phenomena due to discrete electronic perturbations are not always detectable. It may be observed occasionally while repeating the same measurement of the same cell, as mentioned above. In addition, if it is seen as an electric current change, it may accompany the continuous current with no relation to the electronic perturbations. A little ingenuity is necessary to study it.

There is a possible situation which is able to distinguish discrete electronic perturbations from a continuous current. In Figure 8, we illustrate stacked oxides with an interface layer (IL) sandwiched by two electrodes under a moderate electric field across the stacked oxides. The z-axis expands horizontally, and the conduction band edge is drawn on the vertical axis. The left electrode is the cathode that emits electrons to the right one (anode) due to direct tunneling (DT) along the z-axis. The IL expands over the X-Y plane perpendicular to the z-axis between the stacked oxides. In this exemplary illustration, the IL is composed of multiple interface traps having shallow trap levels. In (a), many electrons are stored in the IL, and thus the trap levels are piled up in the energy diagram. This increases the tunnel barrier and then suppresses the DT from the cathode to the anode, which is depicted by the red-dotted arrow. In this way, the tunnel barrier is modulated by stored charge in the traps [14]. The electric field across the oxide is enhanced on the anode side and weakened on the cathode side. The electrons likely are emitted from the IL to the anode in some way (e.g., Poole-Frenkel process [15]), while a sufficient amount of electrons are stored in the interface layer. In (b), the amount of electron charge in the IL was reduced as a result of the leakage, so that the trap levels became lower. The electric field was weakened on the anode side. In other words, the emission current reduces over time due to the emission itself, which is depicted by the red-bulk arrow. The transient current is composed of the emissions from the IL. The saturation current may be due to the DT with the reduced pile-up of the trap levels.

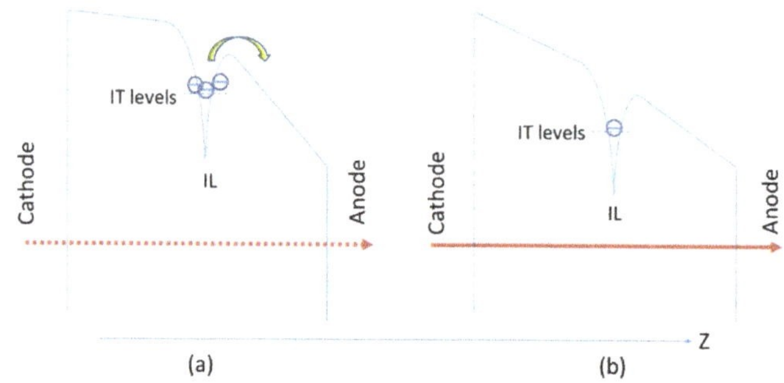

Figure 8. Emission and direct tunneling from an interface between the two oxide layers to be stacked. The horizontal axis depicts the z-axis, which is perpendicular to the interface (IL), and the energy band is drawn on the vertical axis. The IL has shallow trap levels, which are near to the oxide conduction band edge. In (**a**), the conduction band is piled up due to the charge of electrons captured in the IL. It suppresses the direct tunneling depicted by the red-dotted arrow. The Poole–Frenkel process may cause the leakage of the electrons from the IL. In (**b**), the reduction of electron charge in the IL recovers the direct tunneling depicted by the red-bulk arrow.

In Figure 9a, we can observe that the trap levels are lower than in Figure 8b. In such a case, trap-assisted tunneling (TAT) may likely occur. TAT is composed of the capture of electrons from the cathode by the IL due to DT, and the emission from the IL to the anode due to DT or FNT. The emission may be accompanied with energy loss if the TAT process is inelastic [16,17] (See ΔE in (2)). While the number of stored electrons in the IL is small, the emission is suppressed, and the capture is enhanced. When the number of stored electrons in the IL is large, the capture is suppressed, and the emission is enhanced. However, the emission may increase with the number of stored electrons and decrease with the decrease of stored electrons. This leads to a transient current due to the TAT. The saturation of TAT is achieved by the balance of the capture and the emission. The saturated TAT current, if it exists, must be much higher than the saturated DT current. The TAT can likely occur if the trap levels lower as the electric field increases.

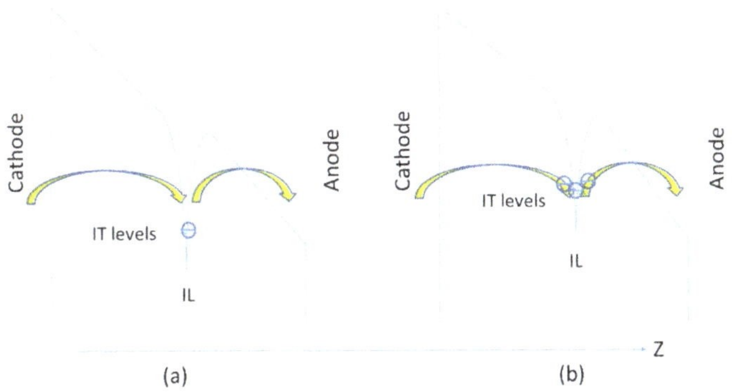

Figure 9. Trap-assisted tunneling (TAT). The *z*-axis expands horizontally, which is perpendicular to the interface (IL), and the conduction band edge is drawn in the vertical axis. The IL has a deeper trap level, which is far from the oxide conduction band edge. In (**a**), the IL has a smaller number of electrons. In (**b**), the IL has a larger number of electrons.

6. Power Spectrum Analysis of Discrete Electronic Perturbations

As we can understand from the discussion of Figures 8 and 9, the emission from the cathode or the IL is the key aspect that we have to analyze. The emission rate must be dependent of the electric field across the oxides, however, in this chapter, we a-priori assume that the emission rate is not explicitly dependent on time. The time dependence of the leakage current is assumed to be attributable only to the time-varying charge in the IL. In other words, we assume that the emission process is a stationary process. The autocorrelation of transient leakage current, $J_g(t)J_g{}^*(t-u)$, can be thus regarded as independent of time (t), where u is the lag. We can depict it as $r_{JJ}(u)$. We apply the Wiener–Khinchin theorem (6a) to analyze the power spectrum of the measured transient current, $S(f)$, where f is the frequency, $E[\cdots]$ depicts the expectation value of $\cdots$ [18–20]. In (6b), $S(f)$ is proportional to $1/f^{1+\alpha}$, where $1+\alpha$ is the spectral index, and ranges from 0.8 to 1.45 [21,22]. It was observed that the spectral index is sensitive to the location of electronic perturbations; at the interface or inside oxides [21,23].

$$S(f) = \int_{-\infty}^{\infty} E[r_{JJ}(u)]exp(-2\pi ifu)du \quad (6a)$$

$$S(f) \sim f^{-(1+\alpha)} \quad (6b)$$

When $\alpha = 0$, it is the $1/f$-fluctuation, i.e., Flicker noise [24–28]. When $\alpha = 1$, it is the $1/f^2$-fluctuation, i.e., Brownian noise [29]. In general, the transient current varies over the samples to be measured. Therefore, it is indispensable to exclude the sample variations from the measured transient leakage current [30]. In Figure 10a, there is an illustration of the measured sample of a metal/high-K stack/metal (MIM) capacitor. The top and bottom electrodes are composed of metal (TiN). The top electrode is applied with a gate voltage (V_g) to measure the current, while the bottom electrode is grounded. The high-K oxide stack is composed of a 5 nm TiO_2 layer and 10 nm ZrO_2 layer from the bottom. There may be an interface between the 10 nm ZrO_2 and the 5 nm TiO_2, which corresponds to the IL in Figures 8 and 9. The gate area is 2500 μm^2. The open probe fluctuation is about 10^{-10} A, which is equivalent to the measuring limit in Figures 4–7. Both the 10 nm ZrO_2 and the

5 nm TiO_2 are polycrystalline, as shown in the TEM image of Figure 10b. It was found that ZrO_2 is cubic or tetragonal and TiO_2 is rutile (tetragonal). Some of the grains inside the polycrystalline may face the IL. The distributions of grain orientations, grain sizes, and grain boundary geometry inside the oxide stack may cause the sample variations to be an obstacle for the power spectrum analysis.

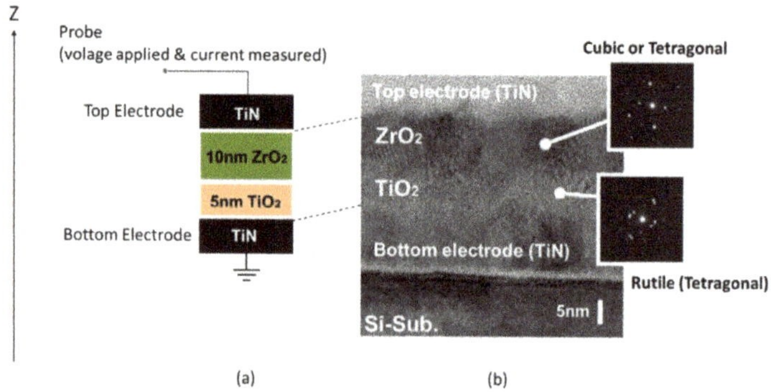

Figure 10. A sample of an MIM capacitor. See Figure 1 in [30]. The z-axis expands perpendicularly to the interface (IL) of 10 nm ZrO_2 having K = 40 and 5 nm TiO_2 having K = 130. The bottom and top electrodes are TiN. In (**a**), the cross-sectional view is illustrated along the z-axis. In (**b**), the TEM image corresponding to (**a**) is shown. We can observe that both ZrO_2 and TiO_2 are polycrystalline.

In Figure 11a,b, we plotted the transient current measured using nine samples (labeled #1–#9) having the same geometry as Figure 11a. The vertical axis is the measured current density and the horizontal is time. In (a), the transient current of the samples was measured under a positive fixed electric field (+0.5 MV/cm). In (b), the transient current of the samples was measured under a negative fixed electric field (−0.5 MV/cm). First of all, the plot shows that the sample-to-sample variation (named, the sample variation) was quite large. Next, we can see an uneven fluctuation (discrete electronic perturbations) in the data of the sample #1 under −0.5 MV/cm and so forth.

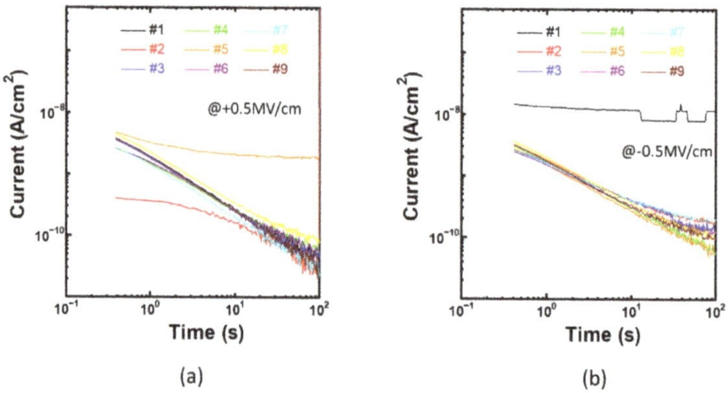

Figure 11. Measured transient leakage current using nine samples under +0.5 MV/cm and −0.5 MV/cm in (**a**,**b**), respectively. See Figure 2 in [30]. The vertical axis is the measured current density, and the horizontal is time. In (**b**), we have an uneven component (slow fluctuation due to electronic perturbation) in the sample #1. We can distinguish the measured data with the uneven component from those without the uneven component.

To analyze the electronic perturbation appropriately, we have to exclude the sample variation. In Figure 12, we illustrate the band diagram of Figure 10 with a moderate positive V_g applied, which corresponds to +0.5 MV/cm. The horizontal axis is the *z*-axis (depth direction) and the vertical axis is the conduction band edge of the stacking film sandwiched by the electrodes. From the bottom, there are three interface layers (IL-1, IL-2, and IL-3) at the interfaces between the bottom electrode (TiN) and the 5 nm TiO_2, between the 5 nm TiO_2 and the 10 nm ZrO_2, and between the 10 nm ZrO_2 and the top electrode (TiN), respectively. In this figure, we assumed no electrons were stored in the interface layers (i.e., the IL-1, the IL-2, and the IL-3). The barrier height of the 5 nm TiO_2 is 0.35 eV from the bottom electrode [31–34]. The barrier height of the 10 nm ZrO_2 from the top electrode is 0.94 eV [31–34]. In general, the dielectric constant decreases with the band gap (or tunnel barrier). Therefore, the electric field across the 10 nm ZrO_2 is higher than the 5 nm TiO_2, while we can ignore the stored electrons in the interface layers.

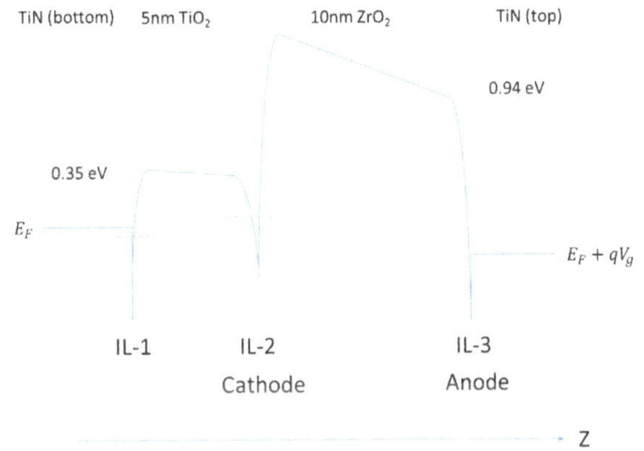

Figure 12. Band diagram with no electrons in the interface layers (IL-1, IL-2, IL-3) under a positive electric field. The horizontal axis is along the *z*-axis (depth direction) and the vertical axis is the conduction band edge of the stacking film sandwiched by the electrodes. There are three interface layers. Among them, IL-2 and IL-3 sandwich the ZrO_2, which dominates the tunnel resistance.

In a usual analysis, the IL-1 and IL-3 may be ignored by supposing that the electrons come from an electrode to the other electrode passing through homogenous oxides (having no sample variations). However, both the oxide stack and the interface layers must have sample variations due to grain-related distributions. In other words, for +0.5 MV/cm, an electron can be emitted from the IL-1 to the IL-2 through the 5 nm TiO_2 with sample variations and then be emitted to the IL-3 through the 10 nm ZrO_2 with sample variations. Due to the higher barrier and the thicker width, the 10 nm ZrO_2 dominates the serial resistance of this oxide stack. Therefore, for a positive electric field, to analyze the transient current, we must consider the emission of electrons from the IL-2 (i.e., the cathode interface) to the IL-3 (i.e., the anode interface). For a negative electric field, we must consider the emission of electrons from the IL-3 (i.e., the cathode interface) to the IL-2 (i.e., the anode interface). That is, the IL-2 and IL-3 are the cathode interfaces for a positive and negative electric field, respectively. The IL-2 and IL-3 are the anode interfaces for a negative and positive electric field, respectively. The DT may likely occur between the IL-2 and the IL-3 due to the high tunnel barrier under a moderate electric field. Thus, we have the sample variations (see Figure 11) that may be attributable to the distribution of grains inside the 10 nm ZrO_2 sandwiched by the IL-2 and IL-3. If we have a sufficient number of stored electrons in the IL-2, the band diagram is modulated to Figure 13. This corresponds to Figure 9b.

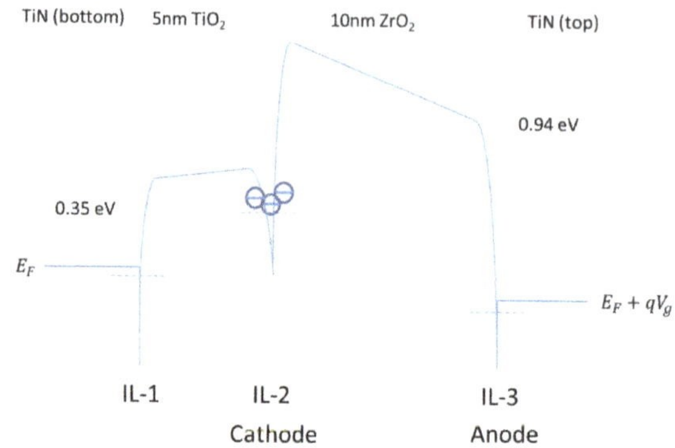

Figure 13. Band diagram with sufficient electrons stored in the cathode under a positive electric field. The horizontal axis is along the *z*-axis (depth direction) and the vertical axis is for the conduction band edge of the stacking film sandwiched by the electrodes.

The 10 nm ZrO_2 is a polycrystalline having variations of grain orientation, grain sizes, and the grain boundary patterns inside. The IL-2 and IL-3 may also be affected by the orientations, the sizes, and the boundary patterns of the grains facing the interfaces. This may be the source of the sample variations. In [30], we regulated these complex sources of sample variation and then obtained a formula to analyze the transient DT current density, as follows: where $J_{g,\infty}$ is the stationary DT current density between the IL-2 and the IL-3.

$$J_{DT}(t) \approx J_{g,\infty}\left(1 - s\,exp\left(-\frac{t}{\tau}\right)\right)^{-1} \tag{7}$$

We, a-priori, ignore the influence of the anode interface on the sample variations and then consider the grain-related variations in the cathode interface and the 10 nm ZrO_2. However, s is the average of surface roughness, which varies over grains facing the cathode interface, and τ is the average of the longest dwell time of electrons which are emitted from the cathode interface over the grains. The s is always positive and less than unity. As s is close to unity, the surface roughness reduces at the cathode interface.

Subsequently, suppose there is a trapping site inside the 10 nm ZrO_2, as illustrated in Figure 14. The horizontal axis is the *z*-axis (depth direction) and the vertical axis is the conduction band edge of the stacking film sandwiched by the electrodes. If the energy levels of the trapping sites are low enough, then the TAT may likely occur from the cathode interface (IL-2) to the anode interface (IL-3) via the trapping site. This may dominate the direct tunneling between the IL-2 and the IL-3. In a similar way to the derivation of (7), we obtained the formula to analyze the transient TAT current density, as follows: where $J'_{g,\infty}$ is the stationary TAT current density, τ' is the dwell time of electrons emitted from the trapping sites for the TAT process, and s' is a positive number.

$$J_{TAT}(t) \approx J'_{g,\infty}\left(1 - s'\,exp\left(-\frac{t}{\tau'}\right)\right)^{-1} \tag{8}$$

However, the trapping sites may also vary in depth in the energy diagram between samples. In one case, the electrons are stuck in the trapping sites. This cannot contribute to the measured leakage current. If s' is smaller than unity, the TAT dominates. Otherwise, the electron is stuck as a fixed charge there.

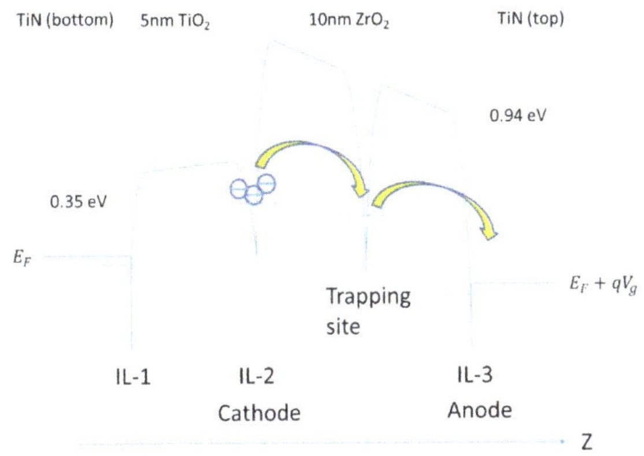

Figure 14. Band diagram with a trapping site inside the 10 nm ZrO_2 under a positive electric field. The horizontal axis is along the z-axis (depth direction) and the vertical axis is the conduction band edge of the stacking film sandwiched by the electrodes.

The parameters ($J_{g,\infty}$, τ, s, $J'_{g,\infty}$, τ', s') characterize the sample variations and not the discrete electronic perturbations. By tuning these parameters, we obtained an excellent agreement with the measured data, without uneven fluctuations (i.e., discrete electronic perturbations). An example is sample No. 6 (#6), shown in Figure 15. In (a), the vertical axis is the current density and the horizontal one is time. The measured data under the electric fields (0.5 MV/cm, 0.7 MV/cm, and 1 MV/cm) are depicted by black open circles. The lowest current can be excellently reproduced only by using (7) (i.e., DT only). The others can be excellently reproduced by using both (7) and (8) (i.e., both DT and TAT). The TAT can become involved as the electric field increases. In (b), the vertical axis is for the power spectrum densities and the horizontal one is for frequency. The black, red and blue open circles depict the power spectrum densities converted from the measured transient current densities at +0.5 MV/cm, +0.7 MV/cm, and +1.0 MV/cm using (6a), respectively. The calibrated α is 0.224, 0.321, and 0.408 at +0.5 MV/cm, +0.7 MV/cm, and +1.0 MV/cm using (6b), respectively. It increases as TAT becomes involved. The calibrated parameters are shown in Table 1. In Figure 16a, the measured transient current of samples No. 2 (#2), No. 6 (#6), and No. 8 (#8) under +0.5 MV/cm are depicted by black open circles. The vertical axis is the current density and the horizontal one is time. We obtained an excellent agreement by using only (7) (i.e., DT). In (b), the vertical axis is the power spectrum densities and the horizontal one is frequency. The red, black, and blue open circles depict the power spectrum densities converted from the measured transient current densities of samples #2, #6, and #8 at +0.5 MV/cm using (6a), respectively. The black open circles are the same as in Figure 15b. In addition, we observed that $\alpha = 0.813$ and 0.108 for #2 and #8 at +0.5 MV/cm using (6b), respectively. The calibrated parameters are also shown in Table 1.

Table 1. Calibrated fitting parameters for Figures 15 and 16. See Table 1 in [30]. In the lowest electric field (+0.5 MV/cm), the leakage current is dominated by the direct tunneling (DT). When increasing the electric field, trap-assisted tunneling (TAT) can become involved.

Sample No.	Field (MV/cm)	$J_{g,\infty}$ (pA/cm^2)	τ (s)	s	$J'_{g,\infty}$ (pA/cm^2)	τ' (s)	s'	Note	α
#2	0.5	17.5	120	0.9580	N/A	N/A	N/A	DT	0.813
#6	0.5	51	48	0.9885	N/A	N/A	N/A	DT	0.224
	0.7	56	75	0.9962	390	0.5	0.5	DT + TAT	0.321
	1.0	140	70	0.9920	860	0.1	0.1	DT + TAT	0.408
#8	0.5	87.5	32	0.9920	N/A	N/A	N/A	DT	0.108

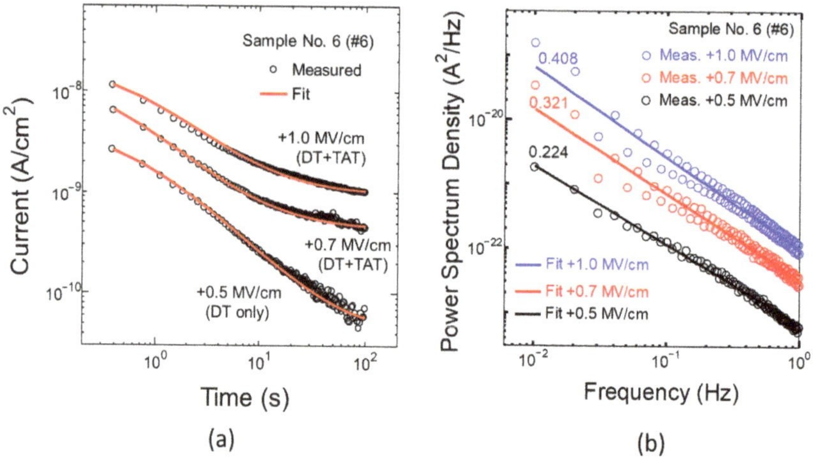

Figure 15. Fitting of the sample No. 6 (#6) under moderate electric fields (+0.5, +0.7, and +1.0 MV/cm). See Figure 5 in [30]. In (**a**), the vertical axis is the current density and the horizontal one is time. The marks depict the measured current and the red lines depict the fit using (7) and/or (7). In (**b**), the marks depict the power spectrum densities converted from the measurement using (6a). The lines fit the measurement for each electric field using (6b) and then we obtained $\alpha = 0.224,\ 0.321$, and 0.408 for +0.5 MV/cm, +0.7 MV/cm, and +1.0 MV/cm, respectively.

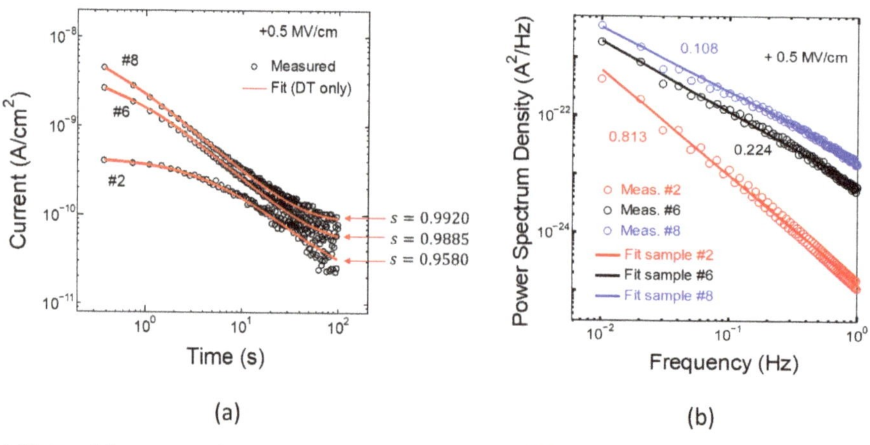

Figure 16. Fitting of the transient direct tunnel current. See Figure 6 in [30]. In (**a**), the vertical axis is the current density and the horizontal one is time. The marks depict the measured transient leakage current densities using sample #2, #6, and #8. The red lines depict the fit using (6a) for each sample. The calibrated s is 0.9580, 0.9885, and 0992 for #2, #6, and #8, respectively. In (**b**), the vertical axis is the power spectrum distribution and the horizontal one is the frequency. The marks were converted from measured data using (61). The red line is to fit α using (6b). The calibrated α is 0.813, 0.224, and 0.108, respectively.

In Figure 17, the vertical axis is α and the horizontal is the surface roughness index (s). The open dots depict those calibrated using sample #6 at +0.5 MV/cm, +0.7 MV/cm, and +1.0 MV/cm in Figure 16a,b, respectively. Since these transient currents can only be fit using (7) (i.e., DT only), the number of fitting parameters is three ($J_{g,}$, τ, and s). See Table 1. Let us choose $J_{g,\infty} = 51$ pA/cm^2 and $\tau = 48$ s from the sample #6 at +0.5 MV/cm in Table 1 (fit by DT). Regarding s as a variable, we can calculate the transient DT current using (7). Substituting it for (6a), we can calibrate α using (6b) for a given s, as plotted using the line in Figure 17. It is thus found that α increases as s decreases from unity (i.e., no surface roughness). We can thus find that the surface roughness violates the power law of the Flicker noise ($\alpha = 0$). See Table 1 again. In sample #2 at +0.5 MV/cm, the transient current was reproduced using only (7) (i.e., only DT) and $\alpha = 0.813$, which is very high. However, $s = 0.9580$, which is very small compared with the others. This means that #2 has a larger surface roughness. According to the trend in Figure 17, we can assume that α would reduce from 0.813, as the surface roughness could have been suppressed in sample #2. If we can reduce the surface roughness, then α may also have decreased in samples #6 and #8 at +0.5 MV/cm from 0.224 and 0.108, respectively. Thus, we can observe that $\alpha \cong 0$ (i.e., the spectral index is 1) when the electrons are directly emitted from the cathode to the anode interfaces (i.e., DT). The DT turns out the Flicker noise and the surface roughness violates the power of the Flicker noise. In sample #6, furthermore, α increases due to the TAT as the electric field increases. The TAT also violates the power law of the Flicker noise.

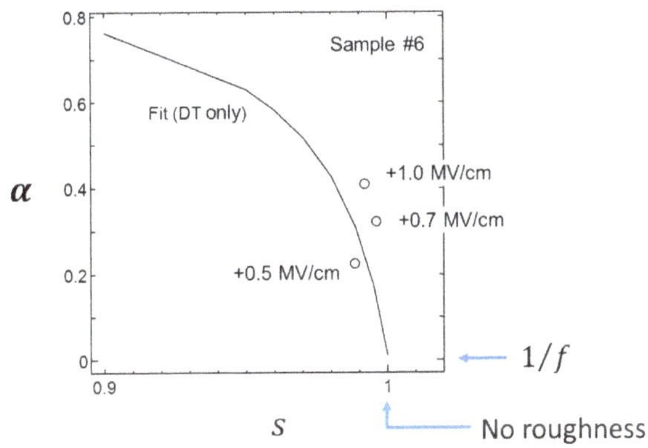

Figure 17. Impact of surface roughness. See Figure 7a in [30]. The parameters (α and s) obtained by fitting the measurement are plotted using a mark for the sample #6 at +0.5 MV/cm, +0.7 MV/cm, and +1.0 MV/cm. The line depicts the theoretical prediction using $J_{g,\infty}$ and τ, calibrated to fit the measured transient current for sample #6 at +0.5 MV/cm. The surface roughness violates the power law of the Flicker noise so that α increases as s decreases.

As mentioned above, we can observe a fluctuation which may be discrete and subject to neither (7) nor (8). For example, there is an uneven transient current for sample No. 1 (#1) under −0.5 MV/cm in Figure 11b. This measured data is replotted using lines and dots in Figure 18a. The vertical axis is the current density and the horizontal one is time. There are three discrete perturbations. In (b) of this figure, by tuning the fitting parameters of (7) and (8), we can obtain the continuous transient current depicted by the red dash line, with the calibrated fitting parameters shown in Table 2. By subtracting it from the measured transient current (depicted dot and line), we obtained the red line. It appears as the summation of a stationary current ($\cong 10^{-8}$ A) and the uneven component of the fluctuating current with no continuous decay. The calibration of the parameters was carried out to exclude the continuous decay from the red line. That is, we can deduce the uneven component from the measured data. We considered the uneven components as random-telegraph noise (RTN) in the transient leakage current [35]. Machlup assumed bi-states (i.e., discrete) and used (6a) and (6b) to study RTN [18]. Its origin has been considered to be related to number fluctuation [24], to mobility fluctuation [36–38], to both number and mobility fluctuations [39–41], and phonon scattering [42].

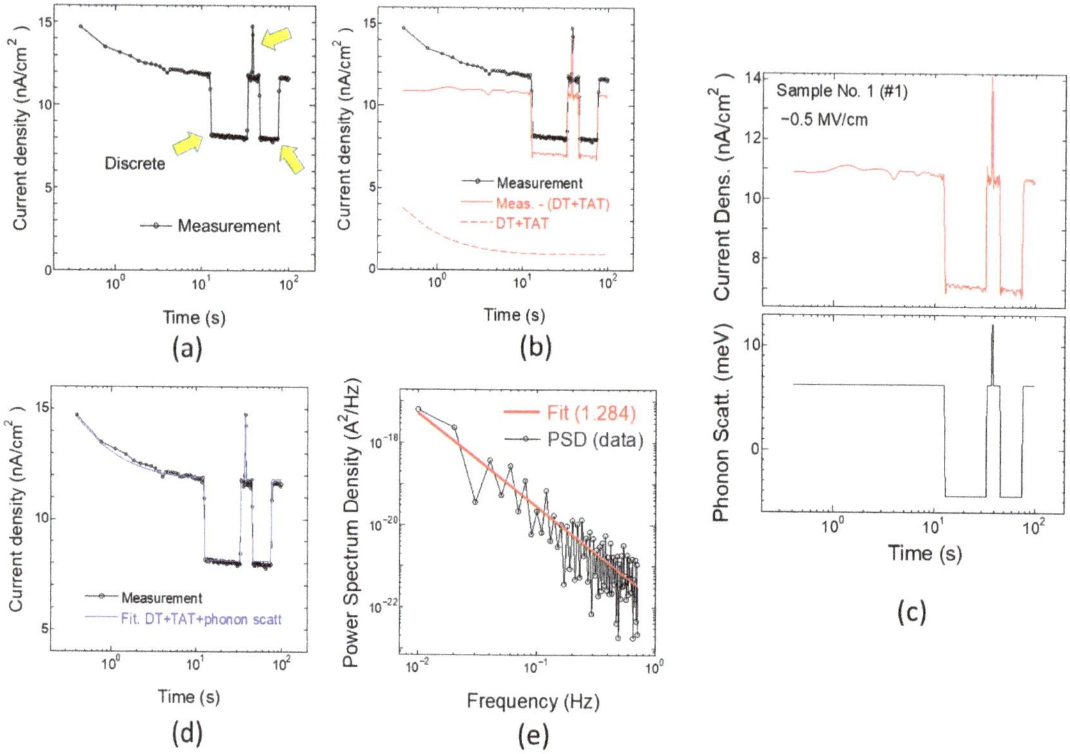

Figure 18. Fit uneven fluctuation with sample No. 1 (#1) under −0.5 MV/cm. See Figure 8 in [30]. In (**a**,**b**,**d**) and the upper of (**c**), the vertical axis is the current density and the horizontal one is time. In the bottom of (**c**), the vertical axis is phonon scattering energy and the horizontal one is time. In (**e**), the vertical axis is the power spectrum density and the horizontal one is the frequency. For the calibration of parameters, we synchronized the occurrence of discrete perturbations and phonon scattering. The marks in (**e**) are converted from measured transient leakage current density using (6a). By fitting the marks, we could extract α using (6b).

Table 2. Calibrated fitting parameters with uneven fluctuation. See Table 2 in [30]. The trap-assisted tunneling (TAT) can be involved in a cell and not in another cell at the same electric field. See #1 and #3 at −0.5 MV/cm. In #1, the TAT is involved, but not in #3. It may be observed that the trap levels of the interface layer are lower in #1 than in #3.

Sample No.	Field (MV/cm)	$J_{g,\infty}$ (pA/cm^2)	τ (s)	s	$J'_{g,\infty}$ (pA/cm^2)	τ' (s)	s'	Note	α
#1	−0.5	160	9.6	0.9880	798	0.1	0.1	DT + TAT	1.284
#3	−0.5	25.5	74.6	0.9950	N/A	N/A	N/A	DT	1.322
#4	1.0	114	50	0.9950	343	0.1	0.1	DT + TAT	1.289
#7	0.7	184	21.8	0.9870	2020	1	0.8	DT + TAT	0.927
#8	−0.7	47.1	85	0.9960	588	12	0.2	DT + TAT	1.367

Let us remember that multiple grains of the 10 nm ZrO_2 may face either of the cathode or anode interfaces. Accordingly, there may be grain boundary patterns at the interfaces. Some traps are on the grain boundaries and others are out of the grain boundaries at the interfaces. We can suppose that the details of trap states are more complicated on the grain boundaries. This may cause the barrier height fluctuations ($\Delta\varphi_c$ and $\Delta\varphi_a$) in the cathode and anode interfaces, respectively. The electrons captured by the trap sites on the grain boundaries may change the trap states by emitting and absorbing the phonon energy (ΔE_p). For the phonon absorption, ΔE_p is positive. For the phonon emission, ΔE_p is negative. Thus, the dwell time of electrons stored in the cathode interface may fluctuate, as follows: where the suffixes of $\pm$ depict the positive and negative electric fields, respectively, and $\tau_{\pm 0}$ is the dwell time without the uneven component.

$$\tau_{\pm} = \tau_{\pm 0} \exp\left(\frac{-\Delta\varphi_c + \Delta E_{\mathrm{p}}}{\Delta_c} + \frac{-\Delta\varphi_a + \Delta E_{\mathrm{p}}}{\Delta_a} \right) \tag{9}$$

The denominators of the terms in the exponent of (9) are written as follows: where m^* is the effective tunnel mass and assumed to be half the rest of the electron mass, t_I is the thickness of the insulator (10 nm in this case), EC_c and EC_a are the conduction band edges in the cathode and anode interfaces, respectively, and u^* is the compensation of u.

$$\Delta_u = -\frac{3}{2} \frac{\hbar}{\sqrt{m^*}} \frac{1}{t_I} \frac{(EC_u - EC_{u^*})^2}{EC_u^{3/2} - 3EC_u^{1/2} EC_{u^*} + EC_{u^*}^{3/2}} \tag{10}$$

If $u^* = a$, then $u = c$ and vice versa. In the bottom plot of Figure 18c, by using these parameters related to phonon scattering, we can extract ΔE_p in the vertical axis over time in the horizontal axis. The upper is the replot of the uneven component depicted by the red line in (b). However, for the calibration of parameters in (9) and (10), we synchronized the occurrence of phonon scattering in our model and that of discrete perturbations in the extracted uneven component (i.e., in the measurement). In (d) of this figure, we obtained an excellent agreement with the measurement by using (7)–(10) with the phonon scattering (i.e., RTN), as depicted by the blue line. The calibrated parameters for phonon scattering are shown in Table 3 (see #1 therein). In (e), furthermore, we can convert this measurement to the power spectrum density in the frequency domain using (6a). We can thus extract $\alpha = 1.284$ using (6b).

Table 3. Calibrated parameters related to phonon scattering. See Table 3 in [30]. The barrier height fluctuation varies over samples around 100 meV.

Sample No.	Field (MV/cm)	$\Delta\varphi_c$ (meV)	$\Delta\varphi_a$ (meV)	Δ_c	Δ_a
#1	−0.5	58.22	104.74	29.47	206.57
#3	−0.5	109.22	74.6	29.47	206.57
#4	1.0	153.87	185.41	30.59	179.50
#7	0.7	85.59	143.39	29.73	205.42
#8	−0.7	134.90	119.21	29.73	205.42

We also observed an uneven transient current with samples #3 at −0.5 MV/cm, #4 at +1.0 MV/cm, #7 at +0.7 MV/cm, and #8 at −0.7 MV/cm. In a similar manner, using (7)–(10), we obtained an excellent agreement for the measured transient current densities, as shown in Figure 19. The vertical axis is current density and the horizontal one is time. In (a), we plot the fitting results of #1 at −0.5 MV/cm (replot of Figure 18d), #3 at −0.5 MV/cm, and #4 at +1.0 MV/cm. In (b), we plot the fitting result of #7 at +0.7 MV/cm and #8 at −0.7 MV/cm. The calibrated parameters are shown in Tables 2 and 3. Subsequently, using the same method to obtain Figure 18c, we extracted phonon energies using these samples, as shown in Figure 20. In the upper plots, we show the discrete perturbation (i.e., uneven component) over time. In the bottom plot, we show the extracted ΔE_p over time. However, for the calibration of parameters in (9) and (10), we synchronized the occurrence of discrete electronic perturbations and phonon scattering. The α extracted using the same method to obtain Figure 18e was around 1.0, as shown in Table 2. Thus, we can find that phonon scattering generates randomness due to the $1/f^2$-fluctuation, that is, Brownian noise.

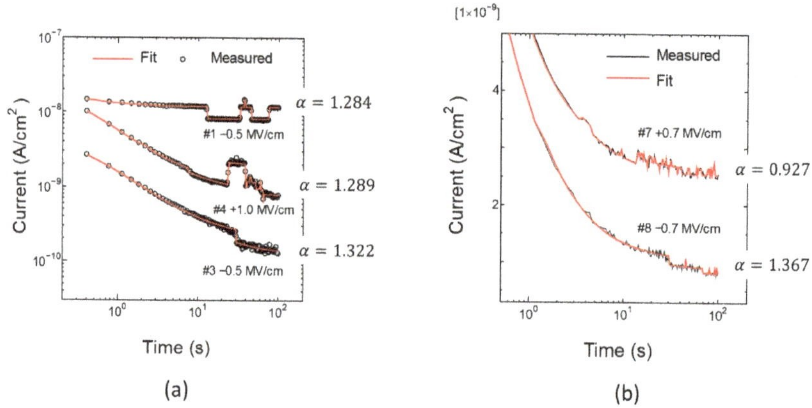

Figure 19. Fit uneven fluctuations. See Figures 8 and 9 in [30]. The vertical axis is the current density and the horizontal one is time. In (**a**), we compare the measured and calculated data for samples #1 at −0.5 MV/cm, #4 at +1.0 MV/cm, #3 at −0.5 MV/cm. In (**b**), we compare the measured and calculated data for samples #7 at +0.7 MV/cm, and #8 at −0.7 MV/cm. Excellent agreement was obtained using (7)–(10) and the calibrated parameters in Tables 2 and 3. The extracted α was 1.284, 1.289, 1.322, 0.927, and 1.367 for samples #1 at −0.5 MV/cm, #4 at +1.0 MV/cm, #3 at −0.5 MV/cm, #7 at +0.7 MV/cm, and #8 at −0.7 MV/cm, respectively.

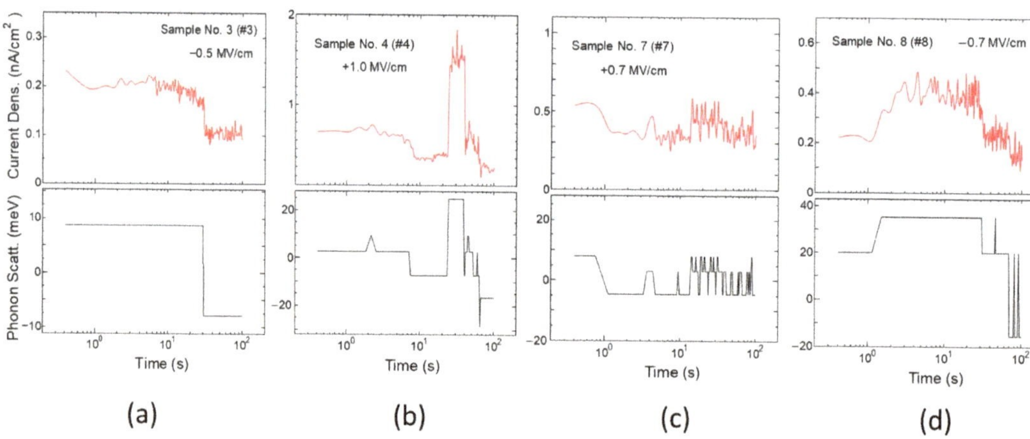

Figure 20. Extracted phonon energies. See Figure 10 in [30]. In the upper plots, the vertical axis is the uneven component of measured transient current density (discrete electronic perturbations) and the horizontal one is time. In the bottom plots, the vertical axis is phonon scattering energies and the horizontal one is time. (**a**–**d**) for the samples #3 at −0.5 MV/cm, #4 at +1.0 MV/cm, #7 at +0.7 MV/cm, and #8 at −0.7 MV/cm, respectively. For the calibration of parameters, we synchronized the occurrence of discrete electronic perturbations and phonon scattering.

By comparing with the theoretical infrared phonon scattering [43], we can suppose that if $\Delta E_p > 10$ meV then grains are cubic or tetragonal, which is consistent with the 10 nm ZrO_2 of the samples used in the present measurement. See Figure 10b. Otherwise, they are monoclinic grains. In this way, we can deduce some physical intuitions from the phonon scattering analysis. By repeating the measurement of the same samples with the same conditions, sometimes we could obtain such uneven fluctuations. On other occasions, we could not obtain it since the discrete electronic perturbations were unstable, as we have shown in Figures 4–7.

However, it may be still an open problem how phonons scatter with a discrete local trap. Grasser et al. reported a time-dependent analysis of metastable defect states [44] and Ielmini et al. performed a structure relaxation due to trapped holes [45]. It would be interesting to know how or if such details are related to phonon scattering, as we could extract some more detailed physical intuitions from phonon scattering energies.

Nowadays, there are many kinds of semiconductor memory. As conventional ones with MIM capacitors, we have static random-access memory (SRAM), dynamic random-access memory (DRAM), NOR flash, and NAND flash. As emerging types, we have phase-change random access memory (PCRAM), magnetic random access memory (MRAM), resistivity random access memory (RRAM), and so forth. The analysis of spectral index will be critically important for investigating various kinds of reliability issues, not only in the conventional ones, but also in the emerging ones. For example, a complex RTN was reported in RRAM [23,46].

7. Conclusions

We consider that sample variation is attributable to grain distributions in stacking oxide films. We thus proposed the formulae to eliminate sample variations from measured transient gate leakage current flowing through stacking oxide films. Thanks to this, we could analyze the discrete electronic perturbations which are seen in measured transient gate leakage current.

The transient gate leakage current has unstable discrete electronic perturbations that are related to local traps existing at the interfaces of the gate stack and inside. By repeatedly measuring the same samples with the same conditions, sometimes we could observe this, but not on other occasions. By assuming that the emission from traps is a stationary process, power spectrum analysis shows that the direct tunneling is subject to the $1/f$-fluctuation, that is, the Flicker noise. However, the trap-assistance and the interface roughness break the power law of the $1/f$-fluctuation. The random telegraph noise due to phonon-scattering shows a $1/f^2$-fluctuation, that is the Brownian noise. Even though it is hard to observe a local trap by physical inspection, we may have a chance to characterize it by using transient analysis.

Funding: The second half of this research was funded by Tokyo Electron Ltd. (Tokyo, Japan).

Acknowledgments: The second half of the present work was performed with respect to the joint development agreement between Tokyo Electron Ltd., and the National Yang Ming Chiao Tung University [30]. The authors would like to thank K. Akiyama, Y. Hirota, Y. Akasaka, G. Nakamura, H. Nagai, T. Morimoto, Y. Morozumi, S. Hsihiya, F. Hayase, K. Harada. The first half of the present work was performed with T. Ushijima, N. Hagiwara, C. Okada, and T. Tanabe [8]. National Chiao Tung University was combined with National Yang Ming University to be National Yang Ming Chiao Tung University since 1 February 2021.

Conflicts of Interest: The authors declare no conflict of interest.

References

1. Dennard, R.H.; Gaensslen, F.H.; Yu, H.-N.; Rideout, V.L.; Bassous, E.; LeBlanc, A.R. Design of ion-implanted MOSFET's with very small physical dimensions. *IEEE J. Solid-State Sicruits* **1974**, *9*, 256–268.
2. Masanet, E.; Shhehabi, A.; Lei, N.; Smith, S.; Koomey, J. Recalibrating global data center energy-use estimates. *Science* **2020**, *367*, 984–986. [CrossRef]
3. Moras, G.; Robayo, D.A.; Lopez, J.M.; Grenouillet, L.; Carabasse, C.; Navarro, G.; Sabbione, C.; Bernard, M.; Cagli, C.; Castellani, N.; et al. Crosspoint memory arrays: Principle, strengths and challenges. In Proceedings of the 2020 IEEE International Memory Workshop, Dresden, Germany, 17–20 May 2020.
4. Dierling, K.; Das, M. GPUs, DPUs, and Storage: Bringing AI and ML to data centers everywhere. In Proceedings of the Flash Memory Summit 2020, Santa Clara, CA, USA, 10–12 November 2020.
5. Ilkbahar, A. Intel®® Optane™ Persistent Memory from vision to reality. In Proceedings of the Flash Memory Summit 2020, Santa Clara, CA, USA, 10–12 November 2020.
6. Burstein, E.; Lundqvist, S. (Eds.) *Tunneling Phenomena in Solids*; Lectures Presented at the 1967 NATO Advanced Study Institute at RiO, Denmark; Plenum: New York, NY, USA, 1969; p. 35.

7. Watanabe, H. Tutorial: Trap-related reliability issues in NAND Flash memory. In Proceedings of the IEEE International Reliability Physics Symposium, Anaheim, CA, USA, 15–19 April 2012.
8. Watanabe, H.; Ushijima, T.; Hagiwara, N.; Okada, C.; Tanabe, T. Integrated Batteryless Electron Timer. *IEEE Trans. Electron Devices* **2010**, *58*, 792–797. [CrossRef]
9. Marichal, O.; Wybo, G.; van Camp, B.; Vanysacker, P.; Keppens, B. SCR-based ESD protection in nanomerter SOI technologies. *Microelectron. Reliab.* **2007**, *47*, 1060–1068. [CrossRef]
10. Naruke, K.; Taguchi, S.; Wada, M. Stress induced leakage current limiting to scale down EEPROM tunnel oxide thickness. In Proceedings of the Technical Digest of the 1988 IEEE International Electron Devices Meeting, San Francisco, CA, USA, 11–14 December 1988; pp. 424–427.
11. Ielmini, D.; Spinelli, A.S.; Lacaita, A.L.; Modelli, A. A new two-trap tunneling model for the anomalous stress-induced leakage current (SILC) in Flash memories. *Microelectron. Eng.* **2001**, *59*, 189–195. [CrossRef]
12. Inatsuka, T.; Kumagai, Y.; Kuroda, R.; Teramoto, A.; Suwa, T.; Ohmi, S.S.T. A test circuit for extremely low gate leakage current measurement of 10 aA for 80,000 MOSFETs in 80s. *IEEE Trans. Semicond. Manuf.* **2013**, *26*, 288. [CrossRef]
13. Köcher, A.; Tohoku University in those days. Hiroshima University at present, Sendai, Japan. Personal communication, 2012.
14. Watanabe, H.; Matsushita, D.; Muraoka, K.; Kato, K. Universal tunnel mass and charge trapping in $[(SiO_2)_{1-x}(Si_3N_4)x]_{1-y}Siy$ film. *IEEE Trans. Electron Devices* **2010**, *57*, 1129–1136. [CrossRef]
15. Simmons, J.G. Poole-Frenkel Effect and Schottky Effect in Metal-Insulator-Metal Systems. *Phys. Rev.* **1967**, *155*, 657–660. [CrossRef]
16. Takagi, S.; Yasuda, N.; Toriumi, A. Experimental evidence of inelastic tunneling and new I-V model for stress-induced leakage current. In Proceedings of the Technical Digest of the 1996 IEEE International Electron Devices Meeting, San Francisco, CA, USA, 8–11 December 1996; pp. 323–326.
17. Wu, J.; Register, L.F.; Rosenbaum, E. Trap-assisted tunneling current through ultra-thin oxide. In Proceedings of the 37th IEEE International Reliability Physics Symposium, San Diego, CA, USA, 23–25 March 1999; pp. 389–395.
18. Machlup, S. Noise in semiconductors: Spectrum of a two-parameter random signal. *J. Appl. Phys.* **1954**, *25*, 341–343. [CrossRef]
19. Dutta, P.; Horn, P.M. Low-frequency fluctuations in solids: 1/f-noise. *Rev. Mod. Phys.* **1981**, *53*, 497–516. [CrossRef]
20. Bak, P.; Tang, C.; Wiesenfeld, K. Self-organized criticality: An explanation of the 1/f noise. *Phys. Rev. Lett.* **1987**, *59*, 381–384. [CrossRef]
21. Weissman, M.B. 1/f noise and other slow, nonexponential kinetics in condensed matter. *Rev. Mod. Phys.* **1988**, *60*, 537–571. [CrossRef]
22. Kogan, S. *Electronic Noise and Fluctuations in Solids*; England Cambridge University Press: Cambridge, UK, 1996; pp. 205–208.
23. Raghavan, N.; Degraeve, R.; Fantini, A.; Goux, L.; Strangio, S.; Govoreanu, B.; Wouters, D.J.; Groeseneken, G.; Jurczak, M. Microscopic origin of random telegraph noise fluctuations in aggressively scaled RRAM and its impact on read disturb variability. In Proceedings of the IEEE IRPS, Monterey, CA, USA, 14–18 April 2013.
24. McWhorter, A.L. *Semiconductor Surface Physics*; Kingston, R.H., Ed.; University of Pennsylvania Press: Philadelphia, PA, USA, 1957.
25. Christensson, S.; Lundström, I.; Svensson, C. Low frequency noise in MOS transistors—I. theory. *Solid-State Electron.* **1968**, *11*, 797–812. [CrossRef]
26. Christensson, S.; Lundström, I. Low frequency noise in MOS transistors—II. Experiments. *Solid-State Electron.* **1968**, *11*, 813–820. [CrossRef]
27. Uren, M.J.; Day, D.J.; Kirton, M.J. $1/f$ and random telegraph noise in silicon metal-oxide-semiconductor field-effect transistors. *Appl. Phys. Lett.* **1985**, *47*, 1195. [CrossRef]
28. Kirton, M.J.; Uren, M.J. Noise in solid-state microstructures: A new perspective on individual defects, interface states and low frequency ($1/f$) noise. *Adv. Phys.* **1989**, *38*, 367. [CrossRef]
29. Boyat, A.; Joshi, B. A review paper: Noise models in digital image processing. *Signal Image Process. Int. J.* **2015**, *6*, 63–75. [CrossRef]
30. Lin, H.-J.; Akiyama, K.; Hirota, Y.; Akasaka, Y.; Nakaumura, G.; Nagai, H.; Morimoto, T.; Watanabe, H. Experimental study of $1/f^{1+\alpha}$ noise in transient leakage current of metal-insulator-metal with stacked high-K polycrystalline films. *IEEE Trans. Electron Devices* **2020**, *67*, 2503–2509. [CrossRef]
31. Robertson, J. Band offsets of high dielectric constant gate oxides on silicon. *J. Non-Cryst. Solids* **2002**, *303*, 994–1000. [CrossRef]
32. Didden, A.; Battjes, H.; Machunze, R.; Dan, B.; van de Krol, R. Titanium nitride: A new ohmic contact material for n-type CdS. *J. Appl. Phys.* **2011**, *110*, 033717. [CrossRef]
33. Williams, R. Photoemission of electrons from silicon to silicon dioxide. *Phys. Rev.* **1965**, *140*, A569–A575. [CrossRef]
34. Watanabe, H. Transient device simulation of floating gate nonvolatile memory cell with a local trap. *IEEE Trans. Electron Devices* **2010**, *57*, 1873–1882. [CrossRef]
35. Chang, C.M.; Chung, S.S.; Hsieh, Y.S.; Cheng, L.W.; Tsai, C.T.; Ma, G.H.; Chien, S.C.; Sun, S.W. The Observation of Trapping and Detrapping Effects in High-k Gate Dielectric MOSFETs by a New Gate Current Random Telegraph Noise (IG-RTN) Approach. In Proceedings of the Technical Digest of the 2008 IEEE International Electron Devices Meeting, San Francisco, CA, USA, 15–17 December 2008.
36. Hooge, F.N. 1/f noise is no surface effect. *Phys. Lett. A* **1969**, *29*, 139–140. [CrossRef]
37. Hooge, F.N. Discussion of recent experiments on 1/f noise. *Physica* **1976**, *60*, 130–144. [CrossRef]
38. Hooge, F.N.; Vandamme, L.K.J. Lattice scattering causes 1/f noise. *Phys. Lett. A* **1978**, *66*, 315–316. [CrossRef]

39. Hung, K.K.; Ko, P.K.; Hu, C.; Cheng, Y.C. A unified model for the flicker noise in metal-oxide-semiconductor field-effect transistors. *IEEE Trans. Electron Devices* **1990**, *37*, 654–665. [CrossRef]
40. Hung, K.K.; Ko, P.K.; Hu, C.; Cheng, Y.C. A physics-based MOSFET noise model for circuit simulators. *IEEE Trans. Electron Devices* **1990**, *37*, 1323–1333. [CrossRef]
41. Hung, K.K.; Ko, P.K.; Hu, C.; Cheng, Y.C. Random telegraph noise of deep-submicrometer MOSFETs. *IEEE Electron Device Lett.* **1990**, *11*, 90–92. [CrossRef]
42. Jindal, R.P.; van der Ziel, A. Carrier fluctuation noise in a MOSFET channel due to traps in the oxide. *Solid-State Electron.* **1978**, *21*, 901–903. [CrossRef]
43. Zhao, X.; Vnanderbilt, D. Phonons and lattice dielectric properties of zirconia. *Phys. Rev. B Condens. Matter* **2002**, *65*, 075105. [CrossRef]
44. Grasser, T.; Reisinger, H.; Wagner, P.-J.; Schanovsky, F.; Goes, W.; Kaczer, B. The time dependent defect spectroscopy (TDDS) for the characterization of the bias temperature instability. In Proceedings of the IEEE International Reliability Physics Symposium, Anaheim, CA, USA, 2–6 May 2010; pp. 16–25.
45. Ielmini, D.; Manigrasso, M.; Gattel, F.; Valentini, M.G. A new NBTI model based on hole trapping and structural relaxation in MOS dielectrics. *IEEE Trans. Electron Devices* **2009**, *56*, 1943–1952. [CrossRef]
46. Ambrogio, S.; Balatti, S.; Cubeta, A.; Calderoni, A.; Ramaswamy, N.; Ielmini, D. Understanding switching variability and random telegraph noise in resistive RAM. In Proceedings of the Technical Digest of the 2013 IEEE International Electron Devices Meeting, Washington, DC, USA, 9–11 December 2013; pp. 782–785.

electronics

MDPI

Review

Evaluation of Low-Frequency Noise in MOSFETs Used as a Key Component in Semiconductor Memory Devices

Akinobu Teramoto

Research Institute for Nanodevice and Bio Systems, Hiroshima University, Higashi-Hiroshima 739-8527, Japan; teramo10@hiroshima-u.ac.jp; Tel.: +81-82-424-6266

Abstract: Methods for evaluating low-frequency noise, such as 1/f noise and random telegraph noise, and evaluation results are described. Variability and fluctuation are critical in miniaturized semiconductor devices because signal voltage must be reduced in such devices. Especially, the signal voltage in multi-bit memories must be small. One of the most serious issues in metal-oxide-semiconductor field-effect-transistors (MOSFETs) is low-frequency noise, which occurs when the signal current flows at the interface of different materials, such as SiO_2/Si. Variability of low-frequency noise increases with MOSFET shrinkage. To assess the effect of this noise on MOSFETs, we must first understand their characteristics statistically, and then, sufficient samples must be accurately evaluated in a short period. This study compares statistical evaluation methods of low-frequency noise to the trend of conventional evaluation methods, and this study's findings are presented.

Keywords: MOSFET; low-frequency noise; random telegraph noise; evaluation method; array test pattern

Citation: Teramoto, A. Evaluation of Low-Frequency Noise in MOSFETs Used as a Key Component in Semiconductor Memory Devices. *Electronics* **2021**, *10*, 1759. https://doi.org/10.3390/electronics10151759

Academic Editor: Changhwan Shin

Received: 22 June 2021
Accepted: 20 July 2021
Published: 22 July 2021

1. Introduction

Semiconductor devices have been basically progressed with the shrinking of MOSFETs (metal-oxide-semiconductor field-effect-transistors), which are used as the key component in them. The shrinkage has been performed following the rule of the constant electric field in MOSFETs, which decreases signal voltage [1,2]. In addition, power consumption of electronic devices has skyrocketed because the amount of digital data generated is growing at a rate faster than Moor's law [3,4]. The reduction of power consumption strongly requires decreasing supply voltage of MOSFETs because power consumption (P) is proportional to the square of the supply voltage (V_{dd}) as follows [5].

$$P = C_L V_{dd}^2 f \tag{1}$$

where C_L represents the load capacitor and f represents the switching frequency of the circuit. The growth of clock frequency in the leading edge logic devices has stopped by exceeding the heat extraction capability. However, the downscaling has been continued to reduce the cost. In the other devices, the downscaling the device size has also been continued to reduce the power consumption and the other reasons. As a result, as the device size is reduced, the signal voltage of MOSFETs decreases. Memory devices' power consumption and supply voltage also have to be reduced [6–9] because of the same reasons as the logic devices and reducing a leakage current. On the other hand, a decrease in signal voltage degrades the reliability of electronic circuits, including analog and digital devices.

The logic (bit) error rate (LER) is given by the following equation.

$$LER = P_0 \int_{-\infty}^{\alpha} f_0(x)dx + P_1 \int_{\alpha}^{\infty} f_1(x)dx \tag{2}$$

where P_0 and P_1 represent the probabilities of signals "0" and "1", respectively, α represents the identification level between "0" and "1", and $f_0(x)$ and $f_1(x)$ represent noise amplitude densities superimposed on "0" and "1", respectively. If f_0 and f_1 are Gaussian noise, the following equations apply.

$$f_0(x) = \frac{1}{\sqrt{2\pi\sigma^2}}e^{-\frac{x^2}{2\sigma^2}},\ f_1(x) = \frac{1}{\sqrt{2\pi\sigma^2}}e^{-\frac{(x-A_S)^2}{2\sigma^2}} \tag{3}$$

where A_S represents the signal amplitude and σ represents the standard deviation of the noise. When $P_0 = P_1 = 1/2$ and $\alpha = A_S/2$ are assumed, the LER is given by the following equation from Equations (2) and (3).

$$\text{LER} = \frac{1}{2}\text{erfc}\left(\frac{A_S}{2\sqrt{2\sigma^2}}\right) \tag{4}$$

$$\text{erfc}(x) = 1 - \text{erf}(x) = 1 - \frac{2}{\sqrt{\pi}}\int_0^x e^{-t^2}dt = \frac{2}{\sqrt{\pi}}\int_x^\infty e^{-t^2}dt \tag{5}$$

When Nyquist transmission rate is the same as the signal band, the signal to noise (S/N) ratio (dB) is given by the following equation.

$$\frac{S}{N} = \frac{E_b}{N_0} = 20\log\left(\frac{A_S}{\sigma}\right) \tag{6}$$

where E_b and N_0 represent signal and noise energy per second, respectively. From Equations (4) and (6), the LER is given by the S/N ratio as follows [10].

$$\text{LER} = \frac{1}{2}\text{erfc}\left\{\frac{1}{2\sqrt{2}}10^{\frac{1}{20}\left(\frac{S}{N}\right)}\right\} \tag{7}$$

Figure 1 shows the LER as a function of S/N (dB) and A/σ ratios [10]. The LER decreases with an increase in S/N (A_S/σ) ratio. On the other hand, to guarantee that a system does not make a mistake even once during the operation period, the LER should be reduced as shown in the following equation.

$$\text{LER} \le \frac{1}{N_L \times F \times T} \tag{8}$$

where N_L, F, and T represent the number of logic gates in a chip, the number of operations per second, and the guarantee period, respectively. For example, the LER should be less than 3×10^{-26} for a circuit with 10^8 logic gates, 10^9 Hz operations, and a 10-year (3×10^8 s) operation period, and then the S/N (A_S/σ) ratio should be greater than 26.5 dB (21.1).

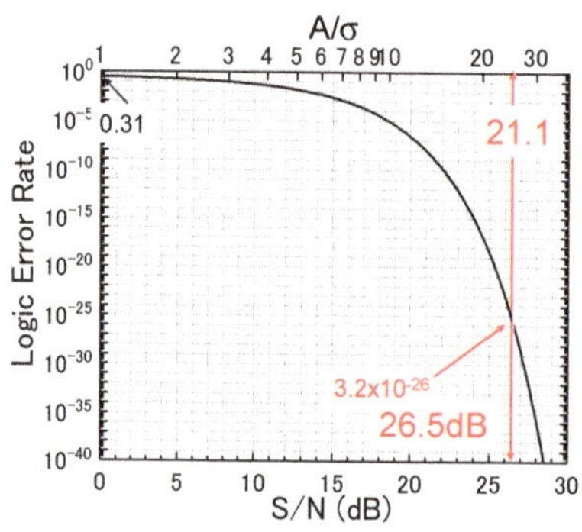

Figure 1. Logic error rate as a function of S/N and A/σ ratios.

For example, we consider signal electrons required from a no error operation. Figure 2 shows (a) Variation in the number of electrons as a function of the number of signal electrons. (b) Logic error rate as a function of the number of signal electrons. When a signal is constructed by a constant number of electrons (N_e), the standard deviation of the number of electrons is $(N_e)^{1/2}$, and then the A/σ ratio is $(N_e)^{1/2}$. As a result, the number of electrons must be greater than 445 to maintain an S/N (A_S/σ) ratio of 26.5 dB (21.1). If the number of signal electrons is 1, the LER must be equal to 0.3. Then, such a system produces the wrong output once every three calculations.

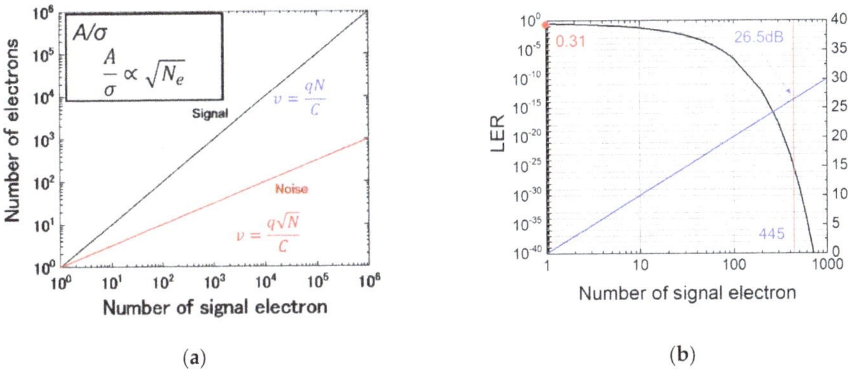

Figure 2. (**a**) Variation in the number of electrons as a function of the number of signal electrons. (**b**) Logic error rate as a function of the number of signal electrons.

The electronic circuit will be influenced by noise, such as thermal noise, quantum noise, and flicker noise. The noise voltage (v_{nf}) in 1/f noise is defined by the following equation.

$$v_{nf} = \sqrt{\frac{K_F}{C_{OX} \times L \times W} \ln\left(\frac{f_H}{f_L}\right)} \quad (9)$$

where K_F represents the flicker noise coefficient, f_L~f_H is the frequency period for device operation, C_{OX}, L, and W represent the gate oxide capacitance, gate length, and gate width of a MOSFET, respectively. Noise increases with device shrinkage because v_{nf} is inversely proportional to $\sqrt{C_{OX}LW}$ [11–14]. It has been pointed out that 1/f noise may influence not only analog devices, but also digital devices when device shrinkage and the decreasing signal voltage are moved on [15]. Random telegraph noise (RTN), another low-frequency noise also affects electronic devices, such as CMOS image sensor [16–21], static random access memory (SRAM) [22–25], dynamic random access memory (DRAM) [25], and flash memory [26–32]. Low-frequency noise, such as 1/f noise and RTN, have high variability [33,34] because they must be statistical phenomena by nature, and statistical analysis is required to fully understand this phenomenon. The conventional evaluations of the noise in MOSFETs have performed with a few sample numbers, and then we could understand only typical noise characteristics of MOSFETs having relatively large noise. However, we need statistical information of the noise for the design of LSI. Then, low-frequency noise statistical evaluation methods and evaluation results are described in this study.

2. Evaluation Methods

2.1. Test Pattern for Noise Evaluation

The test structure is constructed using 0.22 μm, 1-poly 2-metal standard CMOS technology and includes n-MOSFETs of various gate sizes [35–38] as shown in Table 1. The measured MOSFETs are arrayed in 1024 rows and 1776 columns (total number of MOSFETs:

1217856) in a chip at 5 μm intervals. The size of the MOSFETs, and their number, and location in a chip are shown in Table 1. The test chip has an area of 5.5 mm × 14 mm. The gate insulator is formed by pyrogenic oxidation and is 5.8 nm thick.

Table 1. Number of n-MOSFETs for each transistor size.

Gate Length (μm)	Gate Width (μm)	Number of MOSFETs	Supply Voltage (V)
0.22	0.28	131,072 (128 × 1024)	2.5
0.22	0.30	131,072	
0.24	0.30	131,072	
0.24	1.5	131,072	
0.24	15	131,072	
0.4	1.5	32,768 (128 × 256)	
0.4	15	32,768	
1.2	0.3	65,536 (64 × 1024)	
1.2	1.5	65,536	
4.0	0.30	65,536	
4.0	1.5	65,536	
0.24	0.30	32,768 (AR:100)	
0.24	0.30	4096 (16 × 256, AR:1000)	
0.24	0.30	1344 (32 × 42, AR:10000)	
0.4	1.5	131,072	3.3
0.4	15	32,768	
1.2	15	16,384 (64 × 256)	
4.0	15	16,384	

AR: Antenna ratio.

A schematic block diagram of a test pattern is shown in Figure 3a [35,37,39,40]. This is composed of MOSFETs measured in arrayed unit cells, vertical and horizontal shift registers for addressing measured MOSFETs, MOSFETs located on each column for current control of measured MOSFET, analog memories for storing the source voltage of the measured MOSFETs within one line, and a source follower circuit for amplifying the output signal. The drain (V_D) and gate (V_G) voltage in measured MOSFETs and the gate voltage applied to current source MOSFETs (V_{REF}) are supplied from the external voltage source simultaneously. V_{DD} and V_{SS} are the supply voltages in the peripheral circuits and ground voltage, respectively. The measured MOSFET and current source transistor construct a source follower circuit using a select transistor. This test structure uses simple peripheral circuits. Therefore, it can be used to evaluate various MOSFETs with varying gate lengths, gate widths, gate insulator films, thicknesses, and other characteristics. Figure 3b shows the circuit schematic of a unit cell and current source transistor in Figure 1, which is the principle of this measurement. A unit cell is constructed with a measured MOSFET and a select transistor. When the current source transistor operates at a saturation region, I_{REF} is independent of the voltage between the source and drain in the current source transistor (V_{out}). When the gate bias of the select transistor (Φ_x) is applied from a vertical shift register, I_{REF} flows into the measured MOSFET. The output voltage (V_{out}) is indicated as follows.

$$V_{gs} = V_G - V_{out} - I_{REF} \cdot R_{select} \approx V_G - V_{out} \tag{10}$$

where R_{select} is the channel resistance of the select switch transistor. The select transistor must be operated in the linear region to have sufficient high channel conductivity compared with the measured MOSFET, and then, $I_{REF} \cdot R_{select}$ can be neglected. The output signal can be obtained as a source voltage for each measured MOSFET by shift register scanning, and then 1.2 million MOSFETs can be measured within approximately 0.7 s. The electrical characteristics of the measured MOSFETs can be observed as the V_{gs} included in the output voltage V_{out} (Figure 3b). In this frame measurement mode, each MOSFET can be measured every 0.7 s. This test pattern has another measurement mode, which can measure a specific MOSFET every 1 μs.

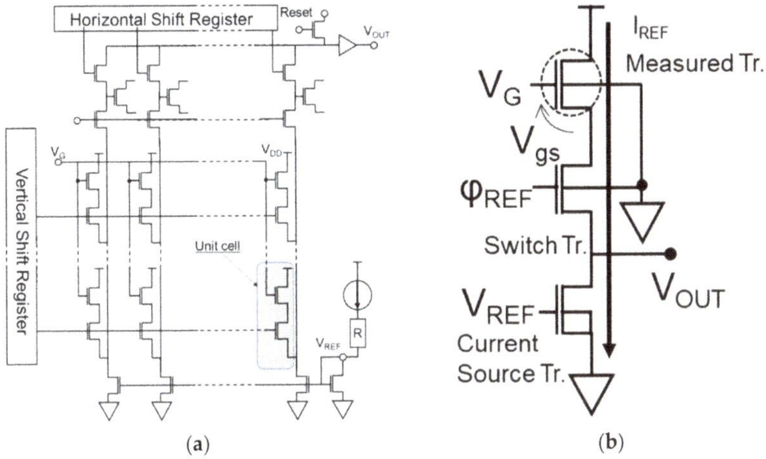

Figure 3. (**a**) Schematic block diagram of a test pattern. (**b**) Unit cell.

2.2. Extraction of Amplitude and Time Constant of RTN

Two-level type RTN is characterized by only three parameters, which are the mean time to capture ($<\tau_c>$), mean time to emission ($<\tau_e>$), and amplitude (ΔV_{gs}). The time constants correspond to two physical states of a trap, that is, τ_c and τ_e represent spans in a low V_{gs} level (carrier trapping state) and high V_{gs} level (carrier emission state), respectively (Figure 4a). The RTN amplitude ΔV_{gs} is defined as the difference between two normal distributions in a voltage histogram (Figure 4b). We extract the time constants by fitting the distributions of τ_c and τ_e to the exponential distribution ($Ae^{-t/<\tau>}$) because the phenomenon is governed by the Poisson process. The time constants can be extracted with μs accuracy using the specific MOSFET measurement mode.

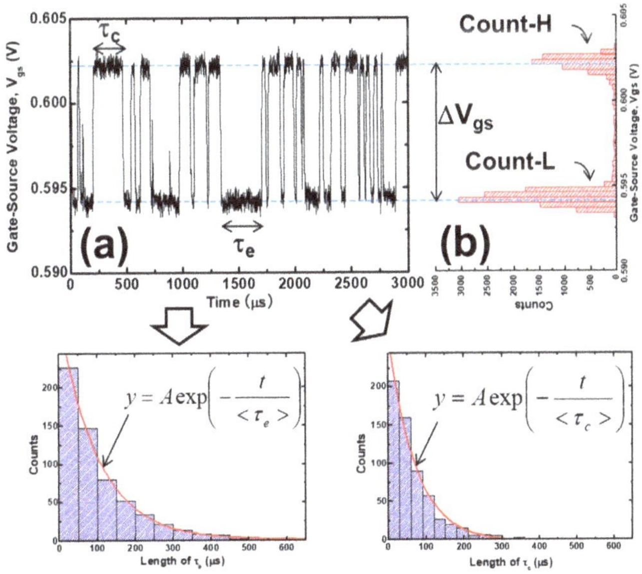

Figure 4. Definitions and extractions of two time constants ($<\tau_c>$, $<\tau_e>$) and amplitude (ΔV_{gs}) from RTN (**a**)waveform data, (**b**)histogram of Vgs.

The time constant ratio <τ_e>/<τ_c> is also an important parameter in RTN because the energy level of a trap that causes RTN is related to the constant ratio as follows [12,41–43].

$$\frac{<\tau_c>}{<\tau_e>} = \mathrm{g}\exp\left(\frac{\mathrm{E_T}-\mathrm{E_F}}{\mathrm{kT}}\right) \tag{11}$$

where E_T and E_F represent the energy of the trap and Fermi energy of the channel, respectively; k, T, and g represent Boltzmann constant, temperature, and degeneracy factor, respectively, where g is assumed to 1. Then, the energy of the trap level is indicated by (12).

$$\mathrm{E_T}-\mathrm{E_F} = \mathrm{kTln}\left(\frac{<\tau_c>}{<\tau_e>}\right) \tag{12}$$

We can use the frame measurement mode to extract the time constant ratio, and the 1.2 million MOSFETs can be measured 10,000 times in 7000 s (sampling period = 0.7 s). An average of the time constant ratio <τ_e>/<τ_c> is the same as Count-L/Count-H (shown in Figure 4b), where Count-L and count-H are the numbers of low and high states, respectively [41–44]. When the time constant is greater than a sampling frequency of 0.7 s, the detected number of transition times is the same as the transition time of RTS characteristics. However, when the time constant is less than 0.7 s, the detected number of transition times is less than the real one; however, it is proportional to the real one because the absolute value of the time constant, which is less than the sampling frequency, cannot be extracted in this measurement. Then, the number of transition times is defined as the detected ones in the sampling frequency of 0.7 s.

2.3. Root Mean Square of RTN Waveform

The root mean square (RMS) of the signal waveform is often used for the representative parameter of noise [45], and the RMS of the output voltage V_{RMS} is defined as follows in this study.

$$\mathrm{V_{RMS}} = \sqrt{\frac{\sum_{i=1}^{N}\left(\mathrm{V_{out,i}}-\overline{\mathrm{V_{out}}}\right)^2}{\mathrm{N}-1}} = \mathrm{A}\frac{\sqrt{<\tau_e><\tau_c>}}{<\tau_e>+<\tau_c>} \tag{13}$$

where $V_{out,i}$, $\overline{V_{out}}$, N, and A are the output voltage at ith sampling, average of V_{out}, sampling numbers, and the amplitude of two-state RTN, respectively. Using V_{RMS}, we can obtain MOSFETs with high noise from many measured MOSFETs. Figure 5 shows the relationship between V_{RMS} and RTN waveform. The waveform with large RTN corresponds to large V_{RMS}.

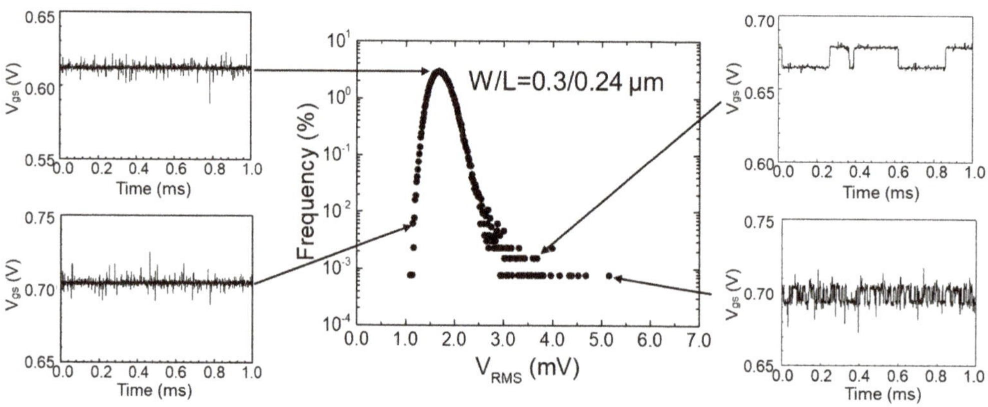

Figure 5. Relation between V_{RMS} and RTN waveform.

3. Results and Discussion

3.1. Statistical Evaluation of RTN Characteristics

The 1/f noise increases with downscaling of MOSFETs, as mentioned above, and RTN also increase with the downscaling [12,36,37]. Figure 6 shows the Gumbel plot of V_{RMS} for the various MOSFET sizes [36,37]. A large V_{RMS} can be observed in small-size MOSFETs (L/W = 0.22/0.28, 0.22/0.3, 0.24/0.3 μm). In this experiment, noise cannot be observed in large MOSFETs because the floor noise is relatively high at ~2.5 mV.

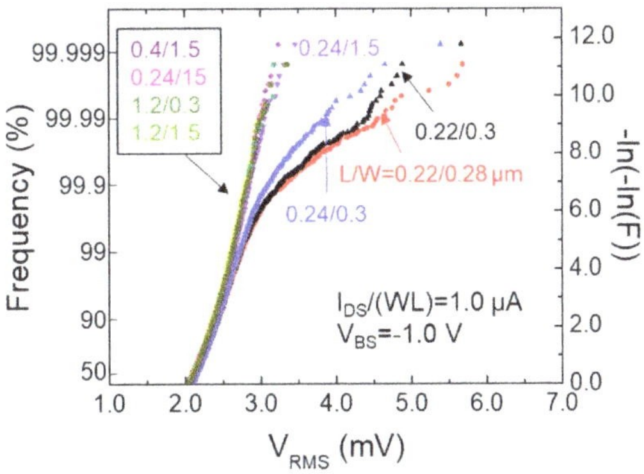

Figure 6. Gumbel plot of V_{RMS} for various MOSFET sizes. The measured sizes of MOSFETs are L/W = 0.22/0.28, 0.22/0.3, 0.24/0.3, 0.24/1.5, 0.4/1.5, 0.24/15, 1.2/0.3, 1.2/1.5 μm.

Figure 7a shows the Gumbel plot of V_{RMS} for the various I_{DS} varied from 0.13 to 12.7 μA. The sizes of MOSFETs are L/W = 0.22/0.28 μm and $V_{BS} = -1.0$ V [39]. The data in (a) and (b) were measured by the frame and specific MOSFET measurement modes, respectively. The number of MOSFETs with large noise increases with decreasing I_{DS}, which is controlled by V_{gs}. This means that the event probability of large noise increases with decreasing V_{gs} because the number of channel electrons decreases with decreasing V_{gs}, and then the effect of a trapped electron charge becomes large with decreasing number of channel electrons. The number of channel electrons also decreases with the shrinkage of transistor size shown in Figure 6, and then, the probability increases with decreasing channel size. Figure 7b shows the waveform of typical MOSFETs for I_{DS} of 0.13, 0.38, and 1.3 μA [39]. The time constants and amplitude are modulated by I_{DS}. With increasing I_{DS} (V_{gs}), amplitude and τ_c decrease, whereas τ_e slightly increases. An increase in V_{gs} decreases E_T-E_F, and then, the time to capture decreases as shown in Equation (11). The difference between the modulation of τ_e and τ_c is discussed later. The modulation of amplitude is caused by a decrease in the number of electrons, as discussed above. It is considered that decreasing the time to capture and increasing amplitude with decreasing V_{gs} increases the event probability of large noise. Figure 8a shows the Gumbel plot of V_{RMS} for the various back bias (V_{BS}). V_{BS} varied from -0.075 to -1.38 V, and Figure 8b shows the waveform of typical MOSFETs for V_{BS} of 0.6 1.0 and 1.3 V [39]. The probability increases with the absolute value of V_{BS} in (a). In this experiment, I_{DS} was constant at 1.0 μA, and this means that the number of electrons was almost the same for each V_{BS}. Increasing V_{BS} caused channel percolation [46–49], making the channel thickness narrow and percolated and increasing electron energy [50]. The probability is increased by channel percolation [46,47], and the varying electron energy modulates the time constants. In MOSFETs with RTN, the amplitude does not increase with increasing V_{BS} because the number of electrons is the same for each V_{BS}. This means that channel percolation increases the probability of RTN generation.

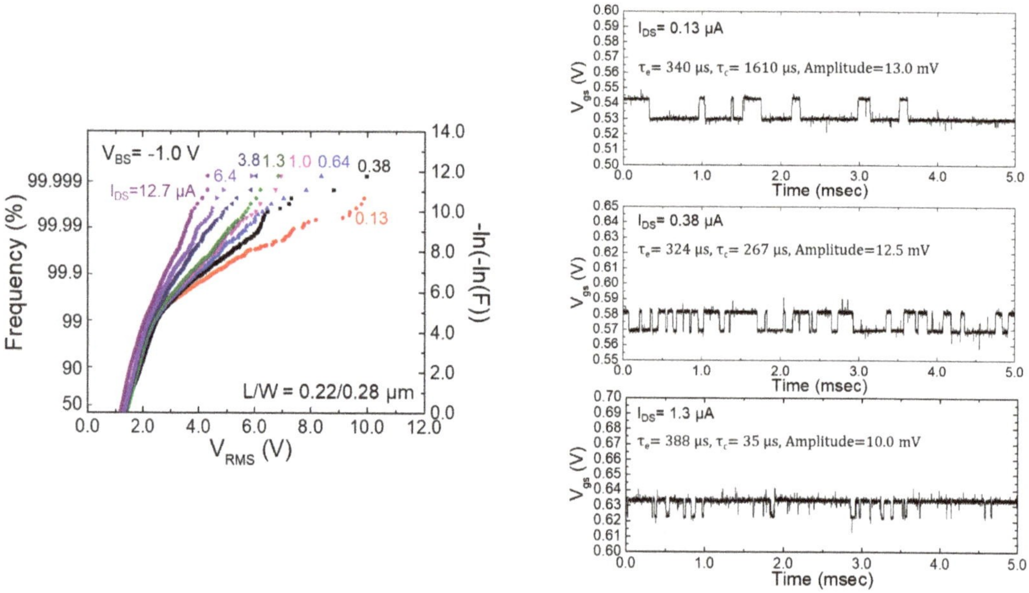

(**a**) (**b**)

Figure 7. (**a**) Gumbel plot of V_{RMS} for the various I_{DS}. I_{DS} varied from 0.13 to 12.7 μA, and (**b**) waveform of typical MOSFETs for I_{DS} of 0.13, 0.38, and 1.3 μA. The sizes of MOSFETs are L/W = 0.22/0.28 μm and V_{BS} = 1.0 V.

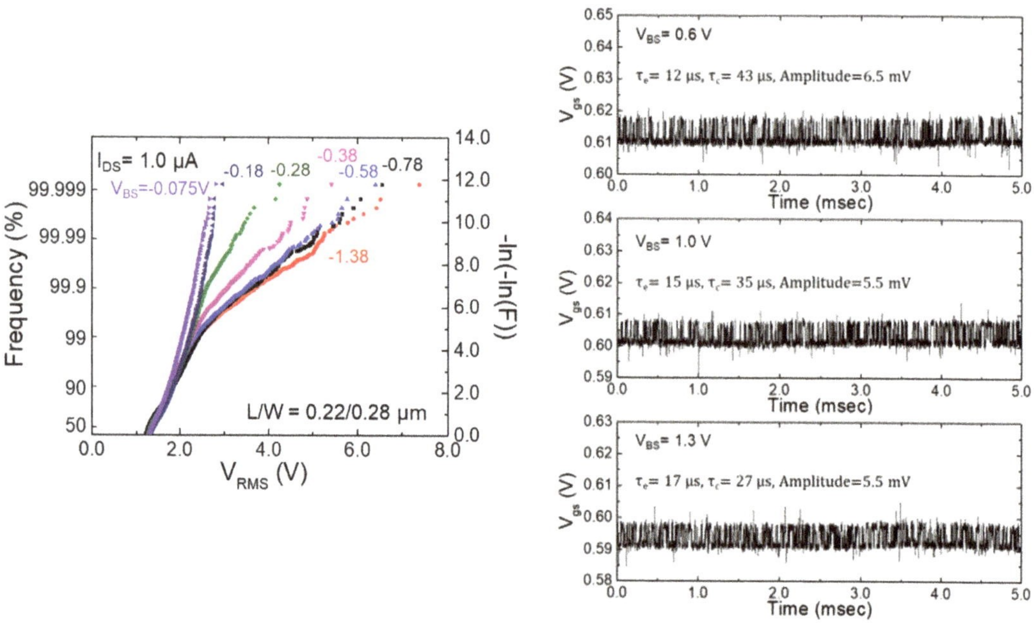

(**a**) (**b**)

Figure 8. (**a**) Gumbel plot of V_{RMS} for various V_{BS}. V_{BS} varied from 0.075 to 1.38 V, and (**b**) waveform of typical MOSFETs for V_{BS} of 0.6 1.0 and 1.3 V. The sizes of MOSFETs are L/W = 0.22/0.28 μm and I_{DS} = 1.0 μA.

Figure 9 shows the Gumbel plot of the RTN amplitude for MOSFETs with varying channel doping [51]. The channel percolation is accelerated by increasing channel doping concentration [46–48]. This figure shows that the probability of the number of MOSFETs with large amplitude increases with doping concentration. RTN is increased by channel doping as well as doping the concentration near the source and drain regions. Figure 10 shows the Gumbel plot of V_{RMS} for various Halo implantation concentrations [52]. The number of MOSFETs with large RTN increases with an increase in Halo implantation concentration. This indicates that the high dose in the channel region or near the source/drain region results in high RTN because of channel percolation enhancement.

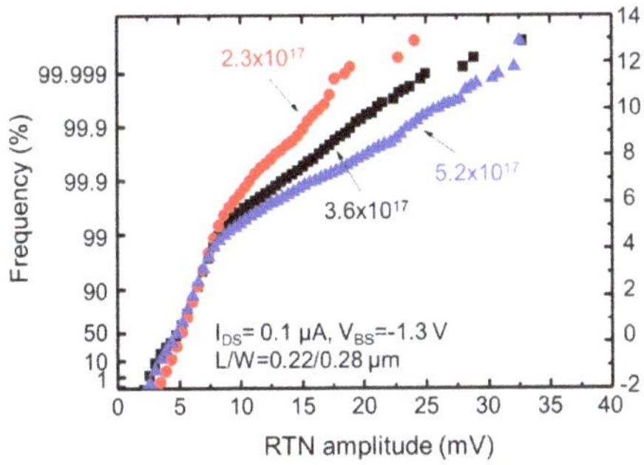

Figure 9. Gumbel plot of the RTN amplitude for MOSFETs with varying channel doping. The channel doping concentrations are varied 2.3×10^{17}, 3.6×10^{17}, 5.2×10^{17} cm^{-3}, respectively.

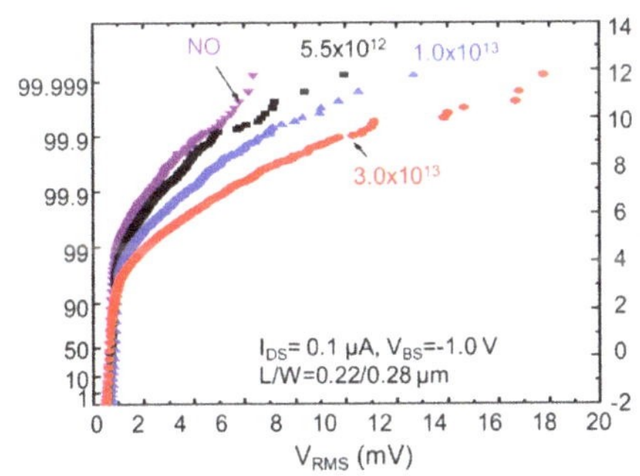

Figure 10. Gumbel plot of the RTN amplitude for MOSFETs with varying channel doping. The channel doping concentrations are varied 2.3×10^{17}, 3.6×10^{17}, 5.2×10^{17} cm^{-2}, respectively.

Figure 11 shows the energy band diagrams and energy distribution of traps causing RTN for V_{gs} of 0.57, 0.53, and 0.46 V, respectively [41,42]. The difference between E_T and E_F for electrons is calculated using Equation (11). The blue shading and red solid bars in Figure 11 show the energy distribution of traps causing RTN in each measurement condition and common traps in all conditions, respectively. Although the shape of the distribution of each V_{gs} is almost the same, and the energy of common traps in all V_{gs} increases with decreasing V_{gs}. The conduction band edge (E_C), the bottom sub-band energy (E_{sub}), and 2nd sub-band energy (E_{2nd}) in the inversion layer and E_F are indicated in Figure 11 [50]. The energy levels of sub-bands were calculated using Equation (14) [50].

$$E_j = \left[\frac{3hqE_s}{4\sqrt{2m_x}}\left(j+\frac{3}{4}\right)\right]^{\frac{2}{3}}, \ j = 0, 1, \tag{14}$$

where E_s is the electric field, h and m_x represent Planck's constant and effective mass of electrons, respectively. E_j is jth sub-band energy, and E_{sub} and E_{2nd} represent E_0 and E_1, respectively. The main energy distribution for each V_{gs} locates higher energy than the conduction band edge. It is considered that the energy level of traps is widely distributed, and the energy of the detected traps is determined by the electron energy in E_{sub} and E_{2nd}. Conversely, the energy of common traps in all V_{gs} increases with decreasing V_{gs} because the influence of trap energy on V_{gs} is larger than that of electron energy in the channel.

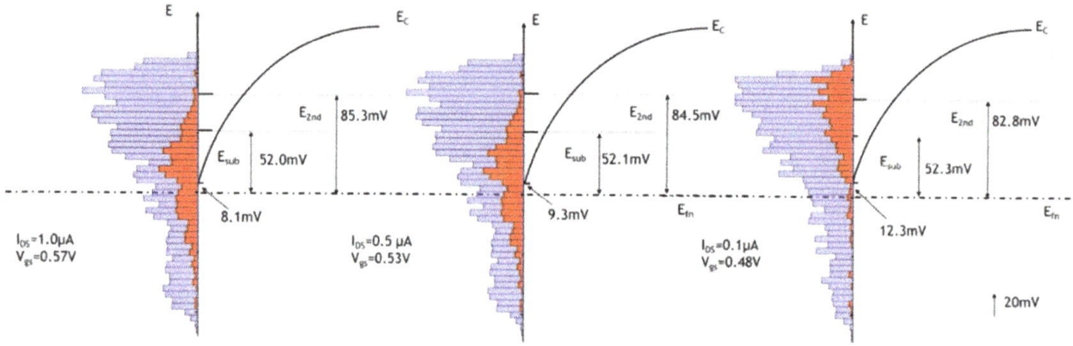

Figure 11. Conduction band diagrams and energy distribution of the traps causing RTN for V_{gs} of 0.57, 0.53, and 0.46 V, respectively.

3.2. Multi-State RTN

Large V_{RMS} RTN includes both two-state and multi-state RTN [34,53–56], which is considered to be generated by multi-traps. The analysis of trap characteristics, such as time constants and amplitude, in multi-state RTN is more difficult than that of two-state. Figure 12 shows the appearance probability of RTN with two, three, four, and more than four states. A large RTN (V_{RMS} > 680 μV) was obtained by a frame measurement mode of the sampling period of 0.7 s/frame in I_{DS} = 1 μA. 131,072 MOSFETs (L/W = 0.22/0.28 μm) were measured, and 2575 MOSFETs with large V_{RMS} can be extracted. Then, we selected MOSFETs with large RTN and measured them by a specific measurement mode of a sampling period of 1 μs and a long sampling time of 10 min (sampling points = 6 × 10^8) for the same bias condition [57,58]. Figures 13–15 show the (a) waveform, (b) time lag plot (TLP), and (c) histogram for typical three-, four-, and six-state RTN. The number of peaks and the transition of each state can be understood via TLP [53,54,56].

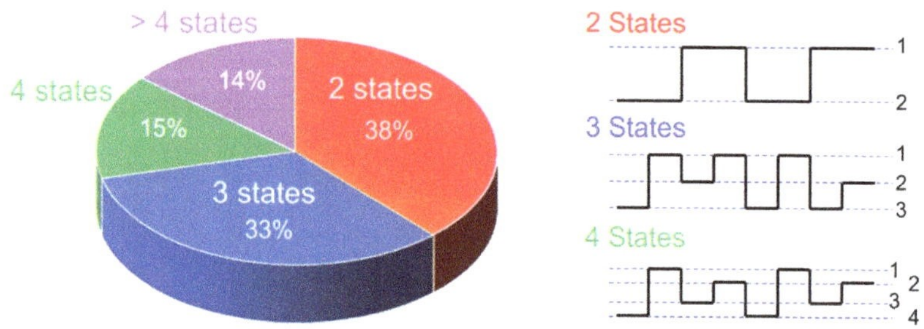

Figure 12. Appearance probability of RTN with two, three, four, and more than four states. 2575 MOSFETs with large V_{RMS} can be extracted from 131,072 MOSFETs (L/W = 0.22/0.28 μm).

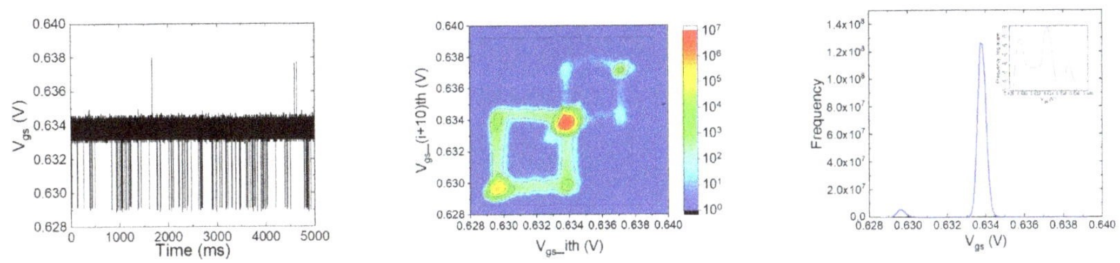

(a) (b) (c)

Figure 13. (**a**) Waveform, (**b**) TLP, and (**c**) histogram for a typical three-state RTN. The inset figure (**c**) shows the same data of the vertical axis of the log (histogram).

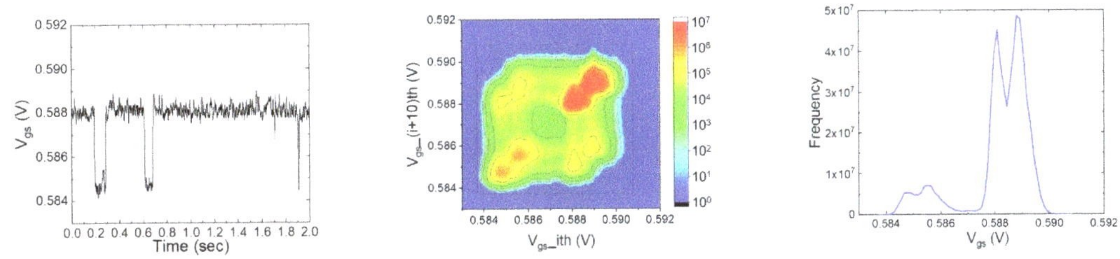

(a) (b) (c)

Figure 14. (**a**) Waveform, (**b**) TLP, and (**c**) histogram for a typical four-state RTN.

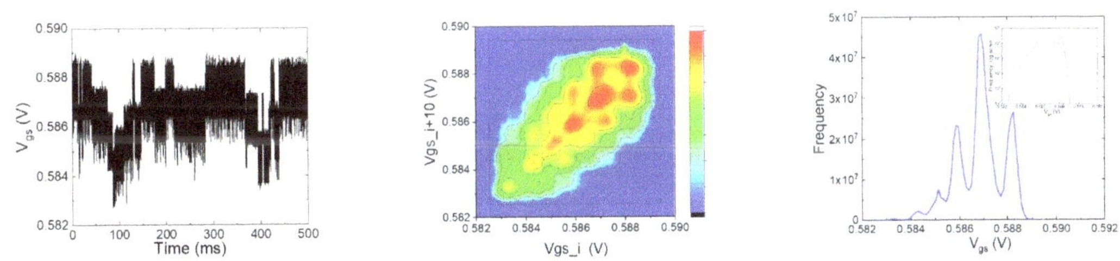

(a) (b) (c)

Figure 15. (**a**) Waveform, (**b**) TLP, and (**c**) histogram for a typical six-state RTN. The insect figure in (**c**) shows the same data of the vertical axis of the log (histogram).

Figures 13–15 show the (a) waveform, (b) time lag plot (TLP), and (c) histogram for typical three-, four-, and six-state RTN. The number of peaks and the transition of each state can be understood via TLP [53,54,56]. Figures 13, 14 and 15b show the relationship of ith and (i + 10)th V_{gs} for the constant IDS. As shown in Figure 13b, transitions occur not from the lowest state to the highest state, but only via the medium state. When the trapping probability of some traps is even, the number of states should be even, and the transition from one position to the next can occur. To begin with, an odd number of states implies that the trapping probability for each trap is not independent of each other. A similar transition phenomenon occurs even in a four-state case. As shown in Figure 14b, transitions did not occur from the lowest state to the highest state or from the second-lowest state to the second-highest state. This also means that there are more than two traps, and the probability of trapping for each trap is not independent of each other. The characteristics of multi-trap RTN can be understood via TLP and waveforms [56]; however, these analyses become more difficult as the number of states increases.

3.3. Time Constants in Individual RTN

As the extraction of multi-trap phenomena is difficult, as discussed in Section 3.2, we discuss the time constants and amplitude only in two-state RTN [59]. Figure 16 shows (a) τ_c and (b) τ_e as a function of I_{DS}, respectively. These data are measured at I_{DS} of 0.1, 0.3, 1.0, 3.0, and 5.0 μA. The data in Figure 16 show how parameters from all two-level RTN can be extracted in common under four or five I_{DS}, and thus, the number of selected data points was 22.

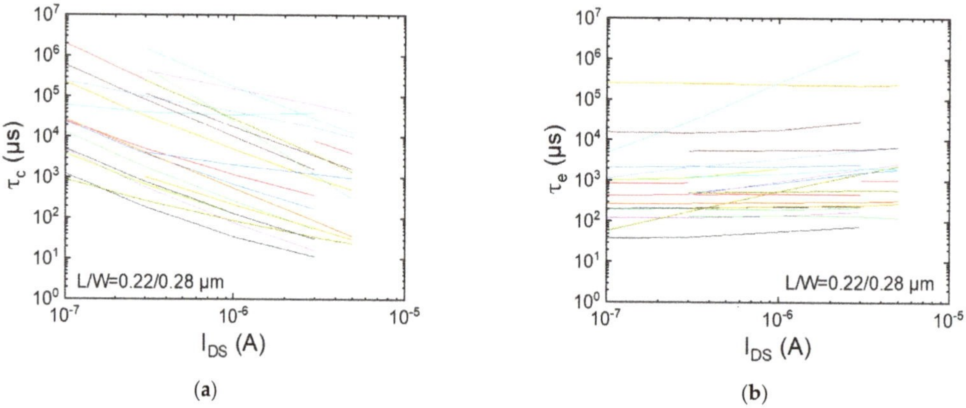

Figure 16. (**a**) τ_c and (**b**) τ_e as a function of I_{DS}. These data are measured at I_{DS} of 0.1, 0.3, 1.0, 3.0, and 5.0 μA. The number of data points is 22.

Figure 17a shows τ_c/τ_e and E_T-E_F as a function of I_{DS} and Figure 17b band diagram of MOS structure for changing V_{GS} (I_{DS}). τ_c/τ_e and E_T-E_F are calculated from the data in Figure 16 and Equation (12), respectively. τ_c and τ_e decrease and increase with an increase in I_{DS} (V_{GS}), and the absolute slope of τ_c is significantly larger than that of τ_e. Large trap energy (E_T) decreases with an increase in V_{GS} than that of channel electron (E_C: bottom energy of conduction band). Then, with increasing V_{GS} (I_{DS}), the energy barrier from the channel electron to the trap decreases and that from the trapped electron to the channel increases. As a result, the time to capture and time to emission decreases and increases, respectively, with increasing V_{GS} (I_{DS}). The transition probability depends on the energy barrier height between a trap and channel. τ_e depends only on the energy barrier because only one electron is captured in a trap. Meanwhile, τ_c depends not only on the energy barrier, but also on the number of electrons in a channel because the number of channel electrons increases as V_{GS} (I_{DS}) increases. Then, the dependency of τ_c on I_{DS} is more significant than that of τ_e. E_T-E_F in Figure 17 changed by approximately 175 mV during I_{DS} (V_{GS}) from 0.1 (0.53V) to 5.0 μA (0.75V). Based on these values, E_T-E_F changes by 0.18 V, whereas V_{GS} changes by 0.22 V. The distance between the traps and the channel was 4.6 nm due to the gate oxide thickness of 5.7 nm. However, τ_c depends not only on the trap energy, but also on the number of channel electrons; thus, E_T-E_F values cannot be calculated using Equation (12). The distance is considered to be shorter than the calculated value. τ_e values for almost all samples monotonically increased with increasing I_{DS}. This suggests that the distance from the trap to the channel is shorter than that to the gate electrode. The distance between the trap and channel is shorter than 2.85 nm, which is the center of the gate oxide thickness. Figure 18a,b show the amplitude and transition frequency as a function of I_{DS} for the same samples as those in Figures 16 and 17, respectively. For a sufficiently long measuring period, the transition frequency (TF) was calculated using the following equation.

$$\mathrm{TF} \approx \frac{N_e}{\tau_e} \approx \frac{N_c}{\tau_c} \quad (15)$$

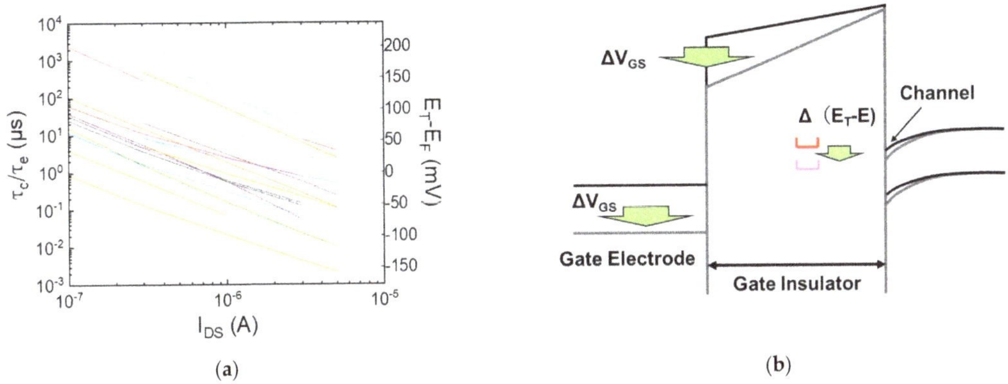

Figure 17. (**a**) shows τ_c/τ_e and E_T-E_F as a function of I_{DS}, and (**b**) band diagram of MOS structure for varying V_{GS} (I_{DS}). τ_c/τ_e and E_T-E_F are calculated from the data in Figure 16 and Equation (12), respectively.

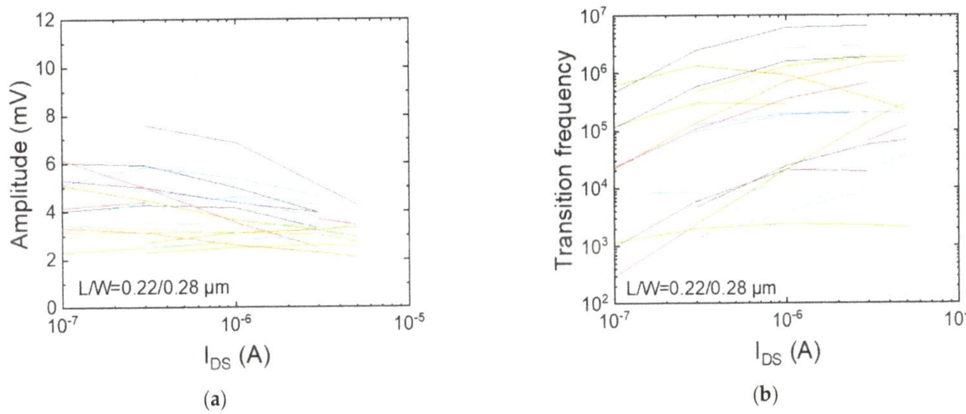

Figure 18. (**a**) Amplitude and (**b**) transition frequency as a function of I_{DS}.

The amplitude decreases and the TF increases with an increase in I_{DS} for almost all samples. This is caused by the increase in the number of channel electrons as I_{DS} increases. However, the amplitude and TF of some samples did not exhibit monotony, which is due to the percolation channel effect [46–49]. The distance between the channel and trap changes as I_{DS} (V_{GS}) changes because of the formation of the percolation channel.

3.4. *Effect of Drain Current on Appearance Probability and Amplitude*

Figure 19 shows the Gumbel plot of the V_{RMS} for 18,048 MOSFETs. I_{DS} varied from 0.1 to 20 µA. The floor noise in this experiment was smaller than the others and was approximately 35 μV_{RMS} [60]. In Figure 7, V_{RMS} decreases with an increase in I_{DS} for all V_{RMS}. In Figure 19, larger V_{RMS} can also be observed in small I_{DS} in relatively large V_{RMS} regions. However, the higher appearance probability in large I_{DS} than that in small I_{DS} for the small V_{RMS} region of less than 500 µV could not be observed in Figure 7 because the floor noise was approximately 1 mV in that experiment. The amplitude characteristics are the same as the V_{RMS} characteristics, and the distribution of the time constants is the same for all conditions [60]. Figure 20 shows the frequency of RTN with two, three, and more than three states in 18,048 MOSFETs. The frequency of all states increases with I_{DS}.

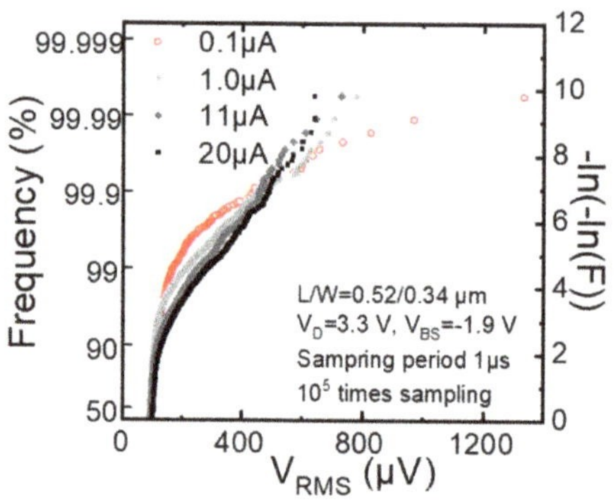

Figure 19. Gumbel plot of the V_{RMS} for 18,048 MOSFETs. I_{DS} varied from 0.1 to 20 μA. The floor noise was a 35-μV_{RMS}.

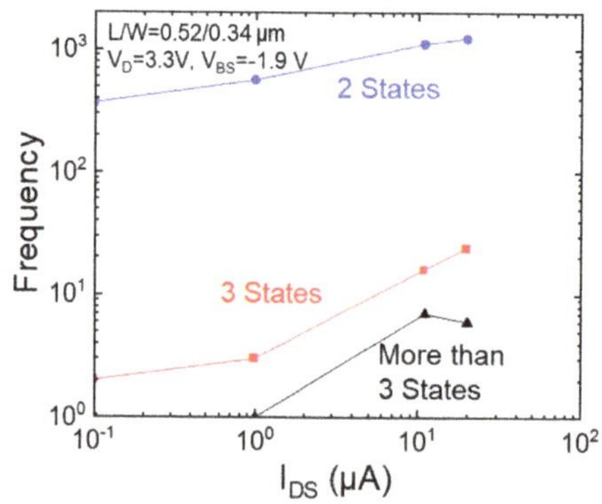

Figure 20. Frequency of RTN with two, three, and more than three states in 18,048 MOSFETs.

It is assumed that the probability that electrons and percolation paths are close to each other increases as I_{DS} increases, increasing the number of electrons in the channel and the number of percolation paths. Figure 21 shows the Gumbel plot for the amplitude of two-state RTN. Notably, frequency sometimes increases with an increase in I_{DS} even though the effect of trapped electrons on the channel decreases with an increase in the electron density in the channel.

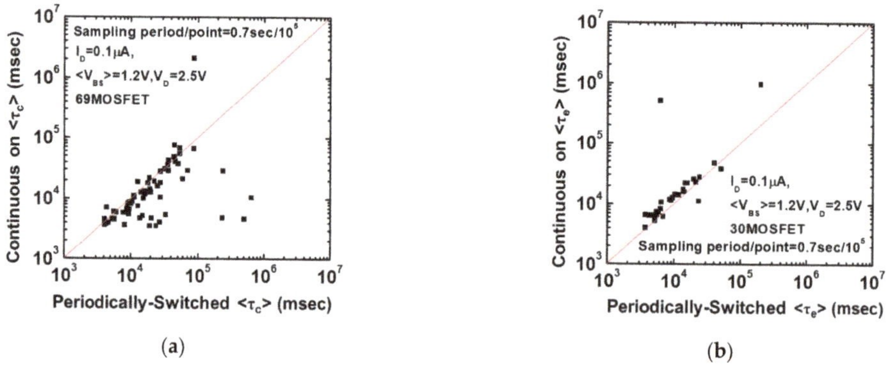

Figure 21. Relationship of time constants between continuous on-state and periodically switched conditions. (**a**) τ_c, (**b**) τ_e.

3.5. Modulation of Time Constants

In Section 3.3, the time constants were measured under constant conditions. The differences between time constants and VRMS in continuous on-state and periodically switched conditions are discussed in this section [61]. As shown in Figure 16, V_{GS} dependance on τ_c is larger than that of τ_e, and this suggests that the cycle period ($\tau_c + \tau_e$) in on-state is longer than that in off-state because the V_{GS} of off-state is smaller than that of on-state. Figure 21a,b show the time constant relationship between continuous on-state and periodically switched conditions. In the periodically switched condition, MOSFETs cycled for 10.6 msec in 700 ms, which was a measurement cycle. Although τ_e of almost all samples are the same in both conditions, τ_c of some samples in the periodically switched condition is larger than that in continuous on-state. Figure 22 shows (a) histogram of the V_{RMS} difference between continuous on-state and periodically off-state (ΔV_{RMS}) and (b) schematic waveform of RTN in continuous on-state and periodically off-state, respectively. Though V_{RMS} of 5% samples was increased, that of 95% was not changed or decreased in the periodically switched condition. As a result, V_{RMS} decreased in the periodically switched condition, with a few exceptions. This suggests that V_{RMS} can be reduced by the modulation of operation conditions, even in the same MOSFET.

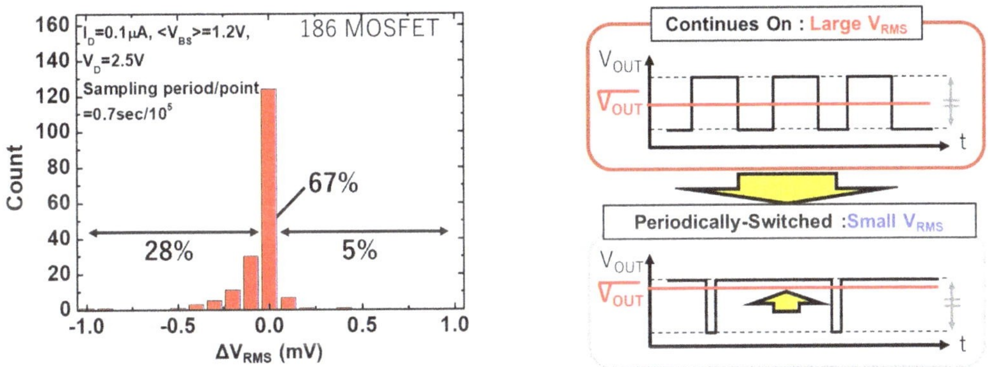

(a) (b)

Figure 22. (a) Histogram of the V_{RMS} difference between continuous on-state and periodically off-state (ΔV_{RMS}), and (b) schematic waveform of RTN in continuous on-state and periodically off-state.

3.6. Device Structure Dependence of RTN

In the above results and discussions in Sections 3.1–3.5, the dependance of RTN characteristics on operation conditions is mainly described. In Section 3.6, the dependance of RTN characteristics on the structure of MOSFETs, such as buried channel MOSFETs and asymmetric source–drain structure MOSFETs, are described [57,62–65].

3.6.1. Buried Channel MOSFETs

Figure 23 shows the structure of buried channel MOSFETs studied in this work. To discuss the effects of n-Si layer widths and the distance between the channel and SiO_2/Si interface, n-Si layer width was varied to be 0, 10, 25, and 60 nm for standard, narrow, middle, and deep samples formed by arsenic ion implantation, respectively, and the high-energy ion implantation created not only a deep channel, but also a wide channel in the vertical direction to the SiO_2/Si interface. Figures 24 and 25 show the Gumbel plots of V_{RMS} for the standard, narrow, middle, and deep samples and the V_{BS} dependance of V_{RMS} for the narrow, middle, and deep samples, respectively [57,65].

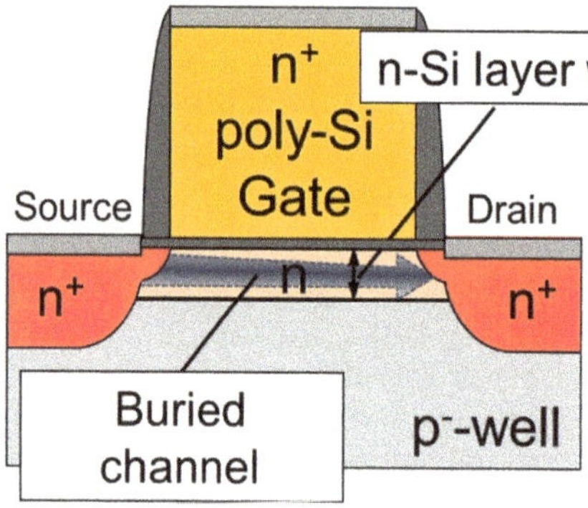

Sample name	n-Si layer width (nm)
Standard	No ion implantation
Narrow	20 nm
Middle	40 nm
Wide	60 nm

Figure 23. Structure of buried channel MOSFETs studied in this work and the channel width for each sample.

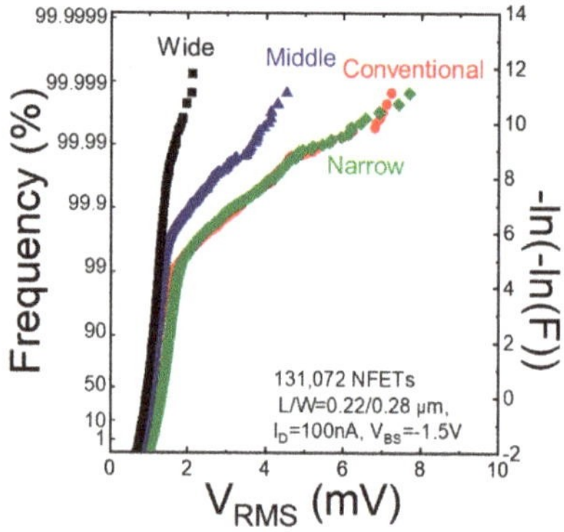

Figure 24. Gumbel plots of V_{RMS} for standard, narrow, middle, and deep samples.

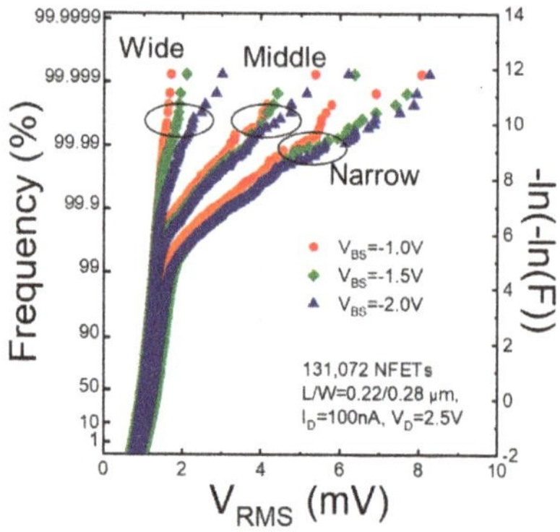

Figure 25. V_{BS} dependance of V_{RMS} for narrow, middle, and deep samples.

The channel length and width were 0.22 and 0.28 µm, respectively, I_{DS} was 100 nA, and V_{BS} in Figure 24 was −1.5 V and varied from −1.0 V to −2.0 V in Figure 25, respectively. The V_{RMS} values for the standard and Narrow samples are the same, and the frequency of large V_{RMS} decreases with an increase in n-Si width and/or depth. This means that RTN cannot be decreased by the 20 nm buried channel, but can be decreased by forming a buried channel of 40 nm and more. By increasing the back bias, V_{RMS} increases for all samples, and the effect of V_{BS} remarkably appeared for the wide sample. By applying the back bias, the channel pushes onto the SiO_2/Si interface, and the channel thickness decreases. The buried channel is extremely effective for decreasing RTN because the channel is separated from SiO_2/Si interface and the wide channel becomes difficult to form the percolation path. Furthermore, the buried channel MOSFET in the isolated well was employed to evaluate V_{BS} dependance [64], and V_{BS} can be varied from 0 V in this structure because the well voltage can be changed freely. The gate length and gate width of the MOSFETs are 0.32 and 0.32 μm, respectively, I_{DS} was 1 μA, vs. was 1.5 V, and V_{well} of the normal well and isolated well were 0 and 1.5 V, respectively; thus, V_{BS} was set at −1.5 and 0 V for the normal well and isolated well, respectively.

Figure 26 shows the Gumbel plot of the V_{RMS} for the buried channel and surface channel MOSFETs with the back bias conditions of 0 and −1.5 V. V_{RMS} of the buried channel MOSFETs at V_{BS} = −1.5 V is significantly less than those of surface channel; however, that at V_{BS} = 0 V is larger than those of the surface channel even though those of the surface channel do not depend on V_{BS}. Figure 27 shows the normal probability plot of the subthreshold swing for the same sample of Figure 26. The subthreshold swing of buried channel MOSFETs with V_{BS} = 0 is much smaller than the others. A strong relationship between the subthreshold swing and V_{RMS} has been reported [64]. The result strongly suggests that the increase in V_{RMS} is enlarged by the physical origin, which increases the subthreshold swing, and the origin is an enhancement of the percolation path formation [64]. Note that RTN has to be enhanced by a minimal small gate control effect on the channel. Furthermore, the variability of the threshold voltage is increased using the buried channel MOSFETs; thus, we cannot introduce buried channel MOSFETs when the fixed pattern noise is critical for device performance.

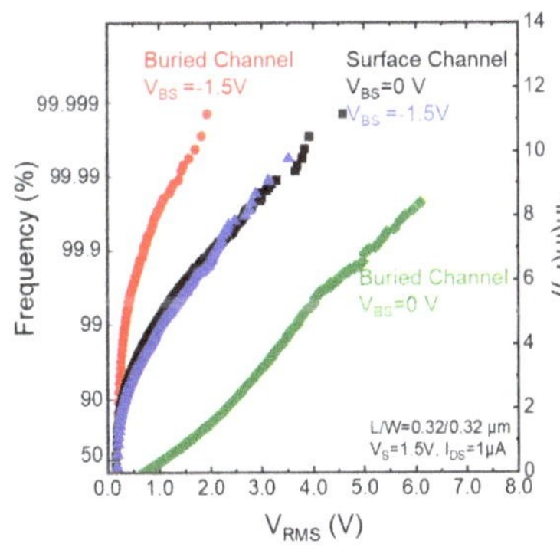

Figure 26. Gumbel plot of V_{RMS} for the buried channel and surface channel MOSFETs with the back bias conditions of 0 and −1.5 V.

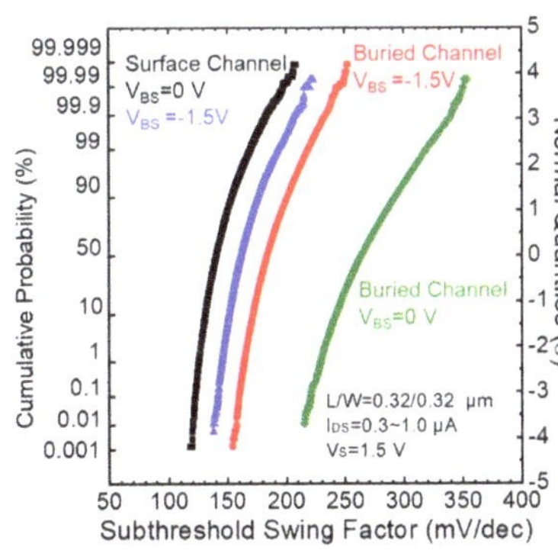

Figure 27. V_{BS} dependance of V_{RMS} for the narrow, middle, and deep samples.

3.6.2. Asymmetry Source and Drain Width MOSFETs

Figure 28 shows the layout structure of rectangular and trapezoidal shape MOSFETs used in this experiment [62,63]. In the trapezoidal MOSFETs, when current flows from the left to right direction, the source has a wide gate width, and when current flows from the right to left direction, the source has a shallow gate width. The Gumbel plots of V_{RMS} for trapezoidal ((a)W_D < W_S and (b) W_S < W_D) and (c) rectangular MOSFETs are shown in Figure 29 [62]. The gate width of rectangular MOSFETs was set as the average of the gate width of trapezoidal MOSFETs. I_{DS} was varied from 0.1 to 11 μA for constant V_{BS} of −1.9 V and V_{DS} of 1.4 V. In (c) rectangular MOSFETs, similar phenomena, as shown in Figure 19, are obtained. In contrast, in the trapezoidal MOSFETs, V_{RMS} increases with an increase in I_{DS} (V_{GS}), and those of W_D < W_S are larger than those of W_S < W_D. V_{DS} is larger than V_{GS} in this experiment. MOSFETs were operated in the saturation region, and the channel formed near the source. Increasing I_{DS} increases electron density in the channel, and the electron density at the source of MOSFETs with W_D < W_S is less than that with W_S < W_D. These characteristics indicate that the influence of a charged trap reduces at a high carrier density condition [39,60,62]. This means that the electron density and the location in the channel are important factors affecting RTN characteristics. RTN characteristics were evaluated for various V_{DS} using rectangular and trapezoidal MOSFETs.

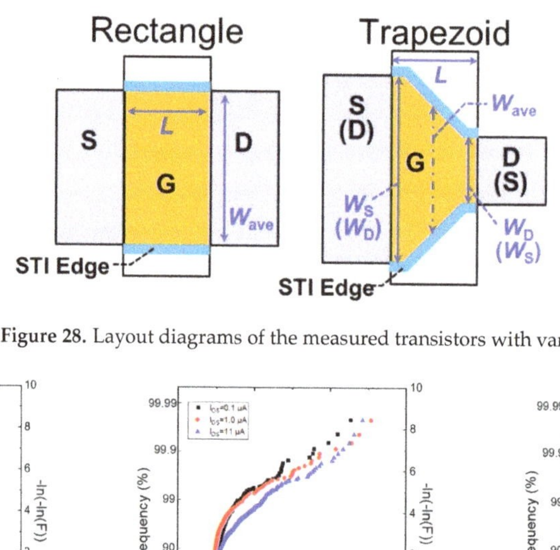

Figure 28. Layout diagrams of the measured transistors with various gate shapes.

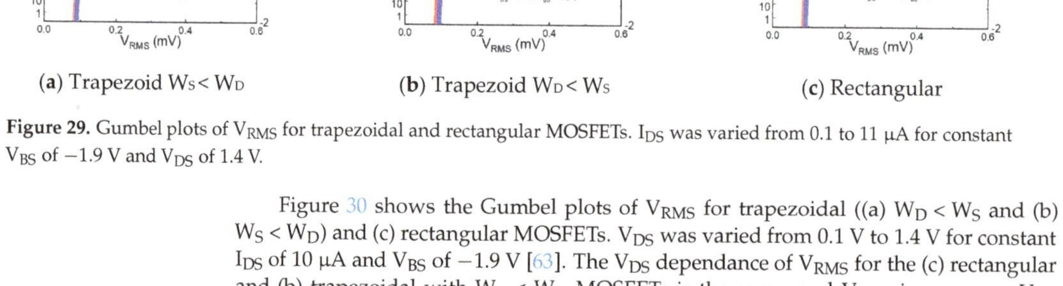

(**a**) Trapezoid W_S < W_D (**b**) Trapezoid W_D < W_S (**c**) Rectangular

Figure 29. Gumbel plots of V_{RMS} for trapezoidal and rectangular MOSFETs. I_{DS} was varied from 0.1 to 11 μA for constant V_{BS} of −1.9 V and V_{DS} of 1.4 V.

Figure 30 shows the Gumbel plots of V_{RMS} for trapezoidal ((a) W_D < W_S and (b) W_S < W_D) and (c) rectangular MOSFETs. V_{DS} was varied from 0.1 V to 1.4 V for constant I_{DS} of 10 μA and V_{BS} of −1.9 V [63]. The V_{DS} dependance of V_{RMS} for the (c) rectangular and (b) trapezoidal with W_S < W_D MOSFETs is the same, and V_{RMS} increases as V_{DS} increases, monotonically. The dependance of (b) trapezoidal with W_S < W_D on V_{DS} is larger than that of rectangular MOSFETs. In contrast, V_{RMS} increases with an increase in V_{DS} at less than 0.3 V; however, V_{RMS} decreases with an increase in V_{DS} at >0.3V for trapezoidal

MOSFETs with $W_S > W_D$. When V_{DS} is smaller than the pinch-off voltage (V_{GS}-V_{TH} = 0.3 V in this experiment), the channel is uniformly formed under the gate oxide, and the channel vanishes at the drain edge at the pinch-off voltage. The vanished region expands with increasing V_{DS}. On the other hand, I_{DS} was set at a constant of 10 μA, and this means that V_{GS} decreases as V_{DS} increases at less than a pinch-off voltage of 0.3V. In rectangular MOSFETs, V_{RMS} increases with a decrease in V_{GS}, which is the same effect as shown in Figure 7. In trapezoidal MOSFETs with $W_S < W_D$, V_{DS} dependance was enhanced by reducing the channel width. In trapezoidal MOSFETs with $W_D < W_S$, V_{DS} dependance is the same as others at less V_{DS} than a pinch-off voltage of 0.3 V. However, the opposite dependency is obtained at larger V_{DS} than the pinch-off voltage. It is considered that the apparent gate width of MOSFETs ($W_D < W_S$) increases when the pinch-off point reaches the source, and then, the size effect of V_{RMS} shown in Figure 6 is obtained. These data imply that the noise strength depends heavily on operation conditions, which means that the location and electron density in a channel are critical for RTN generation.

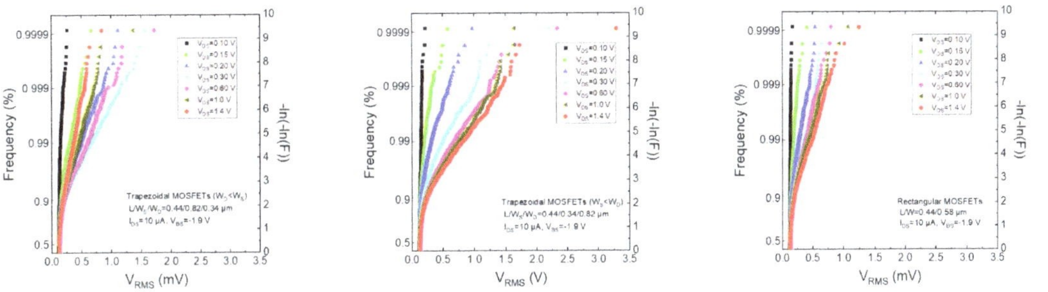

(**a**) Trapezoid $W_S < W_D$ (**b**) Trapezoid $W_D < W_S$ (**c**) Rectangular

Figure 30. Gumbel plots of V_{RMS} for trapezoidal and rectangular MOSFETs. V_{DS} was varied from 0.1 V to 1.4V for constant I_{DS} of 10 μA and V_{BS} of −1.9 V.

Although using trapezoidal MOSFETs in real electronic devices is difficult, changing the shape of MOSFETs is very useful to obtain much information about RTN characteristics. For example, the effect of trap at the isolation edge can be evaluated using octagonal MOSFETs, which have only a gate edge and no shallow trench isolation edge [62].

3.6.3. MOSFETs with Atomically Flat Gate Insulator/Si Interface

The roughness of the interface between the gate insulator and Si is essential for MOSFETs. The interface roughness degrades not only electron mobility [66–71] and gate dielectric reliability [72–74], but also noise generation [71,75,76]. An atomically flat interface [77–84] is effective for reducing low-frequency noise [79,83–87].

Figure 31 shows images of an atomically flat surface and as-received Si(100) measured by atomic force microscopy (AFM). The atomically flat surface was formed in the active region with shallow trench isolation and was measured after the gate oxide formation and following oxide stripping [84]. The average roughness (Ra) of the conventional surface is 0.12 nm, which is the same as the initial surface of Si(100). In an atomically flat surface, a step and terrace structure can be obtained, and the step height is the same as the monoatomic step length of Si(100) of 0.135 nm. The terrace width (L) is defined by the following equation using the off-angle (θ) to the just (100) orientation. The average roughness in the trace of the atomically flat interface was less than 0.04 nm, which is the detection limit of our AFM system.

$$L = \frac{0.135}{\tan\theta}\ (\text{nm}) \quad (16)$$

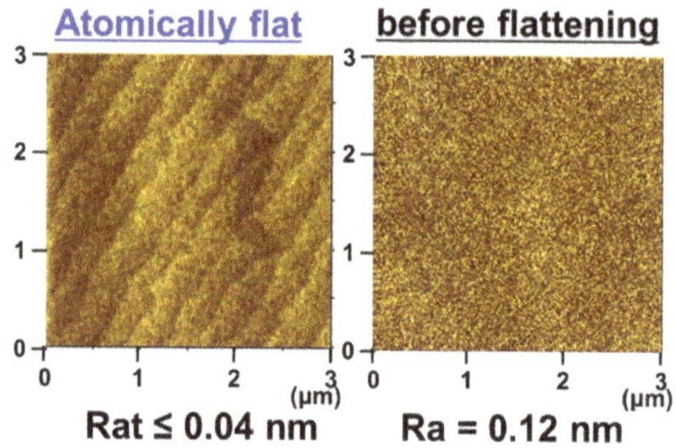

Figure 31. Surface images of atomically flat and as-received Si(100) (before flattening) surfaces measured by atomic force microscopy.

Figure 32 shows the Gumbel plot of V_{RMS} for the atomically flat and conventional SiO_2/Si interface [84]. The noise of the atomically flat interface is less than that of the conventional interface. This means that introducing the atomically flat interface is extremely effective for reducing RTN as well as 1/f noise [86–88]. The atomically flat surface was formed before gate oxidation in this experiment [84].

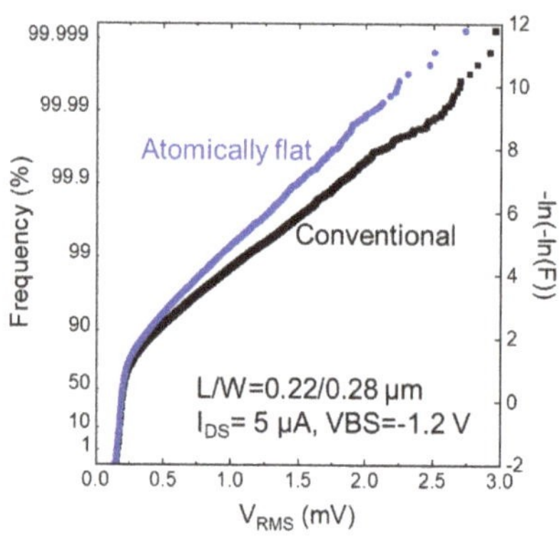

Figure 32. Gumbel plot of V_{RMS} for atomically flat and conventional SiO_2/Si interface.

To implement the surface flattening process, a low temperature of less than 900 °C and low oxidation species, such as O_2 and H_2O, must be required [81,82,85]. There is another method for flattening the surface first and keeping it during the process steps preceding gate oxidation [85,87,88]. Other problems exist, such as STI edge shape and dopant segregation, and the solutions to these problems may affect not only MOSFET characteristics, but also noise [84]. The flattening process just before the gate oxidation is superior to the flattening process in the first step for introducing interface flattening between SiO_2 and Si, and this can be obtained by low temperature Ar annealing by reducing oxidation species.

4. Conclusions

The importance of low-frequency noise in LSI, and various effects on RTN, such as MOSFETs' size, bias and operation conditions, and device structures, are described. The measurement technique using the array test circuit and the extraction of important parameters (time constants and amplitude) in RTN characteristics are also described. Time constants can be extracted essentially using classical equations; however, it is not as simple when downsizing MOSFETs and reducing the number of channel electrons, and the percolation path is formed. Variability of low-frequency noise increases with shrinkage of MOSFETs. In this paper, we evaluated relatively large planer MOSFETs (L = 0.22~0.4 μm), unfortunately. The size of MOSFETs has been downscaled to less than 10 nm and the structure has changed the planer to FinFET, recently. We have to continue the evaluation of such miniaturized and new structure devices. To assess the effect of this noise on MOSFETs, we have to understand their characteristics statistically, and then, sufficient samples must be accurately evaluated in a short period.

Funding: This research received no external funding.

Acknowledgments: The author acknowledges Shigetoshi Sugawa, Rihito Kuroda, Tetsuya Goto, Tomoyuki Suwa, Keninchi Abe, Philippe Gaubert, Yuki Kumagai, Takafumi Fujisawa, Hiroyuki Suzuki, Akihiro Yonezawa, Toshiki Obara, Takezo Mawaki, Shinya Ichino, Ryo Akimoto, and Tatsuki Ueta for their measurements and useful discussions and Yutaka Kamata and Katsuhiko Shibusawa for fabrication of the array test circuit and useful discussions.

Conflicts of Interest: The authors declare no conflict of interest.

References

1. Dennard, R.H.; Gaensslen, F.H.; Kuhn, L.; Yu, H.N. Design of micron MOS switching devices. In Proceedings of the 1972 International Electron Devices Meeting, Washington, DC, USA, 4–6 December 1972; pp. 168–170.
2. Dennard, R.H.; Gaensslen, F.H.; Rideout, V.L.; Bassous, E.; LeBlanc, A.R. Design of ion-implanted MOSFET's with very small physical dimensions. *IEEE J. Solid-State Circuits* **1974**, *9*, 256–268. [CrossRef]
3. Moore, G.E. Cramming more components onto integrated circuits. *Electronics* **1965**, *38*, 114. [CrossRef]
4. Boukhbza, J.; Plivier, P. *Flash Memory Integration -Performance and Energy Considerations*; Elsevier Ltd.: Oxford, UK, 2017.
5. Parameswaran, S.; Hui, G. Power consumption in CMOS combinational logic blocks at high frequencies. In Proceedings of the ASP-DAC '97: Asia and South Pacific Design Automation Conference, Chiba, Japan, 28–31 January 1997; pp. 195–200.
6. Appuswamy, R.; Olma, M.; Ailamaki, A. Scaling the Memory Power Wall With DRAM-Aware Data Management. In Proceedings of the 11th International Workshop on Data Management on New Hardware, Melbourne, VIC, Australia, 31 May–4 July 2015.
7. Poess, M.; Nambiar, R.O. Energy cost, the key challenge of today's data centers: A power consumption analysis of TPC-C results. In Proceedings of the 34th International Conference on Very Large Data Bases, Auckland, New Zealand, 23–28 August 2008; pp. 1229–1240.
8. Karyakin, A.; Salem, K. An analysis of memory power consumption in database systems. In Proceedings of the 13th International Workshop on Data Management on New Hardware, Chicago, IL, USA, 15 May 2017.
9. Carter, J.; Rajamani, K. Designing Energy-Efficient Servers and Data Centers. *Computer* **2010**, *43*, 76–78. [CrossRef]
10. Ohmi, T.; Hirayama, M.; Teramoto, A. New era of silicon technologies due to radical reaction based semiconductor manufacturing. *J. Phys. D Appl. Phys.* **2006**, *39*, R1–R17. [CrossRef]
11. Uren, M.J.; Day, D.J.; Kirton, M.J. 1/f and random telegraph noise in silicon metal-oxide-semiconductor field-effect transistors. *Appl. Phys Lett.* **1985**, *47*, 1195–1197. [CrossRef]
12. Kirton, M.J.; Uren, M.J. Noise in solid-state microstructures: A new perspective on individual defects, interface states and low-frequency (1/f) noise. *Adv. Phys.* **1989**, *38*, 367–468. [CrossRef]
13. Christensson, S.; Lundström, I.; Svensson, C. Low frequency noise in MOS transistors—I Theory. *Solid State Electron.* **1968**, *11*, 797–812. [CrossRef]
14. Toita, M.; Vandamme, L.K.J.; Sugawa, S.; Teramoto, A.; Ohmi, T. Geometry and bias dependence of low-frequency random telegraph signal and 1/f noise levels in mosfets. *Fluct. Noise Lett.* **2005**, *5*, L539–L548. [CrossRef]
15. Sugawa, S. The Advanced Technology of the Semiconductor Integrated Circuits Needs the "Analog Intelligence" Again. *DENSO Tech. Rev.* **2005**, *10*, 3–9. (In Japanese)
16. Leyris, C.; Martinez, F.; Valenza, M.; Hoffmann, A.; Vildeuil, J.C.; Roy, F. Impact of Random Telegraph Signal in CMOS Image Sensors for Low-Light Levels. In Proceedings of the 32nd European Solid-State Circuits Conference, Montreux, Swizerland, 19–21 September 2006; pp. 376–379.
17. Wang, X.; Rao, P.R.; Mierop, A.; Theuwissen, A.J.P. Random Telegraph Signal in CMOS Image Sensor Pixels. In Proceedings of the International Electron Devices Meeting, San Francisco, CA, USA, 11–13 December 2006; pp. 115–118.
18. Chao, C.Y.-P.; Tu, H.; Wu, T.M.-H.; Chou, K.-Y.; Yeh, S.-F.; Yin, C.; Lee, C.-L. Statistical Analysis of the Random Telegraph Noise in a 1.1 μm Pixel, 8.3 MP CMOS Image Sensor Using On-Chip Time Constant Extraction Method. *Sensors* **2017**, *17*, 2704. [CrossRef] [PubMed]
19. Chao, C.Y.-P.; Yeh, S.-F.; Wu, M.-H.; Chou, K.-Y.; Tu, H.; Lee, C.-L.; Yin, C.; Paillet, P.; Goiffon, V. Random Telegraph Noises from the Source Follower, the Photodiode Dark Current, and the Gate-Induced Sense Node Leakage in CMOS Image Sensors. *Sensors* **2019**, *19*, 5447. [CrossRef]

20. Kuroda, R.; Teramoto, A.; Sugawa, S. Impacts of Random Telegraph Noise with Various Time Constants and Number of States in Temporal Noise of CMOS Image Sensors. *ITE Trans. Media Technol. Appl.* **2018**, *6*, 171–179. [CrossRef]
21. Wang, X.; Snoeij, M.F.; Rao, P.R.; Mierop, A.; Theuwissen, A.J.P. A CMOS Image Sensor with a Buried-Channel Source Follower. In Proceedings of the IEEE International Solid-State Circuits Conference, San Francisco, CA, USA, 3–7 February 2008; pp. 62–595.
22. Toh, S.O.; Tsukamoto, Y.; Zheng, G.; Jones, L.; Tsu-Jae King, L.; Nikolic, B. Impact of random telegraph signals on V_{min} in 45 nm SRAM. In Proceedings of the IEEE International Electron Devices Meeting, Baltimore, MD, USA, 24 July 2020; pp. 767–770.
23. Takeuchi, K.; Nagumo, T.; Takeda, K.; Asayama, S.; Yokogawa, S.; Imai, K.; Hayashi, Y. Direct observation of RTN-induced SRAM failure by accelerated testing and its application to product reliability assessment. In Proceedings of the Symposium on VLSI Technology, 15–17 June 2010; pp. 189–190.
24. Tanizawa, M.; Ohbayashi, S.; Okagaki, T.; Sonoda, K.; Eikyu, K.; Hirano, Y.; Ishikawa, K.; Tsuchiya, O.; Inoue, Y. Application of a Statistical Compact Model for Random Telegraph Noise to Scaled-SRAM Vmin Analysis. In Proceedings of the Symposium on VLSI Technology, Honolulu, HI, USA, 15–17 June 2010; pp. 95–96.
25. Aadithya, K.V.; Demir, A.; Venugopalan, S.; Roychowdhury, J. Accurate Prediction of Random Telegraph Noise Effects in SRAMs and DRAMs. *IEEE Trans. Computer-Aided Des. Integr. Circuits Syst.* **2013**, *32*, 73–86. [CrossRef]
26. Compagnoni, C.M.; Gusmeroli, R.; Spinelli, A.S.; Lacaita, A.L.; Bonanomi, M.; Visconti, A. Statistical Investigation of Random Telegraph Noise ID Instabilities in Flash Cells at Different Initial Trap-filling Conditions. In Proceedings of the International Reliability physics symposium, Phoenix, AZ, USA, 15–19 April 2007; pp. 161–166.
27. Tega, N.; Miki, H.; Osabe, T.; Kotabe, A.; Otsuga, K.; Kurata, H.; Kamohara, S.; Tokami, K.; Ikeda, Y.; Yamada, R. Anomalously Large Threshold Voltage Fluctuation by Complex Random Telegraph Signal in Floating Gate Flash Memory. In Proceedings of the International Electron Devices Meeting, San Francisco, CA, USA, 11–13 December 1995; pp. 491–494.
28. Kurata, H.; Otsuga, K.; Kotabe, A.; Kajiyama, S.; Osabe, T.; Sasago, Y.; Narumi, S.; Tokami, K.; Kamohara, S.; Tsuchiya, O. The Impact of Random Telegraph Signals on the Scaling of Multilevel Flash Memories. In Proceedings of the Symposium on VLSI Circuits, Honolulu, HI, USA, 15–17 June 2006; pp. 112–113.
29. Kurata, H.; Otsuga, K.; Kotabe, A.; Kajiyama, S.; Osabe, T.; Sasago, Y.; Narumi, S.; Tokami, K.; Kamohara, S.; Tsuchiya, O. Random Telegraph Signal in Flash Memory: Its Impact on Scaling of Multilevel Flash Memory Beyond the 90-nm Node. *IEEE J. Olid-State Circuits* **2007**, *42*, 1362–1369. [CrossRef]
30. Sing-Rong, L.; McMahon, W.; Lu, Y.L.R.; Yung-Huei, L. RTS Noise Characterization in Flash Cells. *IEEE Electron Device Lett.* **2008**, *29*, 106–108.
31. Ghetti, A.; Compagnoni, C.M.; Spinelli, A.S.; Visconti, A. Comprehensive Analysis of Random Telegraph Noise Instability and Its Scaling in Deca-Nanometer Flash Memories. *IEEE Trans. Electron Devices* **2009**, *56*, 1746–1752. [CrossRef]
32. Cai, Y.; Song, Y.H.; Kwon, W.-H.; Lee, B.Y.; Park, C.-K. The Impact of Random Telegraph Signals on the Threshold Voltage Variation of 65 nm Multilevel NOR Flash Memory. *Jpn. J. Appl. Phys.* **2008**, *47*, 2733. [CrossRef]
33. Toita, M.; Sugawa, S.; Teramoto, A.; Ohmi, T. Sub-Micron MOSFETs Technology Characterization by Low-Frequency Noise. In Proceedings of the 3rd European Microelectronics and Packaging Symposium, Prague, Czech Republic, 16–18 June 2004; pp. 19–24.
34. Abe, K.; Fujisawa, T.; Teramoto, A.; Watabe, S.; Sugawa, S.; Ohmi, T. Anomalous Random Telegraph Signal Extractions from a Very Large Number of n-Metal Oxide Semiconductor Field-Effect Transistors Using Test Element Groups with 0.47 Hz–3.0 MHz Sampling Frequency. *Jpn. J. Appl. Phys.* **2009**, *48*. [CrossRef]
35. Watabe, S.; Sugawa, S.; Teramoto, A.; Ohmi, T. New Statistical Evaluation Method for the Variation of Metal-Oxide-Semiconductor Field-Effect Transistors. *Jpn. J. Appl. Phys.* **2007**, *46*, 2054–2057. [CrossRef]
36. Abe, K.; Sugawa, S.; Watabe, S.; Miyamoto, N.; Teramoto, A.; Toita, M.; Kamata, Y.; Shibusawa, K.; Ohmi, T. Statistical Analysis of RTS Noise and Low Frequency Noise in 1M MOSFETs Using an Advanced TEG. In Proceedings of the 19th International Conference on Noise and Fluctuations, Tokyo, Japan, 9–14 September 2007; pp. 115–118.
37. Abe, K.; Sugawa, S.; Watabe, S.; Miyamoto, N.; Teramoto, A.; Kamata, Y.; Shibusawa, K.; Toita, M.; Ohmi, T. Random Telegraph Signal Statistical Analysis using a Very Large-scale Array TEG with 1M MOSFETs. In Proceedings of the IEEE Symposium on VLSI Technology, Kyoto, Japan, 12–14 June 2007; pp. 210–211.
38. Kumagai, Y.; Abe, K.; Fujisawa, T.; Watabe, S.; Kuroda, R.; Miyamoto, N.; Suwa, T.; Teramoto, A.; Sugawa, S.; Ohmi, T. Large-Scale Test Circuits for High-Speed and Highly Accurate Evaluation of Variability and Noise in Metal-Oxide-Semiconductor Field-Effect Transistor Electrical Characteristics. *Jpn. J. Appl. Phys.* **2011**, *50*. [CrossRef]
39. Abe, K.; Sugawa, S.; Kuroda, R.; Watabe, S.; Miyamoto, N.; Teramoto, A.; Ohmi, T.; Kamata, Y.; Shibusawa, K. Analysis of Source Follower Random Telegraph Signal Using nMOS and pMOS Array TEG. In Proceedings of the International Image Sensor Workshop, Ogunquit, MA, USA, 7–10 June 2007; pp. 62–65.
40. Watabe, S.; Teramoto, A.; Abe, K.; Fujisawa, T.; Miyamoto, N.; Sugawa, S.; Ohmi, T. A Simple Test Structure for Evaluating the Variability in Key Characteristics of a Large Number of MOSFETs. *IEEE Trans. Semicond. Manuf.* **2012**, *25*, 145–154. [CrossRef]
41. Teramoto, A.; Fujisawa, T.; Abe, K.; Sugawa, S.; Ohmi, T. Statistical evaluation for trap energy level of RTS characteristics. In Proceedings of the Symposium on VLSI Technology, Honolulu, HI, USA, 15–17 June 2016; pp. 99–100.
42. Fujisawa, T.; Abe, K.; Watabe, S.; Miyamoto, N.; Teramoto, A.; Sugawa, S.; Ohmi, T. Analysis of Hundreds of Time Constant Ratios and Amplitudes of Random Telegraph Signal with Very Large Scale Array Test Pattern. *Jpn. J. Appl. Phys.* **2010**, *49*. [CrossRef]

43. Fujisawa, T.; Abe, K.; Watabe, S.; Miyamoto, N.; Teramoto, A.; Sugawa, S.; Ohmi, T. Accurate Time Constant of Random Telegraph Signal Extracted by a Sufficient Long Time Measurement in Very Large-Scale Array TEG. In Proceedings of the IEEE International Conference on Microelectronic Test Structures, Oxnard, CA, USA, 30 March–2 April 2009; pp. 19–24.
44. Abe, K.; Teramoto, A.; Sugawa, S.; Ohmi, T. Understanding of Traps Causing Random Telegraph Noise Based on Experimentally Extracted Time Constants and Amplitude. In Proceedings of the International Reliability physics symposium, Monterey, CA, USA, 10–14 April 2011; pp. 381–386.
45. Haartman, M.V.; Östling, M. *LOW-Frequency Noise in Advanced Mos Devices*, 1 ed.; Springer Netherlands: Dordrecht, The Netherland, 2007. [CrossRef]
46. Slavcheva, G.; Davies, J.H.; Brown, A.R.; Asenov, A. Potential fluctuations in metal–oxide–semiconductor field-effect transistors generated by random impurities in the depletion layer. *J. Appl. Phys.* **2002**, *91*, 4326–4334. [CrossRef]
47. Asenov, A.; Balasubramaniam, R.; Brown, A.R.; Davies, J.H. RTS amplitudes in decananometer MOSFETs: 3-D simulation study. *IEEE Trans. Electron Devices* **2003**, *50*, 839–845. [CrossRef]
48. Sano, N.; Yoshida, K.; Yao, C.-W.; Watanabe, H. Physics of Discrete Impurities under the Framework of Device Simulations for Nanostructure Devices. *Materials* **2018**, *11*, 2559. [CrossRef]
49. Sonoda, K.; Ishikawa, K.; Eimori, T.; Tsuchiya, O. Discrete Dopant Effects on Statistical Variation of Random Telegraph Signal Magnitude. *IEEE Trans. Electron Devices* **2007**, *54*, 1918–1925. [CrossRef]
50. Taur, Y.; Ning, T.H. *Fundamentals of Modern VLSI Devices*, 2nd ed.; Cambridge University Press: Cambridge, UK, 2009; pp. 234–239.
51. Abe, K.; Teramoto, A.; Watabe, S.; Fujisawa, T.; Sugawa, S.; Kamata, Y.; Shibusawa, K.; Ohmi, T. Experimental Investigation of Effect of Channel Doping Concentration on Random Telegraph Signal Noise. *Jpn. J. Appl. Phys.* **2010**, *49*. [CrossRef]
52. Hauser, J.R. *Handbook of Semiconductor Manufacturing Technology*, 2 ed.; CRC Press: Boca Raton, FL, USA, 2007.
53. Obara, T.; Teramoto, A.; Yonezawa, A.; Kuroda, R.; Sugawa, S.; Ohmi, T. Analyzing Correlation between Multiple Traps in RTN Characteristics. In Proceedings of the International Reliability Physics Symposium, Waikoloa, HI, USA, 1–5 June 2014.
54. Nagumo, T.; Takeuchi, K.; Yokogawa, S.; Imai, K.; Hayashi, Y. New Analysis Methods for Comprehensive Understanding of Random Telegraph Noise. In Proceedings of the Technical Digest IEEE International Electron Devices Meeting, Baltimore, MD, USA, 7–9 December 2009; pp. 759–762.
55. Nagumo, T.; Takeuchi, K.; Hase, T.; Hayashi, Y. Statistical characterization of trap position, energy, amplitude and time constants by RTN measurement of multiple individual traps. In Proceedings of the 2010 IEEE International Electron Devices Meeting, San Francisco, CA, USA, 6–8 December 2010; pp. 628–631.
56. Tsuchiya, T.; Tamura, N.; Sakakidani, A.; Sonoda, K.; Kamei, M.; Yamakawa, S.; Kuwabara, S. Characterization of Oxide Traps Participating in Random Telegraph Noise Using Charging History Effects in Nano-Scaled MOSFETs. *ECS Trans.* **2013**, *58*, 265–279. [CrossRef]
57. Yonezawa, A.; Teramoto, A.; Kuroda, R.; Suzuki, H.; Sugawa, S.; Ohmi, T. Statistical analysis of Random Telegraph Noise reduction effect by separating channel from the interface. In Proceedings of the IRPS, Anaheim, CA, USA, 15–19 April 2012.
58. Obara, T.; Yonezawa, A.; Teramoto, A.; Kuroda, R.; Sugawa, S.; Ohmi, T. Extraction of time constants ratio over nine orders of magnitude for understanding random telegraph noise in metal–oxide–semiconductor field-effect transistors. *Jpn. J. Appl. Phys.* **2014**, *53*. [CrossRef]
59. Yonezawa, A.; Teramoto, A.; Obara, T.; Kuroda, R.; Sugawa, S.; Ohmi, T. The study of time constant analysis in random telegraph noise at the sub-threshold voltage region. In Proceedings of the International Reliability Physics Symposium Monterey, Anaheim, CA, USA, 14–18 April 2012; p. XT11.
60. Ichino, S.; Mawaki, T.; Teramoto, A.; Kuroda, R.; Park, H.; Wakashima, S.; Goto, T.; Suwa, T.; Sugawa, S. Effect of drain current on appearance probability and amplitude of random telegraph noise in low-noise CMOS image sensors. *Jpn. J. Appl. Phys.* **2018**, *57*. [CrossRef]
61. Yonezawa, A.; Kuroda, R.; Teramoto, A.; Obara, T.; Sugawa, S. A statistical evaluation of effective time constants of random telegraph noise with various operation timings of in-pixel source follower transistors. In Proceedings of the SPIE-IS&T Electronic Imaging, SPIE, San Francisco, CA, USA, 2–6 February 2014; p. 90220F.
62. Ichino, S.; Mawaki, T.; Teramoto, A.; Kuroda, R.; Wakashima, S.; Suwa, T.; Sugawa, S. Statistical Analyses of Random Telegraph Noise in Pixel Source Follower with Various Gate Shapes in CMOS Image Sensor. *ITE Trans. Media Technol. Appl.* **2018**, *6*, 163–170. [CrossRef]
63. Akimoto, R.; Kuroda, R.; Teramoto, A.; Mawaki, T.; Ichino, S.; Suwa, T.; Sugawa, S. Effect of Drain-to-Source Voltage on Random Telegraph Noise Based on Statistical Analysis of MOSFETs with Various Gate Shapes. In Proceedings of the 2020 IEEE International Reliability Physics Symposium, 28 April–30 May 2020; p. 9A2.
64. Kuroda, R.; Yonezawa, A.; Teramoto, A.; Li, T.L.; Tochigi, Y.; Sugawa, S. A Statistical Evaluation of Random Telegraph Noise of In-Pixel Source Follower Equivalent Surface and Buried Channel Transistors. *IEEE Trans. Electron Devices* **2013**, *60*, 3555–3561. [CrossRef]
65. Suzuki, H.; Kuroda, R.; Teramoto, A.; Yonezawa, A.; Sugawa, S.; Ohmi, T. Impact of Random Telegraph Noise Reduction with Buried Channel MOSFET. In Proceedings of the 2011 International Conference on Solid State Devices and Materials, Nagoya, Japan, 28–30 September 2011; pp. 851–852.
66. Yamanaka, T.; Fang, S.J.; Heng-Chih, L.; Snyder, J.P.; Helms, C.R. Correlation between inversion layer mobility and surface roughness measured by AFM. *IEEE Electron Device Lett.* **1996**, *17*, 178–180. [CrossRef]

67. Ohmi, T.; Kotani, K.; Teramoto, A.; Miyashita, M. Dependence of electron channel mobility on Si-SiO_2 interface microroughness. *IEEE Electron Device Lett.* **1991**, *12*, 652–654. [CrossRef]
68. Ishizaka, M.; Iizuka, T.; Ohi, S.; Fukuma, M.; Mikoshiba, H. Advanced electron mobility model of MOS inversion layer considering 2D-degenerate electron gas physics. In Proceedings of the International Electron Devices Meeting, San Francisco, CA, USA, 9–12 December 1990; pp. 763–766.
69. Ferry, D.K. The transport of electrons in quantized inversion and accumulation layers in III–V compounds. *Thin Solid Films* **1979**, *56*, 243–252. [CrossRef]
70. Takagi, S.; Toriumi, A.; Iwase, M.; Tango, H. On the universality of inversion layer mobility in Si MOSFET's: Part I-effects of substrate impurity concentration. *IEEE Trans. Electron Devices* **1994**, *41*, 2357–2362. [CrossRef]
71. Teramoto, A.; Hamada, T.; Yamamoto, M.; Gaubert, P.; Akahori, H.; Nii, K.; Hirayama, M.; Arima, K.; Endo, K.; Sugawa, S.; et al. Very High Carrier Mobility for High-Performance CMOS on a Si(110) Surface. *IEEE Trans. Electron Devices* **2007**, *54*, 1438–1445. [CrossRef]
72. Ohmi, T.; Miyashita, M.; Itano, M.; Imaoka, T.; Kawanabe, I. Dependence of thin-oxide films quality on surface microroughness. *IEEE Trans. Electron Devices* **1992**, *39*, 537–545. [CrossRef]
73. Morita, M.; Teramoto, A.; Makihara, K.; Ohmi, T.; Nakazato, Y.; Uchiyama, A.; Abe, T. Effects of Si Wafer Surface Micro-Roughness on Electrical Properties of Very-Thin Gate Oxide Films. In *ULSI Science and Technology/1991*; Electrochemical Society: Pennington, NJ, USA, 1991; pp. 400–408.
74. Makihara, K.; Teramoto, A.; Nakamura, K.; Kwon, M.Y.; Morita, M.; Ohmi, T. Preoxide-Controlled Oxidation for Very Thin Oxide Films. *Jpn. J. Appl. Phys.* **1993**, *32*, 294–297. [CrossRef]
75. Gaubert, P.; Teramoto, A.; Hamada, T.; Yamamoto, M.; Nii, K.; Akahori, H.; Kotani, K.; Ohmi, T. Impact of interface micro-roughness on low frequency noise in (110) and (100) pMOSFETs. In Proceedings of the 18th International Conference on Noise and Fluctuations, Salamanca, Spain, 19–23 September 2005; pp. 199–202.
76. Gaubert, P.; Teramoto, A.; Hamada, T.; Yamamoto, M.; Kotani, K.; Ohmi, T. 1/f noise suppression of pMOSFETs fabricated on Si(100) and Si(110) using an alkali-free cleaning process. *IEEE Trans. Electron Devices* **2006**, *53*, 851–856. [CrossRef]
77. Matsushita, Y.; Watatsuki, M.; Saito, Y. Improvement of Silicon Surface Quality by H_2 Anneal. In Proceedings of the Conference of Solid State Device and Materials, Tokyo, Japan, 20–22 August 1986; pp. 529–535.
78. Morita, Y.; Tokumoto, H. Atomic scale flattening and hydrogen termination of the Si(001) surface by wet-chemical treatment. In Proceedings of the 42nd national symposium of the American Vacuum Society, Mineapolis, MN, USA, 16–20 October 1995; pp. 854–858.
79. Kuroda, R.; Teramoto, A.; Suwa, T.; Hasebe, R.; Li, X.; Konda, M.; Sugawa, S.; Ohmi, T. Atomically flat gate insulator/silicon (100) interface formation introducing high mobility, ultra-low noise, and small characteristics variation CMOSFET. In Proceedings of the 38th European Solid-State Device Research Conference, Edinburgh, UK, 15–19 September 2008; pp. 83–86.
80. Li, X.; Suwa, T.; Teramoto, A.; Kuroda, R.; Sugawa, S.; Ohmi, T. Atomically Flattening Technology at 850 °C for Si(100) Surface. *ECS Trans.* **2010**, *28*, 299–309. [CrossRef]
81. Li, X.; Teramoto, A.; Suwa, T.; Kuroda, R.; Sugawa, S.; Ohmi, T. Formation speed of atomically flat surface on Si (100) in ultra-pure argon. *Microelectron. Eng.* **2011**, *88*, 3133–3139. [CrossRef]
82. Goto, T.; Kuroda, R.; Suwa, T.; Teramoto, A.; Akagawa, N.; Kimoto, D.; Sugawa, S.; Ohmi, T.; Kamata, Y.; Kumagai, Y.; et al. Low Temperature Atomically Flattening of Si Surface of Shallow Trench Isolation Pattern. *ECS Trans.* **2015**, *66*, 285–292. [CrossRef]
83. Goto, T.; Kuroda, R.; Akagawa, N.; Suwa, T.; Teramoto, A.; Li, X.; Obara, T.; Kimoto, D.; Sugawa, S.; Ohmi, T.; et al. Atomically flattening of Si surface of silicon on insulator and isolation-patterned wafers. *Jpn. J. Appl. Phys.* **2015**, *54*. [CrossRef]
84. Goto, T.; Kuroda, R.; Akagawa, N.; Suwa, T.; Teramoto, A.; Li, X.; Obara, T.; Kimoto, D.; Sugawa, S.; Kamata, Y.; et al. Introduction of Atomically Flattening of Si Surface to Large-Scale Integration Process Employing Shallow Trench Isolation. *ECS J. Solid State Sci. Technol.* **2016**, *5*, P67–P72. [CrossRef]
85. Gaubert, P.; Kircher, A.; Park, H.; Kuroda, R.; Sugawa, S.; Goto, T.; Suwa, T.; Teramoto, A. Atomically flat interface for noise reduction in SOIMOSFETs. In Proceedings of the 24th International Conference on Noise and Fluctuations, Vilnius, Lithuania, 20–23 June 2017.
86. Tanaka, K.; Watanabe, K.; Ishino, H.; Sugawa, S.; Teramoto, A.; Hirayama, M.; Ohmi, T. A Technology for Reducing Flicker Noise for ULSI Applications. *Jpn. J. Appl. Phys.* **2003**, *42*, 2106–2109. [CrossRef]
87. Kuroda, R.; Suwa, T.; Teramoto, A.; Hasebe, R.; Sugawa, S.; Ohmi, T. Atomically Flat Silicon Surface and Silicon/Insulator Interface Formation Technologies for (100) Surface Orientation Large-Diameter Wafers Introducing High Performance and Low-Noise Metal-Insulator-Silicon FETs. *IEEE Trans. Electron Devices* **2009**, *56*, 291–298. [CrossRef]
88. Gaubert, P.; Kuroda, R.; Endo, S.; Kuboyama, Y.; Kitagaki, T.; Nada, H.; Tamura, H.; Teramoto, A.; Ohmi, T. Atomically Flat Interface for the Reduction of the Low Frequency Noise on Si(100) nMOS Transistors. In Proceedings of the 215th ECS Meeting, San Francisco, CA, USA, 24–29 May 2009; p. 916.

electronics

Review

Technique for Profiling the Cycling-Induced Oxide Trapped Charge in NAND Flash Memories

Yung-Yueh Chiu * and Riichiro Shirota

Department of Electrical and Computer Engineering, National Yang Ming Chiao Tung University, Hsinchu 30010, Taiwan; riichiro.shirota@gmail.com
* Correspondence: yungyueh.chiu@gmail.com

Abstract: NAND Flash memories have gained tremendous attention owing to the increasing demand for storage capacity. This implies that NAND cells need to scale continuously to maintain the pace of technological evolution. Even though NAND Flash memory technology has evolved from a traditional 2D concept toward a 3D structure, the traditional reliability problems related to the tunnel oxide continue to persist. In this paper, we review several recent techniques for separating the effects of the oxide charge and tunneling current flow on the endurance characteristics, particularly the transconductance reduction ($\Delta G_{m,max}$) statistics. A detailed analysis allows us to obtain a model based on physical measurements that captures the main features of various endurance testing procedures. The investigated phenomena and results could be useful for the development of both conventional and emerging NAND Flash memories.

Keywords: NAND Flash memory; endurance; reliability; oxide trapped charge

Citation: Chiu, Y.-Y.; Shirota, R. Technique for Profiling the Cycling-Induced Oxide Trapped Charge in NAND Flash Memories. *Electronics* **2021**, *10*, 2492. https://doi.org/10.3390/electronics10202492

Academic Editor: Changhwan Shin

Received: 2 September 2021
Accepted: 7 October 2021
Published: 13 October 2021

1. Introduction

The emergence of NAND Flash memories has revolutionized the data storage industry over the last few decades. NAND Flash devices are used in a wide range of applications in everyday consumer electronics such as laptops, tablets, and smart wearable devices. The first NAND-structured cell was invented in 1987 by Masuoka et al. [1] at Toshiba Corp. Since then, several improvements have been proposed to lower the power consumption of these cells and to enable the contents of the entire chip to be erased at once [2–7]. More recently, its application range has been expanded such that it has become the main storage element, in that solid-state drives (SSDs) are gradually replacing hard disk drives (HDDs) [8,9]. Furthermore, it is increasingly adopted for enterprise-class storage systems. As a result, the size of NAND cells has aggressively shrunk to continuously promote this evolution. However, the ever-shrinking dimensions of the NAND cell create additional challenges in terms of the endurance and retention characteristics, such as random telegraph noise (RTN) fluctuations of the threshold voltage (V_T) [10–12], charge trapping/detrapping mechanisms [13–15], electron injection statistics [16,17], and V_T distribution widening due to parasitic coupling effects [18,19].

Three-dimensional (3D) NAND Flash memories can be considered as a breakthrough to continue to deliver increasing bit density and reduce the bit cost [20]. 3D NAND Flash technology can utilize either floating gate (FG)- or charge trapping (CT)-type cells. Most of the 3D NAND reported to date are CT-type, owing to the simpler fabrication process [21]. The 3D NAND array architecture can be categorized into the following two classes depending on the direction of channel, as schematically shown in Figure 1: vertical gate 3D NAND architecture, which was proposed by Samsung Electronics in 2009 [22]; and vertical channel 3D NAND architecture. There are two main cell structure types that use vertical channels, namely bit cost scalable (BiCS), which was proposed by Toshiba Corp. in 2007 [23,24], and terabit cell array transistor (TCAT), which was developed by Samsung Electronics in 2009 [25]. TCAT subsequently evolved into V-NAND architecture, which has

32-stacked word line (WL) layers [26–28]. The industry has moved beyond 12x-stacked WL layers and achieved a 17x-stacked V-NAND [29,30]. As the memory industry transitions from planar to 3D scaling, traditional device reliability issues must still be considered. The Fowler Nordheim (FN) tunneling mechanism is commonly used in both planar and 3D NAND cells during programming and erasing (P/E) operations [31]. This mechanism leads to the formation of trap states in the tunneling oxide, and thus degrades the oxide reliability. Therefore, overcoming the reliability problems related to the oxide trap is critically important for the development of future advanced NAND Flash memories.

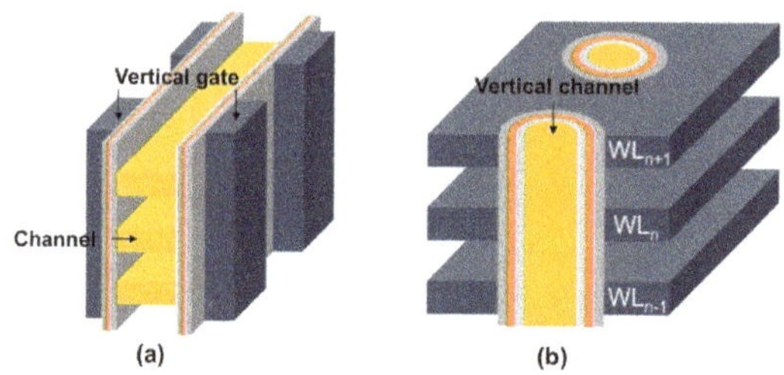

Figure 1. Schematic diagrams of 3D NAND architecture: (**a**) vertical gate and (**b**) vertical channel.

2. Shift in the Midgap Voltage

Generally, the midgap voltage (ΔV_{MG}) during P/E operations is described by a set of two components [32]: the first is the electrostatic shift (ES) that is caused by the creation of oxide trapped charges (Q_T), and the other is the tunneling shift (TS) that is related to the change in the number of floating-gate charges (Q_{FG}). Notably, these two components mutually influence each other. The former deforms the tunneling barrier for P/E operations and thus reduces the number of storage electrons. ΔV_{MG} can be expressed as the sum of these two components.

$$\Delta V_{MG} = \frac{Q_T}{C_i} + \frac{\Delta Q_{FG}}{C_{IPD}} \tag{1}$$

where C_i and C_{IPD} are the tunneling oxide and the interpoly dielectric capacitance, respectively.

Several approaches have been proposed to separate the ES and TS values from ΔV_{MG}. The first category of methods is based on indirect measurements. For example, the ΔV_{MG} in the programming and erasing states combined with tunneling-based modeling is commonly monitored to extract the Q_T distribution from Q_{FG} in NAND Flash memories. Q_T has been presented by a sheet charge located at fixed distance from the channel in the majority of the literature [32–34]. Under this assumption, the tunneling current is calculated straightforwardly along the direction perpendicular to the channel by using the Wentzel–Kramers–Brillouin (WKB) approximation, as schematically shown in Figure 2a. However, as the cell sizes are aggressively shrunk to the nanoscale regime, they are adversely affected by the discrete nature of Q_T. Thus, we must consider all possible tunneling paths across the defective oxide [35,36], as schematically shown in Figure 2b, which increases the computational time and complexity of the method.

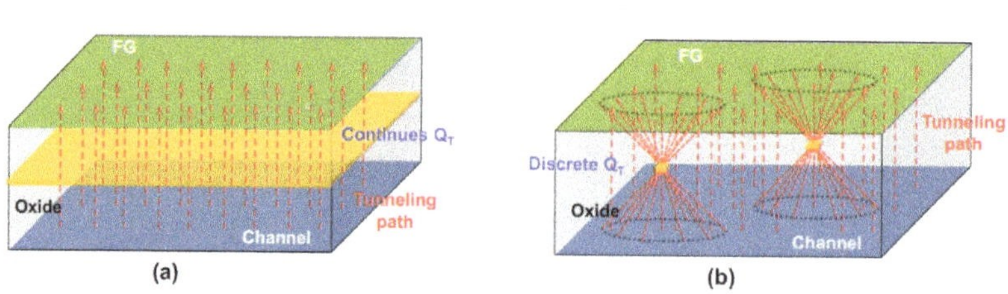

Figure 2. Schematic diagrams of all the possible tunneling paths by the (**a**) continuous and (**b**) discrete Q_T during P/E cycles.

The second category of methods is based on the direct extraction of Q_T and Q_{FG} using a special test device [37,38]. The cross-sectional view and equivalent circuit of the test structure are shown in Figure 3. The device is composed of two memory cells: one with a thick tunneling oxide, referred to as a high-voltage (HV) cell, and the other with a thin tunneling oxide, referred to as a low-voltage (LV) cell. Notably, these two cells have a common FG/common control gate (CG) configuration. During P/E operations, FN tunneling occurs only through the oxide of the LV cell, thus degrading the oxide of this cell. The ES resulting from Q_T is expressed as [38]:

$$\mathrm{ES} = \gamma[\Delta V_T(LV) - \Delta V_T(HV)] = -\frac{1}{\varepsilon_{ox}} \int_0^{T_{OX}} \rho \cdot x dx \tag{2}$$

where γ is the coupling ratio between the FG and CG, ρ is the density of Q_T, and $\Delta V_T(LV)$ and $\Delta V_T(HV)$ are the V_T shifts of the LV and HV cells after P/E cycles, respectively. Unfortunately, the size of the test device (L = 4 µm) is relatively large compared to that of conventional NAND Flash memories, yet it is necessary to continuously evaluate these miniaturized and new device structures. Moreover, this approach can only provide average information for a relatively large sample region rather than statistical information.

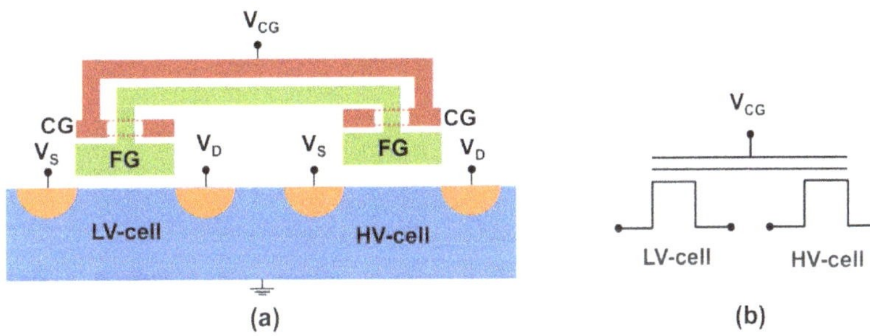

Figure 3. Schematic cross-section view (**a**) and equivalent circuit (**b**) of the test device. Adapted from [37,38].

3. $\Delta G_{m,\,max}$ Statistics

To overcome the limitations of the above-mentioned approaches, we proposed a statistical transconductance reduction ($\Delta G_{m,max}$) method [39], which enables the extraction of Q_T from Q_{FG} in both 2D and 3D NAND memories.

3.1. Experimental Setup

Experiments are carried out in 2D FG-type NAND Flash memory chip. In the NAND array, a string is composed of 32-unit cells, a source-select transistor, and a drain-select transistor, as schematically shown in Figure 4a. The control gates, source-select transistors, and drain-select transistors are connected across different strings to constitute the wordline (WL), source select line (SSL), and drain select (DSL), respectively. The strings are connected to a common sourceline (SL) and bitlines (BLs). The channel length (L) and width (W) are both 42 mm, and the tunneling oxide thickness (T_{ox}) is 8 nm. The measurement scheme was as follows: the program operation is performed by adopting the incremental step pulse programming (ISPP) technique [40] with a starting CG voltage ($V_{CG,0}$) in increments of 0.2 V with a duration of 10 µs, as schematically shown in Figure 4b, driving the selected cells to the desired V_T level. The erase operation is performed on blocks by adopting the incremental step pulse erasing (ISPE; similar to the ISPP) technique. Because it is not possible to apply high negative voltages in NAND chips, a high positive voltage is applied to the p-well. As a result, all cells in the block were erased simultaneously. During the read operation, the CG gate voltage was swept from 0 V to 5 V to harvest the maximum transconductance reduction ($\Delta G_{m,max}$). Figure 5a shows the $I_D - V_{CG}$ characteristics of the

200 randomly selected cells on WL15 in NAND strings before cycling and after 1 k, 3 k, and 30 k P/E cycles, respectively. Then, the corresponding $\Delta G_{m,max}$ distribution can be obtained, as shown in Figure 5b. Notably, the endurance test and $\Delta G_{m,max}$ monitoring were performed at room temperature. The mean value of $\Delta G_{m,max}$ ($\overline{\Delta G}_{m,max}$) is clearly observed to increase, and the distribution to become wider, as the number of cycles increases. This suggests that the $\Delta G_{m,max}$ distribution will be a good parameter for evaluating the oxide degradation.

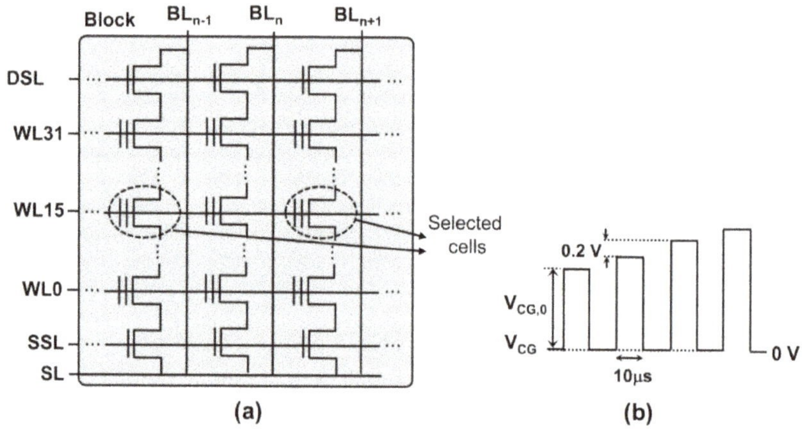

Figure 4. Schematic view of (**a**) NAND Flash array and (**b**) ISPP operation. Adapted from [39].

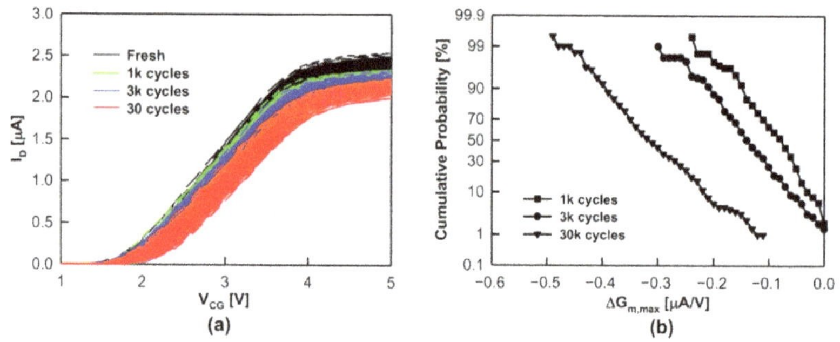

Figure 5. (**a**) $I_D - V_{CG}$ Characteristics and (**b**) cumulative $\Delta G_{m,max}$ statistics of the read cells on WL15 as a function of the number of P/E cycles. Adapted from [39].

3.2. Simulation Methodology

Monte Carlo simulations have been used in an attempt to extract information about Q_T from the measured $\Delta G_{m,max}$ distribution. A NAND string can be modeled to have a selected cell with equivalent source and drain resistances (R_S and R_D), as shown in Figure 6a. The equivalent R_S and R_D can be extracted from the monitoring of the transconductance of the read cells for different positions along the NAND string [41]. The equivalent R_S and R_D are 130 kΩ and 138.2 kΩ, respectively. The TCAD simulations used a 3D drift-diffusion equation and coupled with the Shockley–Read–Hall model for generation/recombination and mobility models (including the electric field dependence, doping-dependent modification, and surface mobility degradation). To determine the $\overline{\Delta G}_{m,max}$ statistics accurately, the simulated $I_D - V_{CG}$ characteristic of the fresh cell is calibrated with experimental data at a probability level $p = 50\%$, as shown in Figure 6b. The simulation was in good agreement with the experimental results. After calibrating the equivalent resistances, the Monte-Carlo-based method was adopted to evaluate the concentration of Q_T (Q_T^C) after P/E cycles, as schematically shown in Figure 7. The step-by-step procedure is as follows: First, discrete Q_T is randomly generated following a uniform distribution in a cuboid volume 420 nm × 840 nm × 8 nm in size (i.e., 20 L × 10 W × T_{ox}), with an equivalent Q_T^C. Notably, the discrete Q_T is treated as a negative point charge corresponding to one electron because the electron mobility is degraded by the Coulomb repulsion.

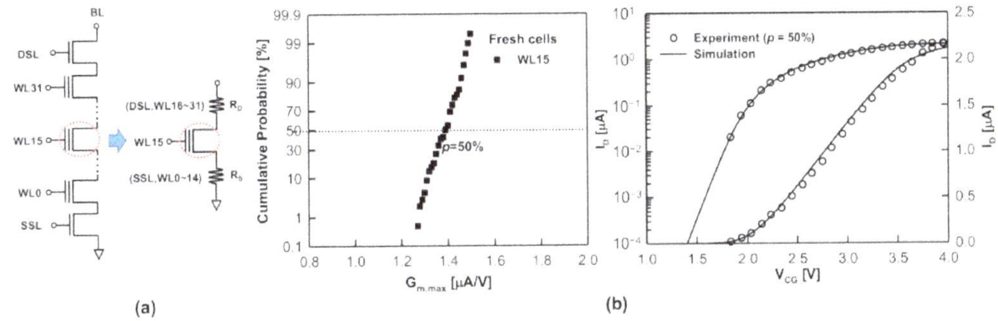

Figure 6. (**a**) Schematic circuit diagram of a NAND string and an equivalent model when cells on WL15 are read. (**b**) Comparison between measured and simulated $I_D - V_{CG}$ curve of cells on WL15 at $p = 50\%$, plotted on the linear and logarithmic scales. Adapted from [39].

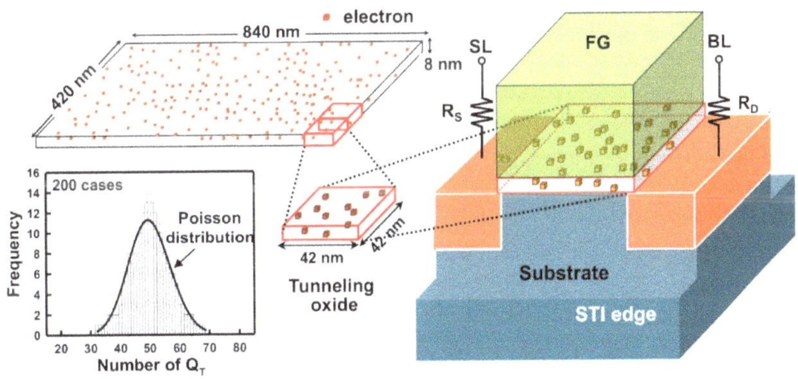

Figure 7. Schematic diagram of the random discrete Q_T generating algorithm. Adapted from [39].

Second, the cuboid is partitioned into 200 sub-cuboids and then mapped into the tunneling oxide region. Thus, the numbers of discrete Q_T in these 200 cases approximately follow a Poisson distribution, as shown in Figure 7. Finally, a comparison of the simulated and measured $\Delta G_{m,max}$ statistics allowed us to evaluate the Q_T^C during P/E cycles.

Moreover, even though the simulation does not directly account for interface trap (D_{it}) generation, the effect thereof is reflected in the model. The measured $I_D - V_{CG}$ characteristics indicated that the transconductance reached a maximum when V_{CG} slightly exceeded V_T; therefore, the occupied D_{it} can be considered as a fixed Q_T located at the silicon/oxide interface because the bending of the surface potential remains almost unchanged [42] (see Figure 8).

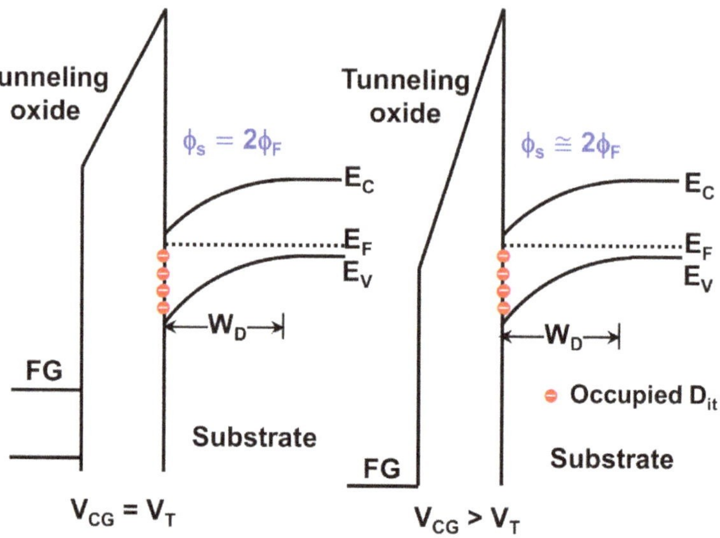

Figure 8. Band diagram and trap occupation of interface trap states at different biases. Reprinted from [39].

4. Endurance Characteristics

4.1. *QT* Extraction

Figure 9 indicates that simulations can reproduce the experimental results satisfactorily, where the extracted equivalent Q_T^C for 1 k, 10 k, and 30 k P/E cycles are 2.6×10^{18} cm^{-3}, 5×10^{18} cm^{-3}, and 1.9×10^{19} cm^{-3}, respectively. Furthermore, the proposed approach can be extended to include the array effect on $\Delta G_{m,max}$ statistics. Figure 10a shows the $\Delta G_{m,max}$ distribution as a function of the position of the WLs in a NAND string after 10 k P/E cycles. The simulations correspond well with the measurements when the following appropriate parameters are adopted: WL_1, $R_S = 16.3$ kΩ and $R_D = 251.9$ kΩ; WL_{15}, $R_S = 130$ kΩ and $R_D = 138.1$ kΩ; WL_{30}, $R_S = 251.7$ kΩ and $R_D = 16.3$ kΩ. Clearly, R_S increases as the position of the WLs changes from WL_0 to WL_{31} owing to the increase in the number of pass cells. Figure 10b shows the $\Delta G_{m,max}$ distribution as a function of V_{pass} after 10 k cycles. Again, a good agreement between the simulation and experimental results is found. The equivalent resistances of the pass cells are 8.2 kΩ/cell, 6.5 kΩ/cell, and 5.1 kΩ/cell for V_{pass} of 4 V, 5 V, and 6 V, respectively. It is clear that the pass-cell bias with a higher V_{pass} has a smaller equivalent resistance but a larger $\Delta G_{m,max}$. As a final verification, the $\Delta G_{m,max}$ statistics of two different V_T levels of the read cells were compared under the same cycling conditions. As shown in Figure 10c, the $\Delta G_{m,max}$ distributions almost overlap, indicating that Q_{FG} causes a simple parallel shift of the $I_D - V_{CG}$ curve.

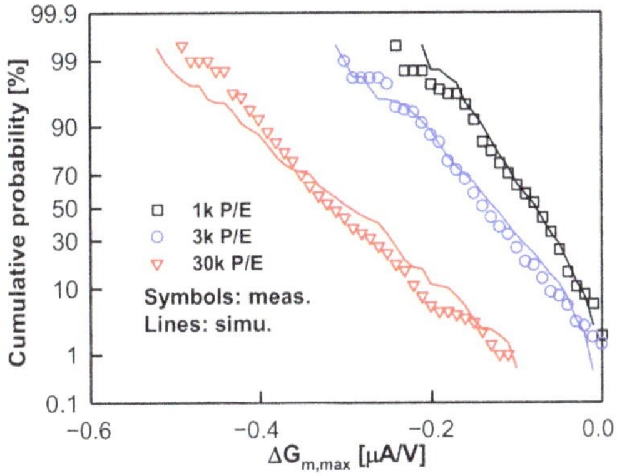

Figure 9. Simulated $\Delta G_{m,max}$ statistics for different number of P/E cycles. The simulations and experimental measurements are in good agreement. Reprinted from [39].

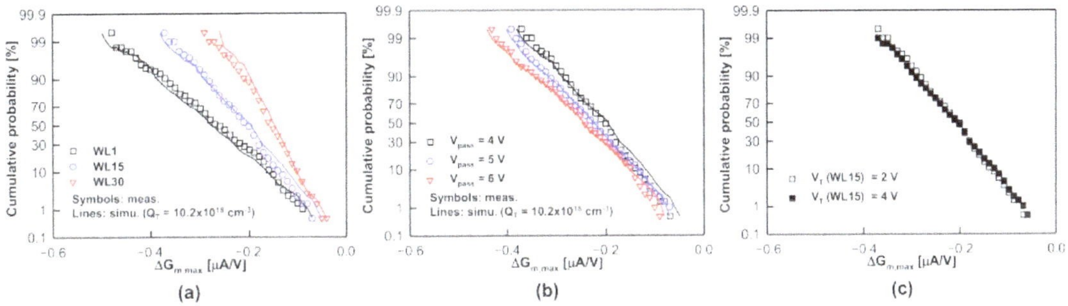

Figure 10. $\Delta G_{m,max}$ distributions for (**a**) different selected WLs in the string. All cells in the string with $V_{pass} = 6\ V$, (**b**) WL15 selected with different values of V_{pass}, and (**c**) WL15 selected with different V_T levels. Except for the read cells, the others are in the erased state. Reprinted from [39].

4.1. Endurance Degradation Model

We start this section with a description of an endurance degradation model that captures the features of the measurement. The evolution of Q_T^C can be conveniently described by the following modified power-law equation [39]:

$$Q_T^C = \frac{Q_0}{1 + (k \cdot N)^{-\alpha}} \tag{3}$$

$$k = k_0 \cdot \exp(-E_{A,G}/k_B T) \tag{4}$$

where Q_0 is the saturated value of Q_T^C, k is the reaction constant, N is the number of P/E cycles, α is the exponential coefficient, $E_{A,G}$ is the activation energy of Q_T creation, and $k_B T$ is the thermal energy. Figure 11 compares the results obtained with the endurance model and with the experimental results. The adopted parameters are as follows: $Q_0 = 1.5 \times 10^{20}$ cm^{-3}, $k = 1.0 \times 10^{-6}$, and $\alpha = 0.58$. When N is small ($\leq$ 30 k cycles), the exponential term in Equation (3) is much greater than 1, and Equation (3) can simply be expressed as a power law. On the other hand, when N is large ($>$ 30 k cycles), Q_T^C gradually approaches the saturated value Q_0. Overall, the proposed model is able to successfully describe the endurance characteristics over a wide range of N (up to 100 k cycles). Notably, Q_0 is supposed to be related to the process condition, which determines the amount of weak Si-O or Si-H bonds that can be broken [43,44].

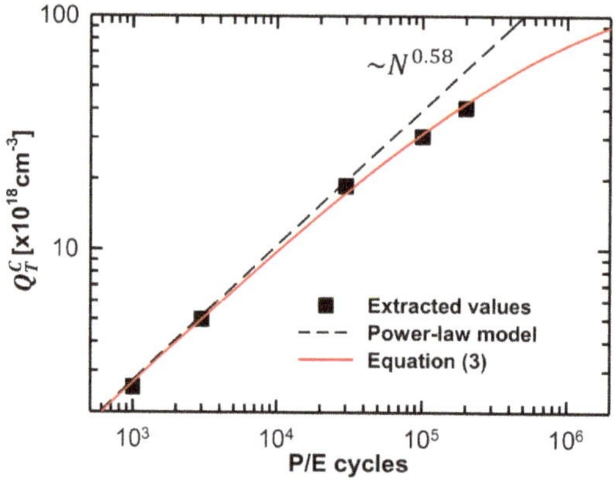

Figure 11. Comparison between extracted values (symbols) and model calculations (lines) according to Equation (3). Reprinted from [39].

To evaluate $E_{A,G}$ in Equation (4), we performed the experiments at various cycling temperatures. Figure 12 shows that $\Delta G_{m,max}$ increases as the cycling temperature (T_{cyc}) increases, suggesting that a higher T_{cyc} causes more oxide damage. The temperature-accelerated Q_T evaluations can be derived by using Equations (3) and (4) as follows [39]:

$$Q_T^C(T_{cyc}) = Q_T^C(T_R)\cdot exp\left[\alpha\cdot E_{A,G}\left(\frac{1}{k_B T_R} - \frac{1}{k_B T_H}\right)\right] \quad (5)$$

where T_R is the cycling performed at room temperature. A good linear relationship between the logarithm of Q_T^C and the reciprocal temperature was observed, and $E_{A,G}$ was evaluated, as shown in Figure 13. Ultimately, the theoretical result fitted the experimental results well, and the yield $E_{A,G}$ was approximately 100 mV, which agreed with that obtained by monitoring the stress-induced gate leakage current [13,45,46].

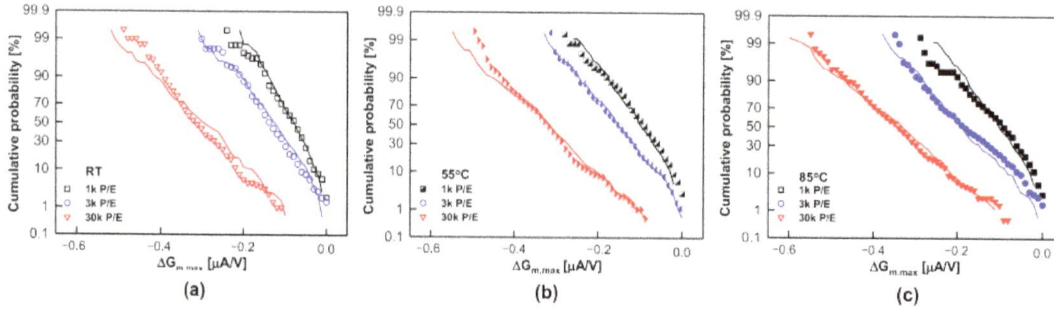

Figure 12. $\Delta G_{m,max}$ distributions for cells on WL15. Selected measured (symbols) and simulated (lines) at T_{cyc} of (**a**) 25 °C, (**b**) 55 °C, and (**c**) 85 °C are shown. Reprinted from [39].

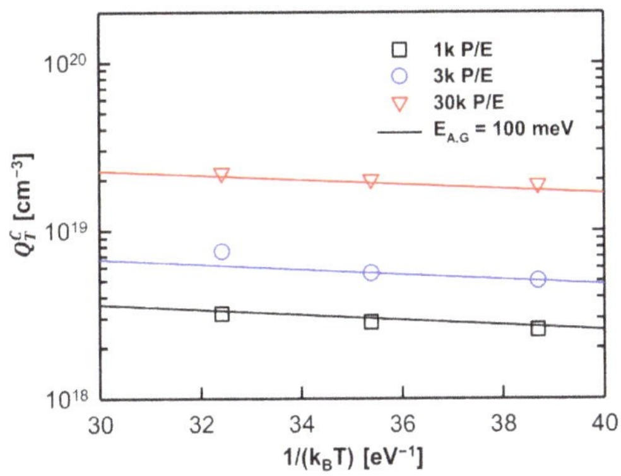

Figure 13. Q_T^C obtained by fitting experimental data using Equation (5) for different values of T_{cyc}. Reprinted from [29].

4.2. Effect of the Time Delay between P/E Cycles

Reportedly, the time delay (t_{wait}) between P/E cycles is an important factor that affects the endurance characteristics [47–51]. Figure 14 shows that the $\Delta G_{m,max}$ statistics become larger as t_{wait} increases when the endurance test is performed at T_R; however, when T_{cyc} increases to 85 °C, the trend is completely the opposite due to recovery from oxide damage through thermal excitation. This suggests that, at high T_{cyc}, the endurance model should not only take into consideration the creation of damage but also the recovery from damage. The time-dependent damage recovery during t_{wait} can be described by a rate equation given by [52,53]

$$f = exp(-t_{wait}/\tau) \tag{6}$$

$$= \tau_0 \cdot \exp(E_{A,R}/k_B T_{cyc}) \tag{7}$$

where f is the occupation function, τ is the time constant, and $E_{A,R}$ is the activation energy for the recovery from the oxide damage. Therefore, Equation (3) can be rewritten as follows [37]:

$$Q_T^C = \frac{Q_0}{1+(k \cdot N)^{-\alpha}} \cdot f \tag{8}$$

when t_{wait} is sufficiently short, for example, 0.1 s, Equation (8) tends to Equation (3) because the mechanism according to which recovery from oxide damage takes place plays a negligible role. Accordingly, we can extract the parameters (i.e., Q_0, α, k_0, and $E_{A,G}$) by using the approach described in Section 4.1. However, when t_{wait} becomes longer, the damage creation and recovery effects are mixed, which complicates the simultaneous extraction of $E_{A,G}$ and $E_{A,R}$. To simplify the situation, we assume that under the condition of T_{cyc} = 25 °C, Equation (8) approaches Equation (3) because the thermal excitation of Q_T is not noticeable. This allows us to evaluate the $E_{A,G}$ for longer t_{wait} values of 2 s and 4 s. Figure 15a shows that, by calibrating the $E_{A,G}$ values, Equation (3) can reproduce the characteristics of the extracted Q_T^C with different t_{wait} values. The relationship between the logarithm of t_{wait} and $E_{A,G}$ is linear, as shown in Figure 15b. Moreover, $E_{A,G}$ is in the approximate range of 60–100 meV, which agrees with the results obtained by monitoring the V_T transients after experiments at different T_{cyc} [13]. Once the value of $E_{A,G}$ is determined for different values of t_{wait}, the remaining parameters $E_{A,R}$ can be determined by using the change rate of the celebrated τ at different T_{cyc}. Figure 16 shows the experimental measurements fit the curve calculated with Equation (8).

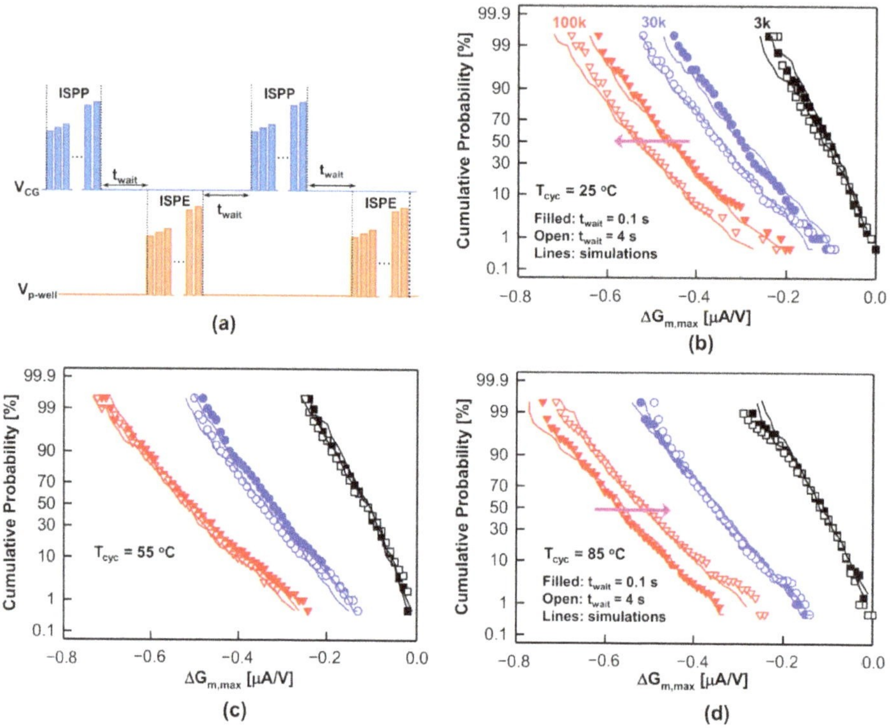

Figure 14. (**a**) Different values of t_{wait} introduced between P/E cycles. $\Delta G_{m,max}$ distributions for cells on WL15 selected measured (symbols) and simulated (lines) for different values of t_{wait} at T_{cyc} of (**b**) 25 °C, (**c**) 55 °C, and (**d**) 85 °C. Reprinted from [47].

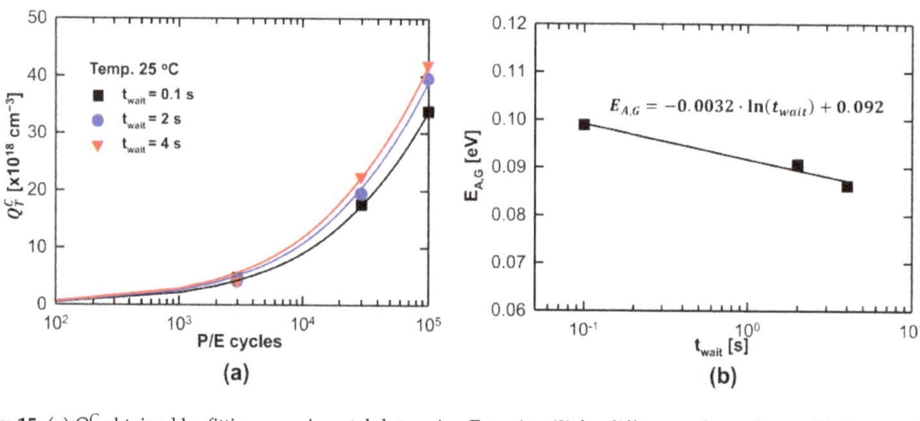

Figure 15. (**a**) Q_T^C obtained by fitting experimental data using Equation (3) for different values of t_{wait}. (**b**) Extracted $E_{A,G}$ for different values of t_{wait}. Reprinted from [47].

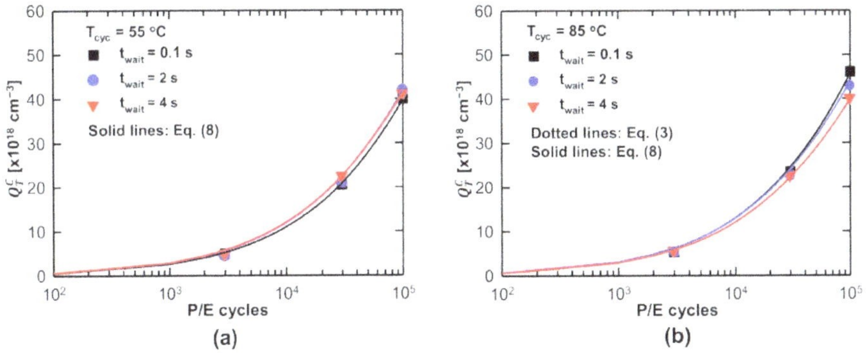

Figure 16. Q_T^C obtained by fitting experimental data using Equation (8) for different values of t_{wait} under T_{cyc} of (**a**) 55 °C and (**b**) 85 °C. Reprinted from [47].

Clearly, if the damage-recovery mechanism is not taken into account, Equation (3) overestimates the experiment, and the discrepancy between them increases with t_{wait}. The optimized τ values were 150 s, 22 s, and 12 s for T_{cyc} of 25 °C, 55 °C, and 85 °C, respectively. Figure 17 shows that the relationship between the logarithm of τ and the reciprocal temperature is linear, and $E_{A,R} \simeq 0.4\ eV$ is obtained. Notably, $E_{A,R} \simeq 0.4\ eV$ is similar to the values reported in the literature for charge detrapping through thermal emission [54]. It is also easily verifiable that the assumption that f approaches one under the condition of T_{cyc} of 25 °C and t_{wait} of 0.1 s is satisfied.

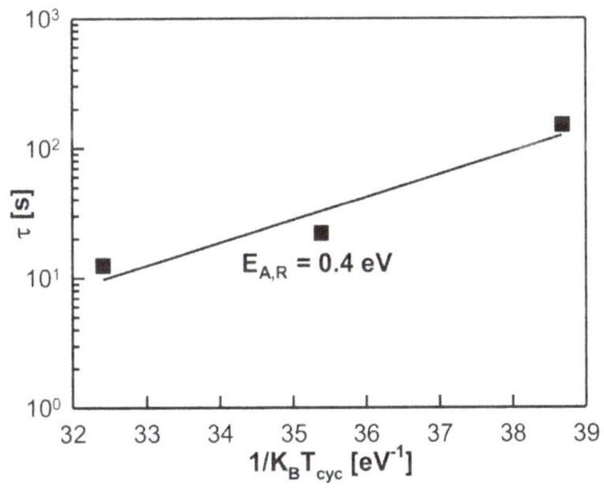

Figure 17. $E_{A,R}$ obtained using Equation (7) for different values of T_{cyc}. Reprinted from [47].

Although the above-mentioned model successfully describes the endurance characteristics, it still does not account for certain features according to more recent research [55]. A comparative analysis of the respective influence of t_{wait} from program to erase (P-to-E) and of that from erase to program (E-to-P) on the median $\Delta G_{m,max}$ ($\overline{\Delta G_{m,max}}$) after 30 k P/E cycles was reported [55] and is plotted in Figure 18, normalized to its initial value ($G_{m0,max}$). The E-to-P t_{wait} clearly had a more significant impact on the normalized $\overline{\Delta G_{m,max}}$ than P-to-E t_{wait}. Moreover, it is also shown that the normalized $\overline{\Delta G_{m,max}}$ increases as $V_{CG,0}$ increases. As a result, adopting ISPP with a lower $V_{CG,0}$ would be better for improving the oxide quality. The physical mechanism is graphically illustrated in Figure 19. Energetic electrons injected from the cathode result in anode hole injection during the P/E operations. During E-to-P t_{wait}, these holes drift near the silicon surface, where they recombine with channel electrons [48,56]. This could create additional trap states, and subsequently, these traps are occupied by electrons.

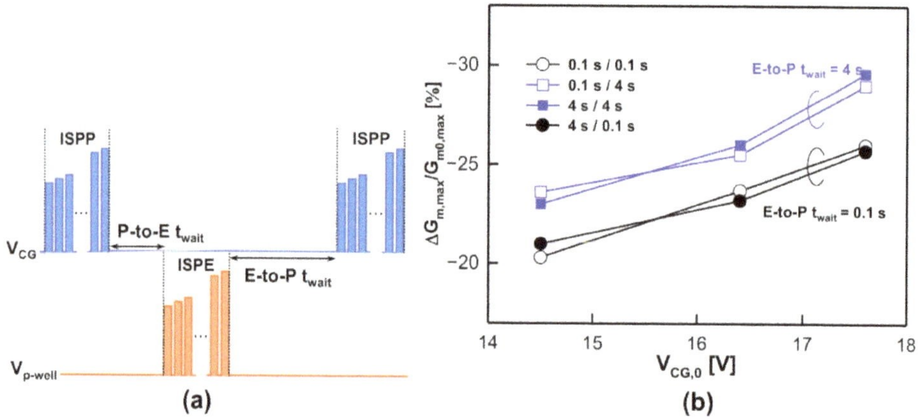

Figure 18. (**a**) Different P-to-E and E-to-P values of t_{wait} are introduced between P/E cycles. (**b**) Dependence of normalized $\Delta G_{m,max}$ on P-to-E and E-to-P values of t_{wait} after 30 k cycles. Reprinted from [55].

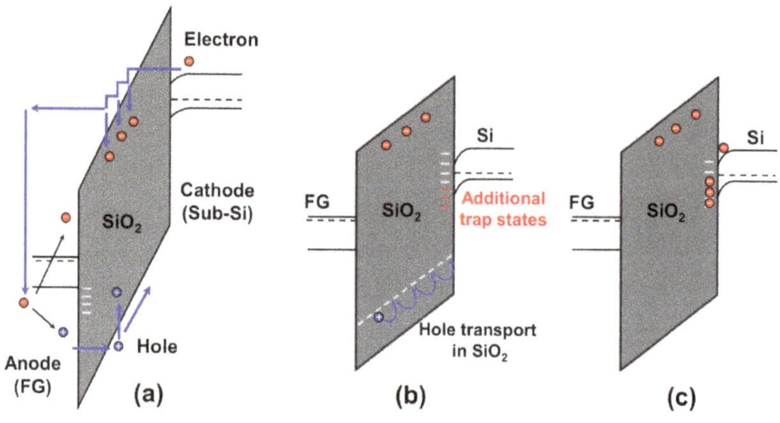

Figure 19. (**a**) Schematic diagram of anode hole injection during P/E cycles. During E-to-P t_{wait}, Q_T is generated via two possible physical mechanisms: (**b**) the holes migrate toward the silicon/oxide interface and recombine with inversion electrons, creating additional trap states. (**c**) Subsequently, electrons are trapped in these trap states. Reprinted from [55].

5. Conclusions

A method for characterizing the endurance characteristics of NAND Flash memories by monitoring the $\Delta G_{m,max}$ statistics is described. The discrete Q_T, gradually generated with P/E cycles, results in the reduction of $\Delta G_{m,max}$, and broadening of the distribution. Based on Monte Carlo simulations, an analytical model for the generation of Q_T, including the effects of T_{cyc} and t_{wait}, is then described. The model represents a powerful tool for the investigation and predictive analysis of next-generation NAND Flash technologies.

Author Contributions: Conceptualization and writing—original draft preparation, Y.-Y.C.; writing—review and editing, R.S. All authors have read and agreed to the published version of the manuscript.

Funding: This research was funded by Winbond Electronics Corp. (Kanagawa, Japan).

Acknowledgments: The present work was performed with respect to the joint development agreement between Winbond Electronics Corp. and the National Yang Ming Chiao Tung University. The authors would like to thank T. Takeshita and M. Yano.

Conflicts of Interest: The authors declare no conflict of interest.

References

1. Masuoka, F.; Momodomi, M.; Iwata, Y.; Shirota, R. New ultra high density EPROM and Flash EEPROM with NAND structure cell. In Proceedings of the 1987 International Electron Devices Meeting, New York, NY, USA, 6–9 December 1987; pp. 552–555.
2. Shirota, R.; Itoh, Y.; Nakayama, R.; Momodomi, M.; Inoue, S.; Kirisawa, R.; Iwata, Y.; Chiba, M.; Masuoka, F. A new NAND cell for ultra high density 5 V-only EEPROMs. In Proceedings of the 1988 Symposium on VLSI Technology—Digest of Technical Papers, San Diego, CA, USA, 10–13 May 1988; pp. 33–34.
3. Momodomi, M.; Kirisawa, R.; Nakayama, R.; Aritome, S.; Endoh, T.; Itoh, Y.; Iwata, Y.; Oodaira, H.; Tanaka, T.; Chiba, M.; et al. New device technologies for 5 Vonly 4 Mb EEPROM with NAND structure cell. In Proceedings of the 1988 International Electron Devices Meeting, San Francisco, CA, USA, 11–14 December 1988; pp. 412–415.
4. Momodomi, M.; Itoh, Y.; Shirota, R.; Iwata, Y.; Nakayama, R.; Kirisawa, R.; Tanaka, T.; Aritome, S.; Endoh, T.; Ohuchi, K.; et al. An experimental 4-Mbit CMOS EEPROM with a NAND structure cell. *IEEE J. Solid-State Circuits* **1989**, *24*, 1238–1243. [CrossRef]
5. Iwata, Y.; Momodomi, M.; Tanaka, T.; Oodaira, H.; Itoh, Y.; Nakayama, R.; Kirisawa, R.; Aritome, S.; Endoh, T.; Shirota, R.; et al. A high-density NAND EEPROM with block-page programming for microcomputer applications. *IEEE J. Solid-State Circuits* **1990**, *25*, 417–424. [CrossRef]
6. Kirisawa, R.; Aritome, S.; Nakayama, R.; Endoh, T.; Shirota, R.; Masuoka, F. A NAND structured cell with a new programming technology for highly reliable SV-only Flash EEPROM. In Proceedings of the 1988 Symposium on VLSI Technology—Digest of Technical Papers, Honololu, HI, USA, 4–7 June 1990; pp. 129–130.
7. Aritome, S.; Shirota, R.; Kirisawa, R.; Endoh, T.; Nakayama, N.; Sakui, K.; Masuoka, F. A reliable bi-polarity write/erase technology in flash EEPROMs. In Proceedings of the 1988 International Electron Devices Meeting, San Francisco, CA, USA, 9–12 December 1990; pp. 111–114.
8. Micheloni, R.; Marelli, A.; Eshghi, K. *Inside Solid State Drives (SSDs)*; Springer: New York, NY, USA, 2013.
9. Spinelli, A.; Compagnoni, C.; Lacaita, A. Reliability of NAND Flash Memories: Planar Cells and Emerging Issues in 3D Devices. *Computers* **2017**, *6*, 16. [CrossRef]
10. Ghetti, A.; Monzio Compagnoni, C.; Spinelli, A.S.; Visconti, A. Comprehensive analysis of random telegraph noise instability and its scaling in deca-nanometer Flash memories. *IEEE Trans. Electron Devices* **2009**, *56*, 1746–1752. [CrossRef]
11. Kurata, H.; Otsuga, K.; Kotabe, A.; Kajiyama, S.; Osabe, T.; Sasago, Y.; Narumi, S.; Tokami, K.; Kamohara, S.; Tsuchiya, O. The impact of random telegraph signals on the scaling of multilevel Flash memories. In Proceedings of the 2006 Symposium on VLSI Technology (VLSI-Technology), Honolulu, HI, USA, 13–15 June 2006; pp. 112–113.
12. Tega, N.; Miki, H.; Osabe, T.; Kotabe, A.; Otsuga, K.; Kurata, H.; Kamohara, S.; Tokami, K.; Ikeda, Y.; Yamada, R. Anomalously large threshold voltage fluctuation by complex random telegraph signal in floating gate Flash memory. In Proceedings of the 2006 International Electron Devices Meeting (IEDM), San Francisco, CA, USA, 11–13 December 2006; pp. 491–494.
13. Resnati, D.; Nicosia, G.; Paolucci, G.M.; Visconti, A.; Monzio Compagnoni, C. Cycling-induced charge trapping/detrapping in Flash memories—Part I: Experimental evidence. *IEEE Trans. Electron Devices* **2016**, *63*, 4753–4760. [CrossRef]
14. Mielke, N.; Belgal, H.; Kalastirsky, I.; Kalavade, P.; Kurtz, A.; Meng, Q.; Righos, N.; Wu, J. Flash EEPROM threshold instabilities due to charge trapping during program/erase cycling. *IEEE Trans. Device Mater. Reliab.* **2004**, *4*, 335–344. [CrossRef]
15. Paolucci, G.M.; Monzio Compagnoni, C.; Miccoli, C.; Spinelli, A.S.; Lacaita, A.L.; Visconti, A. Revisiting charge trapping/detrapping in Flash memories from a discrete and statistical standpoint—Part I: VT instabilities. *IEEE Trans. Electron Devices* **2014**, *61*, 2802–2810. [CrossRef]
16. Monzio Compagnoni, C.; Spinelli, A.S.; Gusmeroli, R.; Beltrami, S.; Ghetti, A.; Visconti, A. Ultimate accuracy for the NAND Flash program algorithm due to the electron injection statistics. *IEEE Trans. Electron Devices* **2008**, *55*, 2695–2702. [CrossRef]
17. Monzio Compagnoni, C.; Gusmeroli, R.; Spinelli, A.S.; Visconti, A. Analytical model for the electron-injection statistics during programming of nanoscale NAND Flash memories. *IEEE Trans. Electron Devices* **2008**, *55*, 3192–3199. [CrossRef]
18. Nishi, Y. (Ed.) *Advances in Non-Volatile Memory and Storage Technology*; Woodhead Publishing: Cambridge, UK, 2014.
19. Lee, J.-D.; Hur, S.-H.; Choi, J.-D. Effects of floating-gate interference on NAND flash memory cell operation. *IEEE Electron Device Lett.* **2002**, *23*, 264–266.
20. Goda, A.; Parat, K. Scaling directions for 2D and 3DNAND cells. In Proceedings of the 2012 IEEE International Electron Devices Meeting (IEDM), San Francisco, CA, USA, 10–13 December 2012.
21. Goda, A. 3-D NAND technology achievements and future scaling perspectives. *IEEE Trans. Electron Devices* **2020**, *67*, 1373–1381. [CrossRef]

22. Kim, W.; Choi, S.; Sung, J.; Lee, T.; Park, C.; Ko, H.; Jung, J.; Yoo, I.; Park, Y. Multi-layered vertical gate NAND Flash overcoming stacking limit for terabit density storage. In Proceedings of the 2009 Symposium on VLSI Technology, Kyoto, Japan, 15–17 June 2009; pp. 188–189.
23. Tanaka, H.; Kido, M.; Yahashi., K.; Oomura, M.; Katsumata, R.; Kito, M.; Fukuzumi, Y.; Sato, M.; Nagata, Y.; Matsuoka, Y.; et al. Bit cost scalable technology with punch and plug process for ultra high density flash memory. In Proceedings of the 2007 Symposium on VLSI Technology, Kyoto, Japan, 12–14 June 2007; pp. 14–15.
24. Fukuzumi, Y.; Katsumata, R.; Kito, M.; Kido, M.; Sato, M.; Tanaka, H.; Nagata, Y.; Matsuoka, Y.; Iwata, Y.; Aochi, H.; et al. Optimal integration and characteristics of vertical array devices for ultra-high density, bit-cost scalable flash memory. In Proceedings of the 2007 International Electron Devices Meeting (IEDM), Washington, DC, USA, 10–12 December 2007; pp. 449–452.
25. Jang, J.; Kim, H.-S.; Cho, W.; Cho, H.; Kim, J.; Shim, S.I.; Jang, Y.; Jeong, J.-H.; Son, B.-K.; Kim, D.-W.; et al. Vertical cell array using TCAT (terabit cell array transistor) technology for ultra high density NAND flash memory. In Proceedings of the 2009 Symposium on VLSI Technology, Kyoto, Japan, 15–17 June 2009; pp. 192–193.
26. Elliott, J.; Jung, E.S. Ushering in the 3D Memory Era with V-NAND. In Proceedings of the Flash Memory Summit, Santa Clara, CA, USA, 13–15 August 2013.
27. Park, K.T.; Byeon, D.S.; Kim, D.H. A world's first product of three-dimensional vertical NAND Flash memory and beyond. In Proceedings of the 2014 14th Annual Non-Volatile Memory Technology Symposium (NVMTS), Jeju Island, Korea, 27–29 October 2014; pp. 1–5.
28. Park, K.T.; Nam, S.; Kim, D.; Kwak, P.; Lee, D.; Choi, Y.H.; Choi, M.H.; Kwak, D.H.; Kim, D.H.; Kim, M.S.; et al. Three-Dimensional 128 Gb MLC Vertical nand Flash Memory With 24-WL Stacked Layers and 50 MB/s High-Speed Programming. *IEEE J. Solid State Circuit* **2015**, *50*, 204–213. [CrossRef]
29. Kang, D.; Kim, M.; Jeon, S.-C.; Jung, W.; Park, J.; Choo, G.; Shim, D.-K.; Kavala, A.; Kim, S.-B.; Kang, K.-M.; et al. A 512Gb 3-bit/cell 3D 6th-generation V-NAND flash memory with 82MB/s write throughput and 1.2Gb/s interface. In Proceedings of the 2019 IEEE International Solid-State Circuits Conference (ISSCC), San Francisco, CA, USA, 17–21 February 2019; pp. 216–218.
30. Cho, J.; Chris Kang, D.; Park, J.; Nam, S.-W.; Song, J.-H.; Jung, B.-K.; Lyu, J.; Lee, H.; Kim, W.-T.; Jeon, H.; et al. A 512Gb 3b/Cell 7th-generation 3D-NAND flash memory with 184MB/s write throughput and 2.0Gb/s interface. In Proceedings of the 2021 IEEE International Solid-State Circuits Conference (ISSCC), San Francisco, CA, USA, 13–22 February 2021; pp. 426–428.
31. Monzio Compagnoni, C.; Goda, A.; Spinelli, A.S.; Feeley, P.; Lacaita, A.L.; Visconti, A. Reviewing the evolution of the NAND Flash technology. *Proc. IEEE* **2017**, *105*, 1609–1633. [CrossRef]
32. Fayrushin, A.; Lee, C.-H.; Park, Y.; Choi, J.-H.; Chung, C. Unified endurance degradation model of floating gate NAND flash memory. *IEEE Trans. Electron Devices* **2013**, *60*, 2031–2037. [CrossRef]
33. Xia, Z.; Kim, D.S.; Jeong, N.; Kim, Y.-G.; Kim, J.-H.; Lee, K.-H.; Park, Y.-K.; Chung, C.; Lee, H.; Han, J. Comprehensive modeling of NAND flash memory reliability: Endurance and data retention. In Proceedings of the IEEE International Reliability Physics Symposium (IRPS), Anaheim, CA, USA, 15–19 April 2012; pp. MY.5.1–MY.5.4.
34. Yang, B.-J.; Wu, Y.-T.; Chiu, Y.-Y.; Kuo, T.-M.; Chang, J.-H.; Wang, P.-Y.; Shirota, R. Evaluation of the role of deep trap state using analytical model in the program/erase cycling of NAND flash memory and its process dependence. *IEEE Trans. Electron Devices* **2018**, *65*, 499–506. [CrossRef]
35. Watanabe, H.; Yao, K.; Lin, J. Numerical study of very small floating islands. *IEEE Trans. Electron. Devices* **2014**, *61*, 1145–1152. [CrossRef]
36. Lin, P.-J.-J.; Lee, C.-A.-A.; Yao, C.-W.-K.; Lin, H.-J.-V.; Watanabe, H. Localized tunneling phenomena of nanometer scaled high-K gate-stack. *IEEE Trans. Electron Devices* **2017**, *64*, 3077–3083. [CrossRef]
37. Shirota, R.; Yang, B.-J.; Chiu, Y.-Y.; Chen, H.-T.; Ng, S.-F.; Wang, P.-Y.; Chang, J.-H.; Kurachi, I. New accurate method to analyze both floating gate charge and tunnel oxide trapped charge profile in NAND flash memory. In Proceedings of the IEEE International Memory Workshop (IMW), Taipei, Taiwan, 18–21 May 2014; pp. 55–58.
38. Shirota, R.; Yang, B.-J.; Chiu, Y.-Y.; Chen, H.-T.; Ng, S.-F.; Wang, P.-Y.; Chang, J.-H.; Kurachi, I. New method to analyze the shift of floating gate charge and generated tunnel oxide trapped charge profile in NAND flash memory by program/erase endurance. *IEEE Trans. Electron Devices* **2015**, *62*, 114–120. [CrossRef]
39. Chiu, Y.-Y.; Lin, I.-C.; Chang, K.-C.; Yang, B.-J.; Takeshita, T.; Yano, M.; Shirota, R. Transconductance distribution in program/erase cycling of NAND flash memory devices: A Statistical Investigation. *IEEE Trans. Electron Devices* **2019**, *66*, 1255–1261. [CrossRef]
40. Hemink, G.J.; Tanaka, T.; Endoh, T.; Aritome, S.; Shirota, R. Fast and accurate programming method for multi-level NAND EEPROMs. In Proceedings of the 1995 Symposium on VLSI Technology—Digest of Technical Papers, Kyoto, Japan, 6–8 June 1995; pp. 129–130.
41. Joe, S.-M.; Yi, J.-H.; Park, S.-K.; Kwon, H.-I.; Lee, J.-H. Position–dependent threshold–voltage variation by random telegraph noise in NAND flash memory strings. *IEEE Trans. Electron Devices* **2010**, *31*, 635–637.
42. van Langevelde, R.; Klaassen, F.M. An explicit surface-potentialbased MOSFET model for circuit simulation. *Solid-State Electron.* **2000**, *44*, 409–418. [CrossRef]
43. Walters, M.; Reisman, A. Radiation-induced neutral electron trap generation in electrically biased insulated gate field effect transistor gate insulators. *J. Electrochem. Soc.* **1991**, *138*, 2756–2762. [CrossRef]
44. Nicklaw, C.J.; Lu, Z.-Y.; Fleetwood, D.M.; Schrimpf, R.D.; Pantelides, S.T. The structure, properties, and dynamics of oxygen vacancies in amorphous SiO_2. *IEEE Trans. Nucl. Sci.* **2002**, *49*, 2667–2673. [CrossRef]

45. Satake, H.; Toriumi, A. Common origin for stress-induced leakage current and electron trap generation in SiO_2. *Appl. Phys. Lett.* **1995**, *67*, 3489–3490. [CrossRef]
46. DiMaria, D.J.; Stasiak, J.W. Trap creation in silicon dioxide produced by hot electrons. *J. Appl. Phys.* **1989**, *65*, 2342–2356. [CrossRef]
47. Chiu, Y.-Y.; Chang, K.-C.; Lin, H.-J.; Tsai, H.-T.-E.; Lin, P.-J.; Li, H.-C.; Takeshita, T.; Yano, M.; Shirota, R. Impact of program/erase Cycling Interval on the transconductance distribution of NAND flash memory devices. *IEEE Trans. Electron Devices* **2020**, *67*, 4897–4903. [CrossRef]
48. Shirota, R.; Yang, B.-J.; Chiu, Y.-Y.; Wu, Y.-T.; Wang, P.-Y.; Chang, J.-H.; Yano, M.; Aoki, M.; Takeshita, T.; Wang, P.-Y.; et al. Improvement of oxide reliability in NAND flash memories using tight endurance cycling with shorter idling period. In Proceedings of the IEEE International Reliability Physics Symposium (IRPS), Monterey, CA, USA, 19–23 April 2015; pp. MY.12.1–MY.12.5.
49. Miccoli, C.; Monzio Compagnoni, C.; Beltrami, S.; Spinelli, A.S.; Visconti, A. Threshold-voltage instability due to damage recovery in nanoscale NAND flash memories. *IEEE Trans. Electron Devices* **2011**, *58*, 2406–2414. [CrossRef]
50. Lee, M.C.; Wong, H.Y. Investigation on the impact of program/erase cycling frequency on data retention of nanoscale charge trap nonvolatile memory. *IEEE Electron Device Lett.* **2014**, *35*, 918–920.
51. Mielke, N.; Belgal, H.P.; Fazio, A.; Meng, Q.; Righos, N. Recovery effects in the distributed cycling of flash memories. In Proceedings of the IEEE International Reliability Physics Symposium (IRPS), San Jose, CA, USA, 26–30 March 2006; pp. 29–35.
52. Wang, Y.; White, M.H. An analytical retention model for SONOS nonvolatile memory devices in the excess electron state. *Solid-State Electron.* **2005**, *49*, 97–107. [CrossRef]
53. Chiu, Y.-Y.; Yang, L.F.-H.; Chang, R.-W.; Sun, W.-T.; Lo, C.-Y.; Hsu, C.-J.; Kuo, C.-W.; Shirota, R. Characterization of the charge trapping properties in pchannel silicon–oxide–nitride–oxide–silicon memory devices including SiO_2/Si_3N_4 interfacial transition layer. *Jpn. J. Appl. Phys.* **2015**, *54*, 104201.1–104201.6. [CrossRef]
54. Thompson, S.E.; Nishida, T. Tunneling and thermal emission of electrons from a distribution of shallow traps in SiO_2. *Appl. Phys. Lett.* **1991**, *58*, 1262–1264. [CrossRef]
55. Chiu, Y.-Y.; Tsai, H.-T.-E.; Chang, K.-C.; Kumari, R.; Li, H.-C.; Takeshita, T.; Yano, M.; Shirota, R. The origin of oxide degradation during time interval between program/erase cycles in NAND Flash memory devices. *Jpn. J. Appl. Phys.* **2021**, *60*, 074004.1–074004.5. [CrossRef]
56. Lai, S.K. Interface trap generation in silicon dioxide when electrons are captured by trapped holes. *J. Appl. Phys.* **1983**, *54*, 2540–2546. [CrossRef]

electronics

MDPI

Review

Recent Progress on 3D NAND Flash Technologies

Akira Goda

Micron Memory Japan, Tokyo 144-0052, Japan; agoda@micron.com

Abstract: Since 3D NAND was introduced to the industry with 24 layers, the areal density has been successfully increased more than ten times, and has exceeded 10 Gb/mm^2 with 176 layers. The physical scaling of XYZ dimensions including layer stacking and footprint scaling enabled the density scaling. Logical scaling has been successfully realized, too. TLC (triple-level cell, 3 bits per cell) is now the mainstream in 3D NAND, while QLC (quad-level cell, 4 bits per cell) is increasing the presence. Several attempts and partial demonstrations were made for PLC (penta-level cell, 5 bits per cell). CMOS under array (CuA) enabled the die size reduction and performance improvements. Program and erase schemes to address the technology challenges such as short-term data retention of the charge-trap cell and the large block size are being investigated.

Keywords: 3D NAND; floating gate cell; charge-trap cell; CMOS under array

Citation: Goda, A. Recent Progress on 3D NAND Flash Technologies. *Electronics* **2021**, *10*, 3156. https://doi.org/10.3390/electronics10243156

Academic Editor: Changhwan Shin

Received: 30 November 2021
Accepted: 13 December 2021
Published: 18 December 2021

1. Introduction

After 2D NAND reached the scaling limit around 15 nm in process node, 3D NAND was proposed as a solution for the continuous NAND scaling [1]. The 3D NAND was introduced into production with 24 layers and MLC technology [2]. The scaling trend of the areal density of 2D NAND and 3D NAND is summarized based on the NAND publications in IEEE ISSCC conferences, (Figure 1). As seen in Figure 1, 3D NAND successfully replaced 2D NAND and has achieved more than 10 Gb/mm^2 areal density [2–31].

At the early stage of the 3D NAND development, various types of 3D NAND technologies were proposed. A comprehensive survey can be found in [32]. After extensive research on various technology options, vertical NAND string architecture with gate-all-around (GAA) cells was introduced to the industry for floating gate (FG) cells and charge-trap cells [2,33].

In this paper, the status of 3D NAND scaling is reviewed and discussed by using the ISSCC conference publications as a reference. This review mainly focuses on the vertical 3D NAND technology adopted in the industry. The physical and logical scaling of 3D NAND will be discussed. The recent progress and topics since the prior review of 3D NAND [34] will be reviewed, including progress on the QLC technology and beyond.

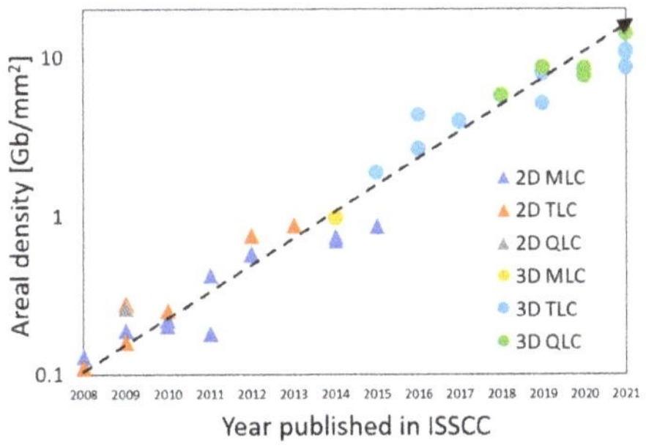

Figure 1. Areal density trend of NAND flash chip presented in ISSCC conferences since 2008 [2–31]. Adapted with permission from ref. [34], Copyright 2020 IEEE.

2. 3D NAND Architecture and Operations

Figure 2 shows the 3D NAND cell array architectures. The strings are placed in the vertical direction. Word lines (WLs) have a plate-like shape and are stacked vertically for 3D cell stacking. There are multiple select gates at the drain side (SGDs) in a block. The channel of the NAND string has a cylinder shape.

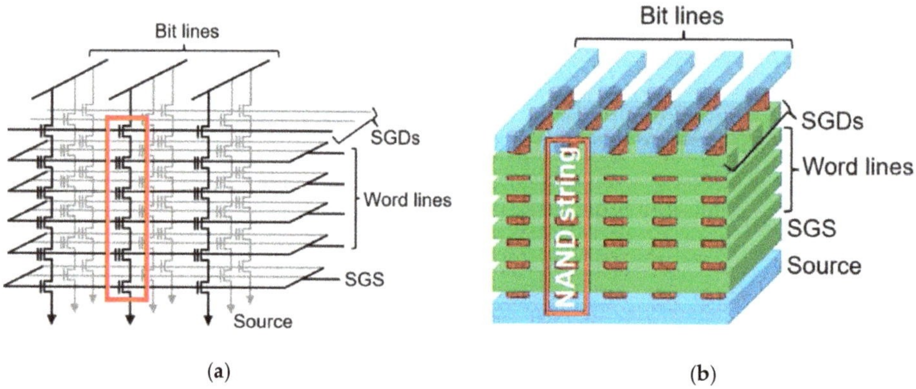

Figure 2. 3D NAND array architecture [35]. (**a**) Schematics of 3 × 3 NAND strings and (**b**) bird's eye view of 3 × 5 NAND strings. Only 4 word lines (WLs) in a string are shown for visibility. The actual 3D NAND has more than 128 WLs in a string.

A block is a unit of the erase operation. As shown in Figure 3, there are two types of erase methods in 3D NAND—the body erase (Figure 3a) and the GIDL erase (Figure 3b) [1,36]. In the body erase, NAND strings are connected to the Si-substrate, and holes are supplied to the NAND string from the Si-substrate, enabling the positive body potential required for erase. In the GIDL erase, the NAND strings are de-decoupled from the Si-substrate and formed on the N+ source layer instead. During erase, the electron-hole pairs are generated at source and drain N+ junctions by GIDL mechanism to supply holes to the NAND strings. The GIDL erase is used for the CMOS under Array (CuA) technology, which will be discussed later [33].

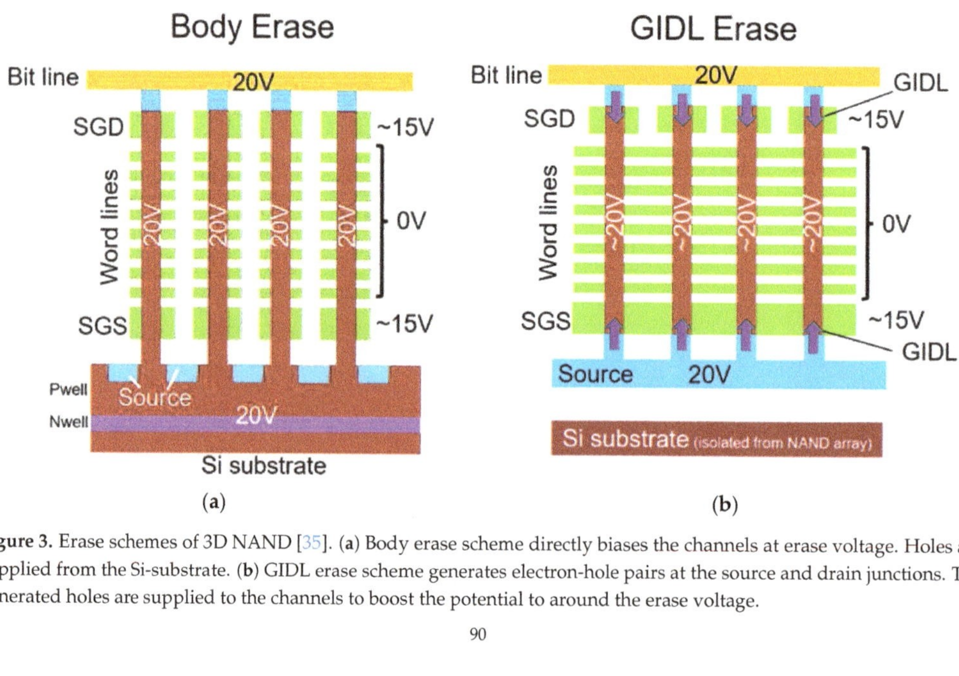

Figure 3. Erase schemes of 3D NAND [35]. (**a**) Body erase scheme directly biases the channels at erase voltage. Holes are supplied from the Si-substrate. (**b**) GIDL erase scheme generates electron-hole pairs at the source and drain junctions. The generated holes are supplied to the channels to boost the potential to around the erase voltage.

During program and read, one of the SGDs is selected in a block so that one NAND string can be selected per bit line (Figure 4).

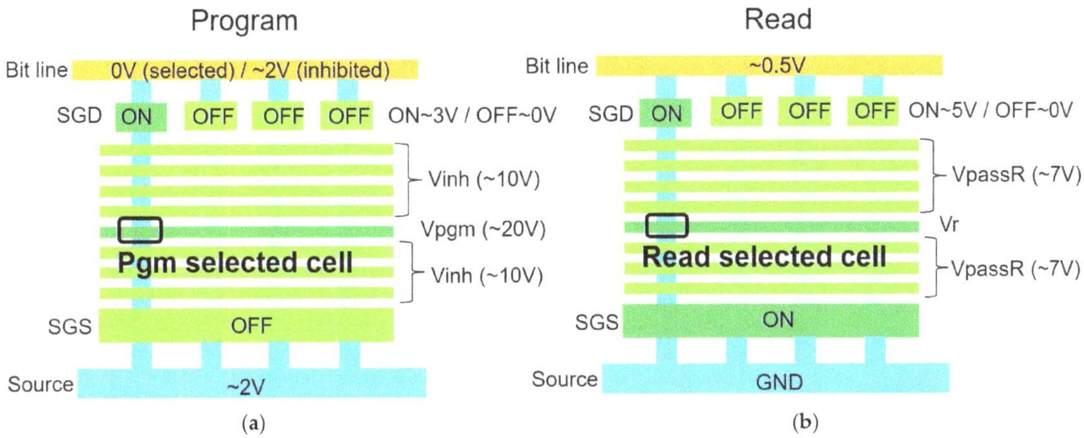

Figure 4. (**a**) Program and (**b**) read operations of 3D NAND. One cell per bit line is selected by selecting one SGD and one WL [35].

3. Floating Gate NAND and Replacement Gate NAND

3.1. Architectures of FG NAND and RG NAND

The floating gate (FG) cell technology was used in 2D NAND. In 3D NAND, in addition to the FG technology (FG NAND), replacement gate cell technology (RG NAND) is also utilized [33,36]. Figure 5 compares the cross-sections of the NAND strings for FG NAND and RG NAND.

In FG NAND, the FG storages are separated between cells because the FG is made of the conductive polysilicon material. The advanced cell integration scheme is adopted to realize the FG separation [33]. In this scheme, the cells are formed outside the pillar holes. Therefore, the diameter of the pillar etching needs to be aligned with the final channel diameter of the NAND string.

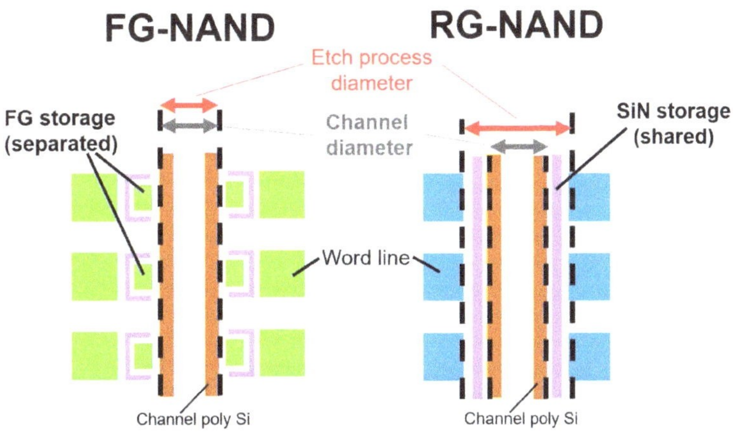

Figure 5. Cross-section comparison of NAND strings between floating gate (FG) NAND and replacement gate (RG) NAND. RG NAND has a larger diameter for the memory hole etching which is advantageous for the pillar height scaling.

In RG NAND, charge-trap cells with a silicon-nitride (SiN) storage layer are employed [2]. Because the SiN storage is a dielectric material which can trap charges, the storage layer can be shared and continuous among the cells. The SiN storage film and other films composing the charge-trap cells are formed after the dry etching of the pillar holes. Therefore, the diameter of the pillar holes for etching can be larger than that of the final channel diameter by the thicknesses of the cell dielectrics. This is advantageous for the pillar etching, especially for the very tall pillar when many cells are stacked vertically. The word line is formed by tungsten metal by replacing the SiN films stacked originally [36]. Therefore, the technology is called replacement gate NAND. RG NAND is the combination of the replacement gate technology and the charge-trap cell technology.

3.2. Band-Engineered Tunneling Dielectrics of the Charge-Trap Cell

In the FG cell, electrons are injected to or emitted from the FG by Fowler–Nordheim (FN) Tunneling (Figure 6a). In the charge-trap cells, the programming is similar to FG cells, where the electrons are injected to the SiN storage by modified FN tunneling. For erase, holes are injected to the SiN storage by direct tunneling (DT) (DT erase) (Figure 6b). In the charge-trap cell, in order to enhance erase efficiently, the cell stack is engineered as shown in Figure 7a. First, the band engineered (BE) tunnel directrices has been introduced [37]. As BE-tunnel dielectrics, an ONO stacked film or a film with an engineered nitrogen profile can be used instead of the SiO_2 tunnel layer. With BE-tunnel dielectrics, holes can tunnel only the thin oxide layer during the erase, while the full ONO thickness can be utilized during the retention. Second, the High-k/Metal gate is used to reduce the unwanted electron injection from the control gate [38]. By combining the BE-tunnel layer and the High-k/Metal gate, good erase characteristics can be achieved with the charge-trap cell (Figure 7b).

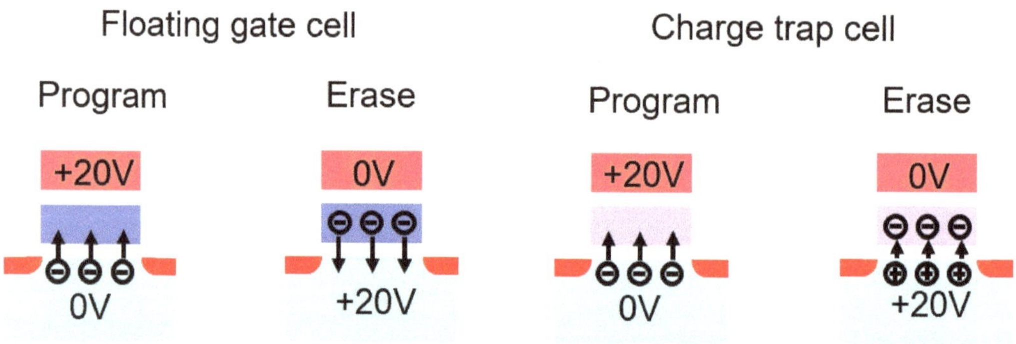

(a) (b)

Figure 6. Program and erase operations for (**a**) floating gate cell and (**b**) charge-trap cell. In the charge-trap cell, holes are injected to the storage layer by direct tunneling.

3.3. Data Retention Mechanisms of the Charge-Trap Cell

The short-term data retention has been a challenge for the charge-trap cell [38]. Figure 8 shows various mechanisms potentially causing the short-term data retention [39]. The first mechanism is charge migration and relaxation. After the programming, the trapped charge in SiN storage could move both laterally (lateral migration, LM) or vertically (vertical relaxation, VR). The second mechanism is detrapping from the SiN storage layer by trap-assisted tunneling (TAT). On top of these, there is trapping at the BE-tunnel oxide as it includes SiN or a nitrogen-rich layer. Due to the very short distance between the trapping sites at BE and the poly Si channel, the detrapping can occur in a very short time.

The impacts on the Vth distribution and programming algorithms solutions for the short-term data retention will be discussed in a later section.

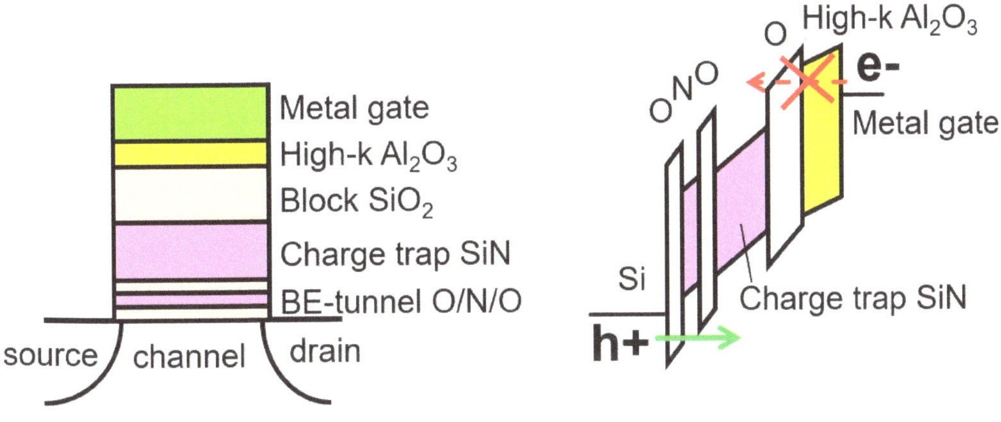

(a) (b)

Figure 7. (**a**) Cell film stack of the charge-trap cell. (**b**) Band diagram during erase of charge-trap cell. Band-engineered (BE) tunnel dielectrics enhances hole injections while High-k/Metal gate prevents electron back tunneling [35].

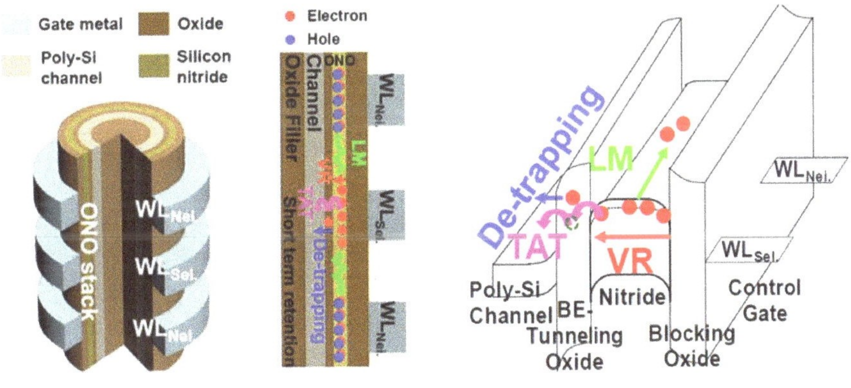

(a) (b) (c)

Figure 8. (**a**) The bird's eye view and (**b**) cross-sectional view of 2D NAND with charge-trap cells. (**c**) Energy band diagram schematic for short-term retention and possible retention mechanisms. Reprinted with permission from ref. [40], copyright 2020 IEEE.

4. 3D NAND Array Physical Scaling by Process Integration

4.1. Conventional Scaling

The physical scaling of the 3D NAND array can be described as XYZ scaling, as shown in Figure 9. XY scaling means reducing the cell footprint. Z scaling means stacking more layers, which is often done together with layer pitch shrink (Z shrink) in order to minimize the increase in the physical height. Z scaling (stacking) has been the main scaling enabler so far. In the ISSCC publications, the steady progress of the layer stacking has been shown, except for in 2020, when the focus was on circuit design technologies for single-level-cells (SLC) and quad-level-cells (QLC) rather than the physical array scaling. In the latest achievement, 176-layer stacked 3D NAND has been demonstrated in both publication and mass production (Figure 10).

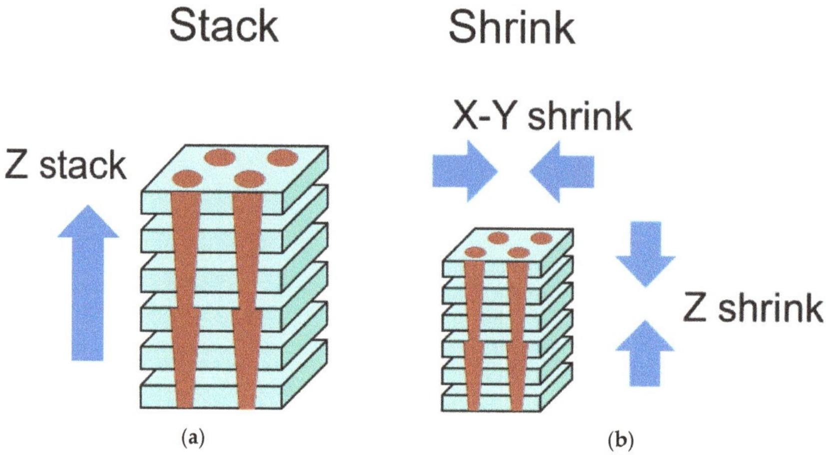

Figure 9. XYZ scaling of 3D NAND. (**a**) Layer stacking in Z-direction. (**b**) Footprint shrink of XY dimensions and layer pitch reduction as Z-shrink. Z-shrink is often combined with Z-stack to minimize the increase in the pillar height.

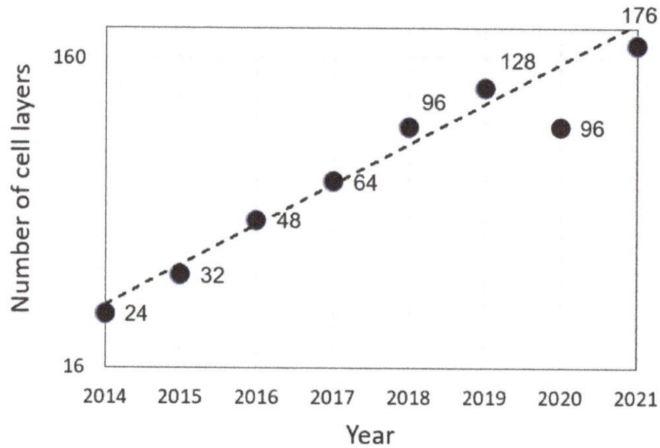

Figure 10. The year-by-year trend of cell layer stacking in 3D NAND published in ISSCC conferences. The reverse trend in 2020 is because the publications are about SLC and QLC technologies on the 96-layer NAND baseline.

The effort for XYZ scaling has been focused on efficiency improvement of the pillar (memory hole) layout [41]. Due to the various layout space requirements such as source line contacts, SGD-to-SGD separations, block-to-block separations, the pillar layout is far from the ideal hexagonal close-packed (HCP) layout. There have been continuous improvements in the array layout so that the pillar arrangement has become closer to HCP, enabling XY scaling (Figure 11).

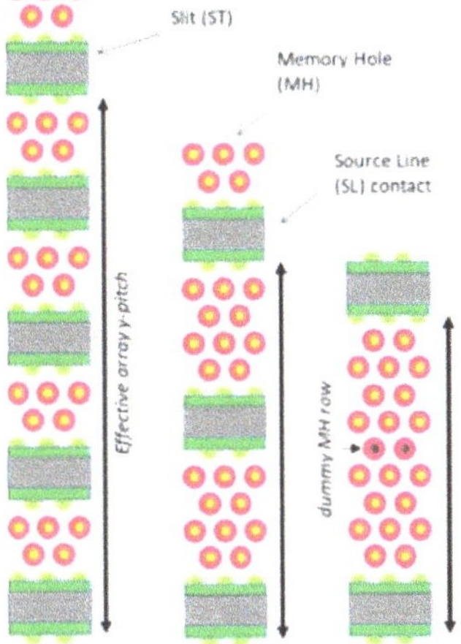

Figure 11. XY scaling by pillar (memory hole) layout efficiency improvement. The pillar layout approaches HPC configuration. Reprinted with permission from ref. [41], Copyright 2020 IEEE.

4.2. Disruptive Scaling

As a disruptive XY scaling, the split cells have been investigated (Figure 12). In the conventional 3D NAND, the cell has a cylinder shape. In the split cell, the cell is split into two parts so that the cell density increases. There are two different types of the split cell proposals. One is a planar-like split cell [42] and the other is a half-cylindrical cell [43,44]. The planar-like cell is similar to the 2D NAND cell, while the half-cylindrical cell can be seen as an evolution of the 3D NAND cylindrical cell. For both split cells, the challenges are the process integration of the cell split, the increased cell-to-cell interference and the reduced gate-coupling ratio.

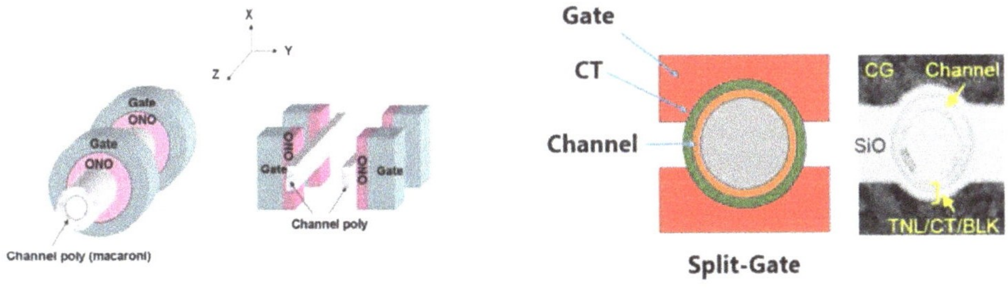

(a) (b)

Figure 12. Split cells with (**a**) planar cell and (**b**) semicylindrical cell. Both of these achieve physical density scaling. The extra cell-to-cell interference needs to be managed to achieve good cell characteristics and high reliability. Reprinted with permission from refs. [42,44], Copyright 2015 and 2019 IEEE, respectively.

As discussed earlier, the pitch shrink of the WL layer is critical to manage the pillar height with layer stacking. In the charge-trap cell, the SiN storage layer is continuous between neighboring cells. With the WL pitch scaling, the trapped charge migration between the neighboring cells would raise the reliability concern. In order to overcome this issue, the SiN storage separation has been proposed (Figure 13) [45]. The process flow is similar to that of 3D FG NAND where the diameter of the pillar etching is smaller compared to the conventional 3D RG NAND. Therefore, the process integration needs to be well-managed for the successful implementation of the SiN storage separation.

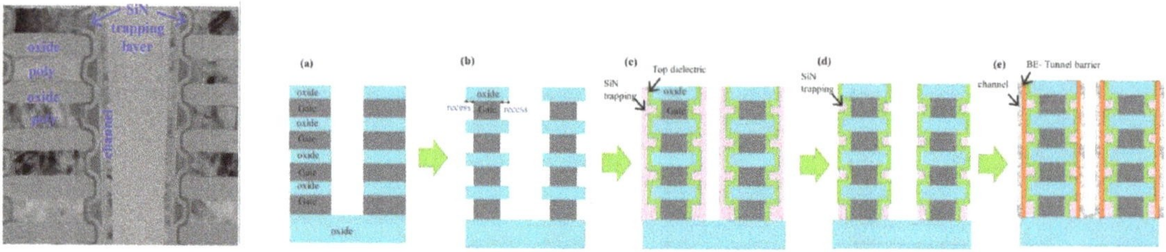

Figure 13. Charge-trap cell with the confined SiN storage layer and the process flow. The SiN storage is successfully separated between cells. The pillar etching diameter needs to be smaller compared to the conventional charge-trap cell. Reprinted with permission from ref. [45], Copyright 2019 IEEE.

5. 3D NAND Array Logical Scaling by Cell Device Engineering

In addition to the physical cell density scaling, the logical density scaling (i.e., more bits per cell scaling) has been actively pursued in 3D NAND.

Figure 14 shows the number of 3D NAND publications in ISSCC conferences. 3D NAND started as MLC technology and rapidly transitioned to TLC, owing to the excellent cell characteristics and reliability. Recently, QLC technology has been introduced to mass production and the presence of the QLC technology has also been increasing in the publications. Currently, TLC is the mainstream for high-performance and high-endurance usages. QLC is becoming mainstream for the high-density and low-cost usages.

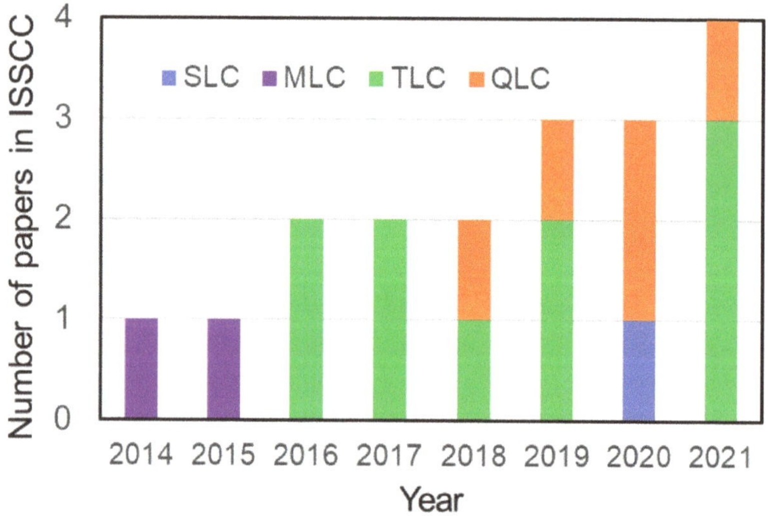

Figure 14. 3D NAND publications in ISSCC for SLC, MLC, TLC and QLC technologies. TLC has been the mainstream for 3D NAND. The interest in QLC is rapidly increasing.

Recently, "beyond QLC" efforts have been reported for both FG and charge-trap cells. PLC (5 bits per cell) distributions were shown with FG cells (Figure 15a) [46]. In this work, the excellent data-retention properties of the FG cell were identified as critical to enable PLC. HLC (6 bits per cell) Vth distributions were experimentally shown at a cryogenic temperature of 77 K for charge-trap cells (Figure 15b) [47]. In that work, it was shown that random telegraph noise (RTN) improves at the 77 K while it degrades at 300 K relative to 358 K. Performance and reliability characteristics as well as the operational conditions warrant further study to enable further bit-per-cell scaling beyond QLC.

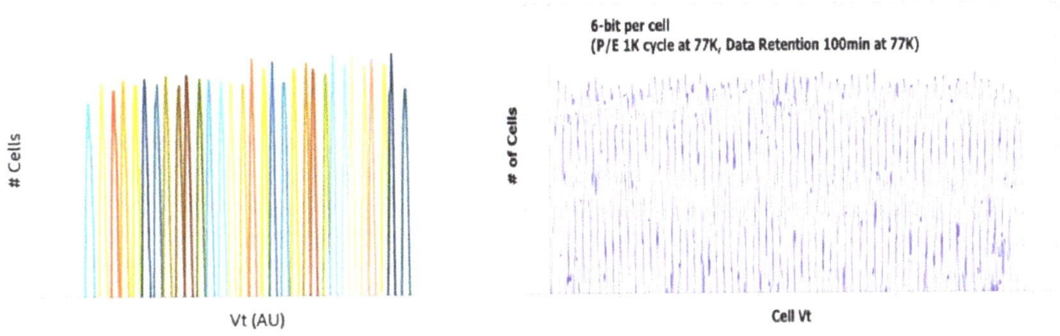

(a) (b)

Figure 15. (**a**) PLC (5 bits per cell) Vth distribution with FG cell. (**b**) HLC (6 bits per cell) Vth placement with charge-trap cell at cryogenic temperature of 77 K. Reprinted with permission from refs. [46,47], Copyright 2020 and 2021 IEEE, respectively.

6. 3D NAND Performance Scaling by Design and Algorithms

6.1. Write Bandwidth for TLC

Write bandwidth is provided by

$$\textit{Page size} \times \textit{number of planes}/\textit{tProg} \tag{1}$$

Therefore, the large write parallelism (= *page size* × *number of planes*) and short *tProg* are critical to achieve high write bandwidth. Figure 16 shows the trend of TLC write bandwidth from the ISSCC publications. There has been incremental improvement over the years and a steep increase in recent years.

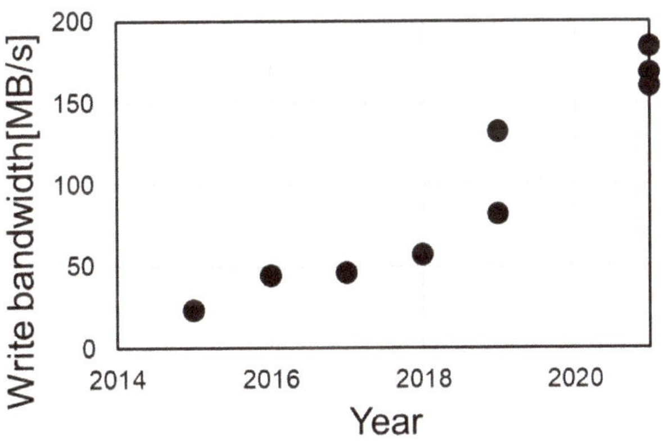

Figure 16. TLC write bandwidth trend reported in ISSCC conferences.

The former is due to the continuous improvement of TLC tProg. Figure 17 is the TLC tProg trend published or estimated from ISSCC publications. In TLC RG NAND, all seven programmed states are programmed in a single programming pass. The improvements in tProg are realized by combinations of WL and BL bias time reduction, fine tuning of the programming voltage compensating cell characteristics' variability across the pillar and the reduction in the program verify operations. Figure 18 shows the center XDEC (WL-driver) architecture which shortens the time for WL loading [31].

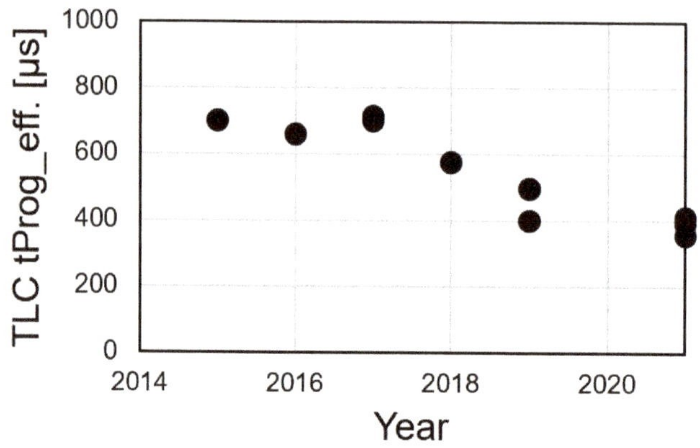

Figure 17. TLC effective tProg reported or calculated from ISSCC publications.

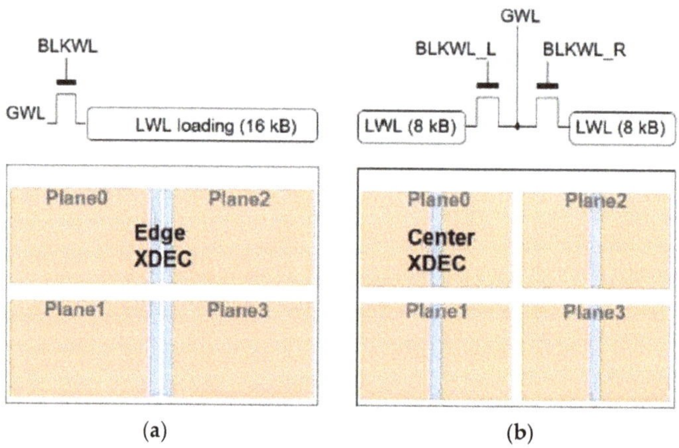

Figure 18. XDEC (WL driver) architecture. (a) XDEC is placed at the edge of the local WL. (b) XDEC is placed at the center of the local WL, reducing the LWL load to a half for faster operations. Reprinted with permission from ref. [31], Copyright 2021 IEEE.

The latter (the steep increase in the write bandwidth) is realized by the increased number of planes owing to the CMOS under Array (CuA) architecture. Figure 19 shows three different CMOS architectures utilized in 3D NAND. In CMOS outside array (CoA), the CMOS circuits are placed next to the array. Therefore, the die size increases if the CMOS area increases. To increase the parallelism, a larger amount of CMOS circuits such as page buffer circuits need to be placed, which results in the increase in the die size. In the CMOS under Array (CuA), the CMOS circuits are placed under the array. More parallelism (more planes) can be realized with CuA because the larger area is available for CMOS circuits. Another variation is to place the CMOS over array by using wafer bonding technologies (Wafer on wafer, WoW) [48]. With this architecture, CMOS is processed separately from the array processing by using a dedicated wafer for CMOS. Therefore, the process flow can be optimized for CMOS devices and interconnect.

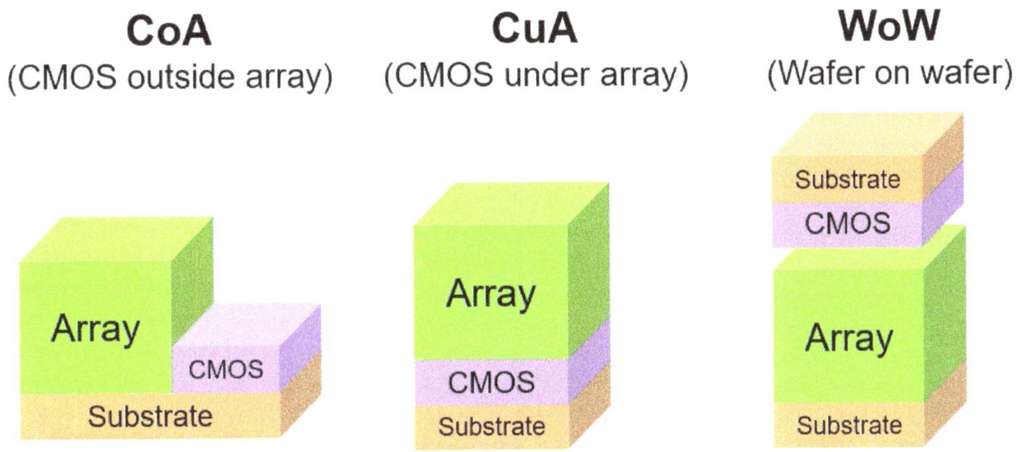

(a) (b) (c)

Figure 19. 3D NAND architectures for CMOS placement. (**a**) CMOS outside array (CoA); (**b**) CMOS under Array (CuA), (**c**) Wafer on wafer bonding (WoW).

Figure 20 shows the number of publications in ISSCC for CNA and CuA. After CuA was presented in 2016 for the first time in an ISSCC conference, publications with CuA have been increasing, and most of the recent 3D NAND publications are for CuA. CuA technology is now the mainstream for 3D NAND.

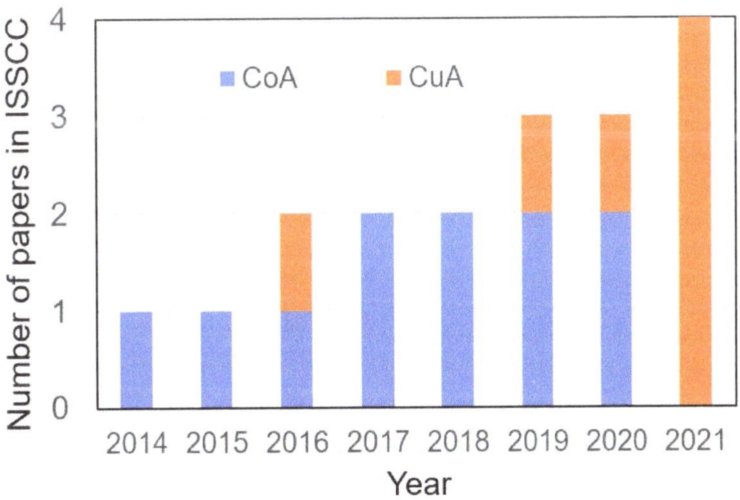

Figure 20. 3D NAND publications in ISSCC for CMOS outside array (CoA) and CMOS under array (CuA) architectures. All publications in 2021 are for CuA.

6.2. QLC Program Schemes

QLC tProg is much longer than TLC tProg (as shown in Figure 21), in order to realize the tight Vth distributions. As discussed earlier, in RG NAND with the charge-trap cells, there is a phenomenon known as short-term data retention. This causes shift and widening of Vth distributions right after programming. To realize the tight Vth distributions for QLC, it is important to manage the short-term data retention effects.

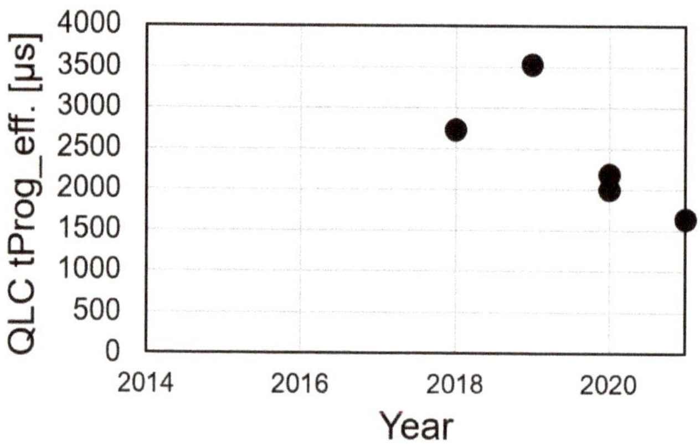

Figure 21. QLC effective tProg reported or calculated from ISSCC publications.

The coarse–fine programming scheme has been introduced for QLC RG NAND [26]. As shown in Figure 22a, all 16 levels are programmed in the first pass with relatively wide distribution widths. Due to the short-term data retention, the Vth distributions shift and widen right after the coarse programming. The fine programming is performed as the second programming path, which tightens the Vth distributions by touching up the distribution tails caused by the short-term data retention. This scheme can be called the 16-16 scheme as the 16-level programming is performed twice. To reduce QLC tProg, an 8-16 scheme was proposed (Figure 22b). With the 8-16 scheme, eight levels are programmed at the coarse programming pass, contributing to the tProg reduction.

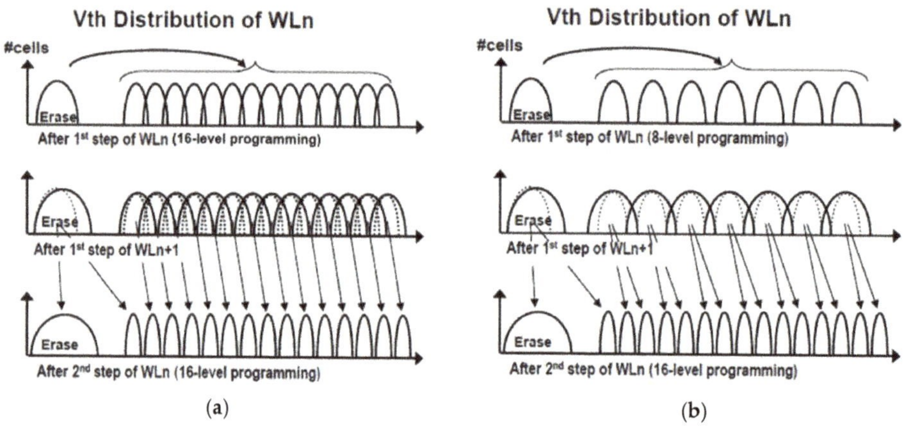

Figure 22. QLC coarse–fine programming schemes with (**a**) 16-16 scheme and (**b**) 8-16 scheme. 16-16 is more efficient to compensate the short-term retention while the 8-16 scheme would have tProg advantage. Reprinted with permission from ref. [26], Copyright 2019 IEEE.

In the FG NAND, the first programming pass only has four levels. The 16 levels are then completed at the second pass [30] because the short-term retention is much smaller in the FG cell and does not require the touch-up operation.

So far, 1.6–2 msec tProg has been reported for QLC and ~0.4 msec tProg has been reported for TLC. For the further enhancement of QLC tProg, solving the short-term data retention by programming scheme and cell improvement is critical.

6.3. Block Size Scaling

With the WL stacking increasing, the block size increases. Figure 23a shows the TLC block size as a function of number of WL stacking. The number of SGDs per block is chosen as a parameter. Typically, FG NAND runs 12–16 SGDs/block, while RG NAND has 4–8 SGDs/blocks. Given that the block size is the minimum granularity of erase, the increase in the block size could increase the system burden of the data management and would degrade system performance. In order to mitigate this problem, the block-by-deck scheme was proposed (Figure 23b) [30]. In this scheme, the NAND string is divided into multiple segments (three segments in this example) and each segment is treated as a different block. During erase operation, the entire pillar is biased to the erase voltage. The WLs of the selected deck block are grounded while the WLs of the unselected deck blocks are ramped to the channel potential. In this way, the cells in the unselected deck blocks can be inhibited for erase.

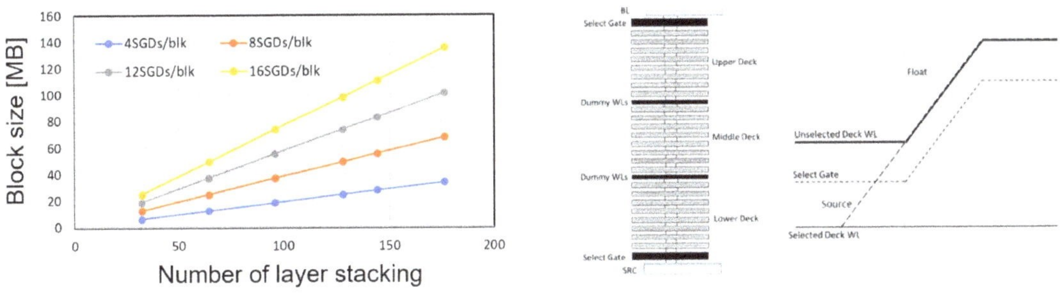

Figure 23. (**a**) Block size scaling as a function of number of layer stacking. (**b**) Block-by-deck erase scheme. The conventional physical block is divided into three logical blocks [30]. Reprinted with permission from ref. [23], Copyright 2021 IEEE.

7. Conclusions

3D NAND scaling has been successfully achieved. The layer stacking has reached 176 layers. QLC has been introduced in both FG NAND and RG NAND. Research on PLC is active, with partial demonstrations for cell capability. CMOS under array (CuA) has been widely adopted and enables performance enhancement by increasing the number of planes. For future scaling, on top of the continuous XYZ physical scaling, disruptive technologies such as split cells are proposed. Program and erase schemes are being developed further to solve cell reliability challenges and block size scaling challenges.

Funding: This research received no external funding.

Conflicts of Interest: The author declares no conflict of interest.

References

1. Tanaka, H.; Kido, M.; Yahashi, K.; Oomura, M.; Katsumata, R.; Kito, M.; Fukuzumi, Y.; Sato, M.; Nagata, Y.; Matsuoka, Y.; et al. Bit Cost Scalable technology with Punch and plug process for ultra high density flash memory. In Proceedings of the 2007 IEEE Symposium on VLSI Technology, Kyoto, Japan, 12–14 June 2007; pp. 14–15. [CrossRef]
2. Park, K.T.; Han, J.M.; Kim, D.; Nam, S.; Choi, K.; Kim, M.S.; Kwak, P.; Lee, D.; Choi, Y.H.; Kang, K.M.; et al. Three-dimensional 128 Gb MLC vertical NAND Flash-memory with 24-WL stacked layers and 50 MB/s high-speed programming. In Proceedings of the 2014 IEEE International Solid-State Circuits Conference Digest of Technical Papers (ISSCC), San Francisco, CA, USA, 9–13 February 2014; Volume 57, pp. 334–335. [CrossRef]
3. Cernea, R.; Pham, L.; Moogat, F.; Chan, S.; Le, B.; Li, Y.; Tsao, S.; Tseng, T.Y.; Nguyen, K.; Li, J.; et al. A 34 MB/s-program-throughput 16 Gb MLC NAND with all-bitline architecture in 56 nm. In Proceedings of the 2008 IEEE International Solid-State Circuits Conference—Digest of Technical Papers, San Francisco, CA, USA, 3–7 February 2008; Volume 51, pp. 420–624. [CrossRef]

4. Li, Y.; Leo, S.; Fong, Y.; Pan, F.; Kuo, T.C.; Park, J.; Samaddar, T.; Nguyen, H.; Mui, M.; Htoo, K.; et al. A 16 Gb 3 b/Cell NAND flash memory in 56 nm with 8 MB/s write rate. In Proceedings of the 2008 IEEE International Solid-State Circuits Conference—Digest of Technical Papers, San Francisco, CA, USA, 3–7 February 2008; Volume 51, pp. 506–508. [CrossRef]
5. Kanda, K.; Koyanagi, M.; Yamamura, T.; Hosono, K.; Yoshihara, M.; Miwa, T.; Kato, Y.; Mak, A.; Chan, S.L.; Tsai, F.; et al. A 120 mm^2 16 Gb 4-MLC NAND flash memory with 43 nm CMOS technology. In Proceedings of the 2008 IEEE International Solid-State Circuits Conference—Digest of Technical Papers, San Francisco, CA, USA, 3–7 February 2008; Volume 51, pp. 430–625. [CrossRef]
6. Zeng, R.; Chalagalla, N.; Chu, D.; Elmhurst, D.; Goldman, M.; Haid, C.; Huq, A.; Ichikawa, T.; Jorgensen, J.; Jungroth, O.; et al. A 172 mm^2 32 Gb MLC NAND flash memory in 34 nm CMOS. In Proceedings of the 2009 IEEE International Solid-State Circuits Conference—Digest of Technical Papers, San Francisco, CA, USA, 8–12 February 2009; pp. 236–237. [CrossRef]
7. Trinh, C.; Shibata, N.; Nakano, T.; Ogawa, M.; Sato, J.; Takeyama, Y.; Isobe, K.; Le, B.; Moogat, F.; Mokhlesi, N.; et al. A 5.6 MB/s 64 Gb 4 b/Cell NAND flash memory in 43 nm CMOS. In Proceedings of the 2009 IEEE International Solid-State Circuits Conference—Digest of Technical Papers, San Francisco, CA, USA, 8–12 February 2009; pp. 246–248. [CrossRef]
8. Futatsuyama, T.; Fujita, N.; Tokiwa, N.; Shindo, Y.; Edahiro, T.; Kamei, T.; Nasu, H.; Iwai, M.; Kato, K.; Fukuda, Y.; et al. A 113 mm^2 32 Gb 3 b/cell NAND flash memory. In Proceedings of the 2009 IEEE International Solid-State Circuits Conference—Digest of Technical Papers, San Francisco, CA, USA, 8–12 February 2009; Volume 1, pp. 242–243. [CrossRef]
9. Lee, C.; Lee, S.K.; Ahn, S.; Lee, J.; Park, W.; Cho, Y.; Jang, C.; Yang, C.; Chung, S.; Yun, I.S.; et al. A 32 Gb MLC NAND-flash memory with Vth-endurance-enhancing schemes in 32 nm CMOS. In Proceedings of the 2010 IEEE International Solid-State Circuits Conference—(ISSCC), San Francisco, CA, USA, 7–11 February 2010; Volume 53, pp. 446–447. [CrossRef]
10. Marotta, G.G.; Macerola, A.; D'Alessandro, A.; Torsi, A.; Cerafogli, C.; Lattaro, C.; Musilli, C.; Rivers, D.; Sirizotti, E.; Paolini, F.; et al. A 3 bit/Cell 32 Gb NAND flash memory at 34 nm with 6 Mb/s program throughput and with dynamic 2 b/cell blocks configuration mode for a program throughput increase up to 13 MB/s. In Proceedings of the 2010 IEEE International Solid-State Circuits Conference—(ISSCC), San Francisco, CA, USA, 7–11 February 2010; Volume 53, pp. 444–445. [CrossRef]
11. Kim, H.; Park, J.H.; Park, K.T.; Kwak, P.; Kwon, O.; Kim, C.; Lee, Y.; Park, S.; Kim, K.; Cho, D.; et al. A 159 mm^2 32 nm 32 Gb MLC NAND-flash memory with 200 MB/s asynchronous DDR interface. In Proceedings of the 2010 IEEE International Solid-State Circuits Conference—(ISSCC), San Francisco, CA, USA, 7–11 February 2010; Volume 53, pp. 442–443. [CrossRef]
12. Kim, T.Y.; Lee, S.D.; Park, J.S.; Cho, H.Y.; You, B.S.; Baek, K.H.; Lee, J.H.; Yang, C.W.; Yun, M.; Kim, M.S.; et al. A 32 Gb MLC NAND flash memory with Vth margin-expanding schemes in 26 nm CMOS. In Proceedings of the 2011 IEEE International Solid-State Circuits Conference, San Francisco, CA, USA, 20–24 February 2011; pp. 202–203. [CrossRef]
13. Park, K.T.; Kwon, O.; Yoon, S.; Choi, M.H.; Kim, I.M.; Kim, B.G.; Kim, M.S.; Choi, Y.H.; Shin, S.H.; Song, Y.; et al. A 7 MB/s 64 Gb 3-bit/cell DDR NAND flash memory in 20 nm-node technology. In Proceedings of the 2011 IEEE International Solid-State Circuits Conference, San Francisco, CA, USA, 20–24 February 2011; Volume 43, pp. 212–213. [CrossRef]
14. Fukuda, K.; Watanabe, Y.; Makino, E.; Kawakami, K.; Sato, J.; Takagiwa, T.; Kanagawa, N.; Shiga, H.; Tokiwa, N.; Shindo, Y.; et al. A 151 mm^2 64 Gb MLC NAND flash memory in 24 nm CMOS technology. In Proceedings of the 2011 IEEE International Solid-State Circuits Conference, San Francisco, CA, USA, 20–24 February 2011; pp. 198–199. [CrossRef]
15. Lee, D.; Chang, I.J.; Yoon, S.Y.; Jang, J.; Jang, D.S.; Hahn, W.G.; Park, J.Y.; Kim, D.G.; Yoon, C.; Lim, B.S.; et al. A 64 Gb 533 Mb/s DDR interface MLC NAND Flash in sub-20 nm technology. In Proceedings of the 2012 IEEE International Solid-State Circuits Conference, San Francisco, CA, USA, 19–23 February 2012; Volume 55, pp. 430–431. [CrossRef]
16. Naso, G.; Botticchio, L.; Castelli, M.; Cerafogli, C.; Cichocki, M.; Conenna, P.; D'Alessandro, A.; De Santis, L.; Di Cicco, D.; Di Francesco, W.; et al. A 128 Gb 3 b/cell NAND flash design using 20 nm planar-cell technology. In Proceedings of the 2013 IEEE International Solid-State Circuits Conference Digest of Technical Papers, San Francisco, CA, USA, 17–21 February 2013; Volume 56, pp. 218–219. [CrossRef]
17. Helm, M.; Park, J.K.; Ghalam, A.; Guo, J.; Ha, C.W.; Hu, C.; Kim, H.; Kavalipurapu, K.; Lee, E.; Mohammadzadeh, A.; et al. A 128 Gb MLC NAND-Flash device using 16 nm planar cell. In Proceedings of the 2014 IEEE International Solid-State Circuits Conference Digest of Technical Papers, San Francisco, CA, USA, 9–13 February 2014; Volume 57, pp. 326–327. [CrossRef]
18. Choi, S.; Kim, D.; Choi, S.; Kim, B.; Jung, S.; Chun, K.; Kim, N.; Lee, W.; Shin, T.; Jin, H.; et al. A 93.4 mm^2 64 Gb MLC NAND-flash memory with 16 nm CMOS technology. In Proceedings of the 2014 IEEE International Solid-State Circuits Conference Digest of Technical Papers, San Francisco, CA, USA, 9–13 February 2014; Volume 57, pp. 328–329. [CrossRef]
19. Sako, M.; Watanabe, Y.; Nakajima, T.; Sato, J.; Muraoka, K.; Fujiu, M.; Kouno, F.; Nakagawa, M.; Masuda, M.; Kato, K.; et al. A low-power 64 Gb MLC NAND-flash memory in 15 nm CMOS technology. In Proceedings of the 2015 IEEE International Solid-State Circuits Conference Digest of Technical Papers, San Francisco, CA, USA, 22–26 February 2015; Volume 58, pp. 128–129. [CrossRef]
20. Im, J.W.; Jeong, W.P.; Kim, D.H.; Nam, S.W.; Shim, D.K.; Choi, M.H.; Yoon, H.J.; Kim, D.H.; Kim, Y.S.; Park, H.W.; et al. A 128 Gb 3 b/cell V-NAND flash memory with 1 Gb/s I/O rate. In Proceedings of the 2015 IEEE International Solid-State Circuits Conference Digest of Technical Papers, San Francisco, CA, USA, 22–26 February 2015; Volume 58, pp. 130–131. [CrossRef]
21. Tanaka, T.; Helm, M.; Vali, T.; Ghodsi, R.; Kawai, K.; Park, J.K.; Yamada, S.; Pan, F.; Einaga, Y.; Ghalam, A.; et al. A 768 Gb 3 b/cell 3D-floating-gate NAND flash memory. In Proceedings of the 2016 IEEE International Solid-State Circuits Conference (ISSCC), San Francisco, CA, USA, 31 January–4 February 2016; Volume 59, pp. 142–144. [CrossRef]

22. Yamashita, R.; Magia, S.; Higuchi, T.; Yoneya, K.; Yamamura, T.; Mizukoshi, H.; Zaitsu, S.; Yamashita, M.; Toyama, S.; Kamae, N.; et al. A 51 2Gb 3 b/cell flash memory on 64-word-line-layer BiCS technology. In Proceedings of the 2017 IEEE International Solid-State Circuits Conference (ISSCC), San Francisco, CA, USA, 5–9 February 2017; Volume 60, pp. 196–197. [CrossRef]
23. Kim, C.; Cho, J.H.; Jeong, W.; Park, I.H.; Park, H.W.; Kim, D.H.; Kang, D.; Lee, S.; Lee, J.S.; Kim, W.; et al. A 512 Gb 3 b/cell 64-stacked WL 3D V-NAND flash memory. In Proceedings of the 2017 IEEE International Solid-State Circuits Conference (ISSCC), San Francisco, CA, USA, 5–9 February 2017; Volume 60, pp. 202–203. [CrossRef]
24. Lee, S.; Kim, C.; Kim, M.; Joe, S.M.; Jang, J.; Kim, S.; Lee, K.; Kim, J.; Park, J.; Lee, H.J.; et al. A 1 Tb 4 b/cell 64-stacked-WL 3D NAND flash memory with 12 MB/s program throughput. In Proceedings of the 2018 IEEE International Solid—State Circuits Conference—(ISSCC), San Francisco, CA, USA, 11–15 February 2018; Volume 61, pp. 340–342. [CrossRef]
25. Maejima, H.; Kanda, K.; Fujimura, S.; Takagiwa, T.; Ozawa, S.; Sato, J.; Shindo, Y.; Sato, M.; Kanagawa, N.; Musha, J.; et al. A 512 Gb 3 b/Cell 3D flash memory on a 96-word-line-layer technology. In Proceedings of the 2018 IEEE International Solid—State Circuits Conference—(ISSCC), San Francisco, CA, USA, 11–15 February 2018; Volume 61, pp. 336–338. [CrossRef]
26. Sugawara, H.; Hosono, K.; Hisada, T.; Kaneko, T.; Nakamura, H. A 1.33 Tb 4-bit/Cell 3D-Flash Memory on a 96-Word-Line-Layer Technology. In Proceedings of the 2019 IEEE International Solid-State Circuits Conference—(ISSCC), San Francisco, CA, USA, 17–21 February 2019; Volume 43, pp. 210–212.
27. Siau, C.; Kim, K.H.; Lee, S.; Isobe, K.; Shibata, N.; Verma, K.; Ariki, T.; Li, J.; Yuh, J.; Amarnath, A.; et al. A 512 Gb 3-bit/Cell 3D Flash Memory on 128-Wordline-Layer with 132 MB/s Write Performance Featuring Circuit-Under-Array Technology. In Proceedings of the 2019 IEEE International Solid-State Circuits Conference—(ISSCC), San Francisco, CA, USA, 17–21 February 2019; pp. 218–220. [CrossRef]
28. Higuchi, T.; Kodama, T.; Kato, K.; Fukuda, R.; Tokiwa, N.; Abe, M.; Takagiwa, T.; Shimizu, Y. A 1 Tb 3 b/Cell 3D-Flash Memory in a 170+ Word-Line-Layer Technology. In Proceedings of the 2021 IEEE International Solid-State Circuits Conference (ISSCC), San Francisco, CA, USA, 13–22 February 2021; pp. 428–430.
29. Huh, H.; Cho, W.; Lee, J.; Noh, Y.; Park, Y.; Ok, S.; Kim, J.; Cho, K.; Lee, H.; Kim, G.; et al. A 1 Tb 4 b/Cell 96-Stacked-WL 3D NAND Flash Memory with 30 MB/s Program Throughput Using Peripheral Circuit under Memory Cell Array Technique. In Proceedings of the 2020 IEEE International Solid-State Circuits Conference—(ISSCC), San Francisco, CA, USA, 16–20 February 2021; pp. 220–221. [CrossRef]
30. Khakifirooz, A.; Balasubrahmanyam, S.; Fastow, R.; Gaewsky, K.H.; Ha, C.W.; Haque, R.; Jungroth, O.W.; Law, S.; Madraswala, A.S.; Ngo, B.; et al. A 1 Tb 4 b/Cell 144-Tier Floating-Gate 3D-NAND Flash Memory with 40 MB/s Program Throughput and 13.8 Gb/mm^2 Bit Density. In Proceedings of the 2021 IEEE International Solid-State Circuits Conference (ISSCC), San Francisco, CA, USA, 16–20 February 2021; Volume 64, pp. 424–426. [CrossRef]
31. Park, J.; Kim, D.; Ok, S.; Park, J.; Kwon, T.; Lee, H.; Lim, S.; Jung, S.; Choi, H.; Kang, T.; et al. A 176-Stacked 512 Gb 3 b/Cell 3D-NAND Flash with 10.8 Gb/mm^2 Density with a Peripheral Circuit Under Cell Array Architecture. In Proceedings of the 2021 IEEE International Solid-State Circuits Conference (ISSCC), San Francisco, CA, USA, 13–22 February 2021; pp. 422–423.
32. Lee, G.H.; Hwang, S.; Yu, J.; Kim, H. Architecture and process integration overview of 3d nand flash technologies. *Appl. Sci.* **2021**, *11*, 6073. [CrossRef]
33. Parat, K.; Dennison, C. A floating gate based 3D NAND technology with CMOS under array. In Proceedings of the 2015 IEEE International Electron Devices Meeting (IEDM), Washington, DC, USA, 7–9 December 2015; pp. 3.3.1–3.3.4. [CrossRef]
34. Goda, A. 3-D NAND Technology Achievements and Future Scaling Perspectives. *IEEE Trans. Electron Devices* **2020**, *67*, 1373–1381. [CrossRef]
35. Hemink, G.; Goda, A. NAND technology status and perspectives. In *Semiconductor Memories and Systems*; Redaelli, A., Pellizzer, F., Eds.; Elsevier: Amsterdam, The Netherlands, 2022; in press.
36. Jang, J.; Kim, H.S.; Cho, W.; Cho, H.; Kim, J.; Sun, I.S.; Jang, Y.; Jeong, J.H.; Son, B.K.; Dong, W.K.; et al. Vertical cell array using TCAT(terabit cell array transistor) technology for ultra high density NAND flash memory. In Proceedings of the 2009 Symposium on VLSI Technology, Kyoto, Japan, 15–17 June 2019; pp. 192–193.
37. Lue, H.T.; Wang, S.Y.; Lai, E.K.; Shih, Y.H.; Lai, S.C.; Yang, L.W.; Chen, K.C.; Ku, J.; Hsieh, K.Y.; Liu, R.; et al. BE-SONOS: A bandgap engineered SONOS with excellent performance and reliability. In Proceedings of the IEEE International Electron Devices Meeting, Washington, DC, USA, 5 December 2005; pp. 547–550. [CrossRef]
38. Park, Y.; Choi, J.; Kang, C.; Lee, C.; Shin, Y.; Choi, B.; Kim, J.; Jeon, S.; Sel, J.; Park, J.; et al. Highly manufacturable 32 Gb multi—Level NAND flash memory with 0.0098 μm^2 cell size using TANOS(Si—Oxide—Al_2O_3—TaN) cell technology. In Proceedings of the 2006 International Electron Devices Meeting, San Francisco, CA, USA, 11–13 December 2006; Volume 2, pp. 5–8. [CrossRef]
39. Chen, C.P.; Lue, H.T.; Hsieh, C.C.; Chang, K.P.; Hsieh, K.Y.; Lu, C.Y. Study of fast initial charge loss and it's impact on the programmed states Vt distribution of charge-trapping NAND flash. In Proceedings of the 2010 International Electron Devices Meeting, San Francisco, CA, USA, 6–8 December 2010; pp. 118–121. [CrossRef]
40. Woo, C.; Kim, S.; Park, J.; Shin, H.; Kim, H.; Choi, G.B.; Seo, M.S.; Noh, K.H. Modeling of Charge Failure Mechanisms during the Short Term Retention Depending on Program/Erase Cycle Counts in 3-D NAND Flash Memories. In Proceedings of the 2020 IEEE International Reliability Physics Symposium (IRPS), Dallas, TX, USA, 28 April–30 May 2020; pp. 3–8. [CrossRef]

41. Alsmeier, J.; Higashitani, M.; Paak, S.S.; Sivaram, S. Past and future of 3D flash. In Proceedings of the 2020 IEEE International Electron Devices Meeting (IEDM), San Francisco, CA, USA, 12–18 December 2020; pp. 6.1.1–6.1.4. [CrossRef]
42. Lue, H.T.; Hsu, T.H.; Wu, C.J.; Chen, W.C.; Yeh, T.H.; Chang, K.P.; Hsieh, C.C.; Du, P.Y.; Hsiao, Y.H.; Jiang, Y.W.; et al. A novel double-density, single-gate vertical channel (SGVC) 3D NAND Flash that is tolerant to deep vertical etching CD variation and possesses robust read-disturb immunity. In Proceedings of the 2015 IEEE International Electron Devices Meeting (IEDM), Washington, DC, USA, 7–9 December 2015; pp. 3.2.1–3.2.4. [CrossRef]
43. Fujiwara, M.; Ishikawa, T.; Arayashiki, Y.; Hirayama, K.; Koyama, Y.; Kashiyama, S.; Cai, W.; Goki, Y.; Sawa, K.; Ikeno, D.; et al. 3D Semicircular Flash Memory Cell: Novel Split-Gate Technology to Boost Bit Density. In Proceedings of the 2019 IEEE International Electron Devices Meeting (IEDM), San Francisco, CA, USA, 7–12 December 2019; pp. 642–645. [CrossRef]
44. Ishimaru, K. Future of Non-Volatile Memory -From Storage to Computing. Procedding of the 2019 IEEE International Electron Devices Meeting (IEDM), San Francisco, CA, USA, 7–12 December 2019; pp. 12–17. [CrossRef]
45. Fu, C.H.; Lue, H.T.; Hsu, T.H.; Chen, W.C.; Lee, G.R.; Chiu, C.J.; Wang, K.C.; Lu, C.Y. A Novel Confined Nitride-Trapping Layer Device for 3D NAND Flash with Robust Retention Performances. In Proceedings of the 2019 Symposium on VLSI Technology, Kyoto, Japan, 9–14 June 2019. [CrossRef]
46. Kalavade, P. 4 bits/cell 96 Layer Floating Gate 3D NAND with CMOS under Array Technology and SSDs. In Proceedings of the 2020 IEEE International Memory Workshop (IMW), Dresden, Germany, 17–20 May 2020; pp. 14–17. [CrossRef]
47. Aiba, Y.; Tanaka, H.; Maeda, T.; Sawa, K.; Kikushima, F.; Miura, M.; Fujisawa, T.; Matsuo, M.; Sanuki, T. Cryogenic Operation of 3D Flash Memory for New Applications and Bit Cost Scaling with 6-Bit per Cell (HLC) and beyond. In Proceedings of the 2021 5th IEEE Electron Devices Technology & Manufacturing Conference (EDTM), Chengdu, China, 8–11 April 2021; pp. 28–30. [CrossRef]
48. Yang, S. Unleashing 3D NAND's Potential with an Innovative Architecture. In Proceedings of the Flash Memory Summit, Santa Clara, CA, USA, 7 August 2018.

electronics

MDPI

Review

Review of Bumpless Build Cube (BBCube) Using Wafer-on-Wafer (WOW) and Chip-on-Wafer (COW) for Tera-Scale Three-Dimensional Integration (3DI)

Takayuki Ohba [1,*], Koji Sakui [1], Shinji Sugatani [1], Hiroyuki Ryoson [1,2] and Norio Chujo [1,3]

[1] Institute of Innovative Research, Tokyo Institute of Technology, 4259 Nagatsuta, Midori-ku, Yokohama 226-8503, Japan; sakui.k.aa@m.titech.ac.jp (K.S.); sugatani.s.aa@m.titech.ac.jp (S.S.); hiroyuki.ryoson@dexerials.com (H.R.); norio.chujo.fj@hitachi.com (N.C.)
[2] Dexerials Co.,Tochigi 323-0194, Japan
[3] Hitachi, Ltd., Tokyo 185-8601, Japan
* Correspondence: ohba.t.ac@m.titech.ac.jp

Abstract: Bumpless Build Cube (BBCube) using Wafer-on-Wafer (WOW) and Chip-on-Wafer (COW) for Tera-Scale Three-Dimensional Integration (3DI) is discussed. Bumpless interconnects between wafers and between chips and wafers are a second-generation alternative to the use of micro-bumps for WOW and COW technologies. WOW and COW technologies for BBCube can be used for homogeneous and heterogeneous 3DI, respectively. Ultra-thinning of wafers down to 4 μm offers the advantage of a small form factor, not only in terms of the total volume of 3D ICs, but also the aspect ratio of Through-Silicon-Vias (TSVs). Bumpless interconnect technology can increase the number of TSVs per chip due to the finer TSV pitch and the lower impedance of bumpless TSV interconnects. In addition, high-density TSV interconnects with a short length provide the highest thermal dissipation from high-temperature devices such as CPUs and GPUs. This paper describes the process platform for BBCube WOW and COW technologies and BBCube DRAMs with high speed and low IO buffer power by enhancing parallelism and increasing yield by using a vertically replaceable memory block architecture, and also presents a comparison of thermal characteristics in 3D structures constructed with micro-bumps and BBCube.

Keywords: bumpless; TSV; WOW; COW; BBCube; bandwidth; yield; power consumption; thermal management

Citation: Ohba, T.; Sakui, K.; Sugatani, S.; Ryoson, H.; Chujo, N. Review of Bumpless Build Cube (BBCube) Using Wafer-on-Wafer (WOW) and Chip-on-Wafer (COW) for Tera-Scale Three-Dimensional Integration (3DI). *Electronics* **2022**, *11*, 236. https://doi.org/10.3390/electronics11020236

Academic Editor: Tae-Yeon Seong

Received: 28 November 2021
Accepted: 7 January 2022
Published: 12 January 2022

1. Introduction

Semiconductor devices and computer systems have evolved as feature sizes have been continuously reduced [1–4]. On the other hand, three-dimensional technology has been considered since the 1980s, mainly from the viewpoint of monolithic ICs [5–11]. From the late 1990s, 3D technology has been widely studied for the hybrid structure, including package from the die-level to wafer-level, e.g., how to stack semiconductor elements and how to connect between stacked dies with the vertical interconnects such as TSVs [12–27].

According to this trend, computer system volumes will reach 50 mm^3, and the power consumption will be 0.5 mW [28,29]. Even in such small computers, high performance and large memory capacity are desired without sacrificing power efficiency and thermal dissipation. Conventional two-dimensional (2D) scaling and three-dimensional (3D) integration methods such as those used in High-Bandwidth-Memory (HBM) [30], however, will inevitably face an economic crisis due to the manufacturing costs and yield required [31–33].

A promising approach to overcome these problems is to combine 3D stacking with high throughput, i.e., co-integration extended into the third dimension (z-direction) using Wafer-on-Wafer (WOW) and Chip-on-Wafer (COW) technologies. In detail, the z-height of a multi-wafer stack must be small, meaning that there should be no bumps between dies, and the dies should be thin. This is the main feature of BBCube, which allows high bandwidth with low power consumption because of the short length of TSVs and high-density signal parallelism [34]. Furthermore, high-density TSVs act as thermal pipes, and, hence, a low temperature, even in a 3D structure, can be expected.

2. Manufacturing Cost Crisis for Two-Dimensional Scaling

Before discussing 3D integration for high-volume manufacturing, it is necessary to investigate the current status and future prospects of semiconductor technology development. Conventional 2D scaling will face a severe economic crisis due to the expensive lithography processes and facilities required. Reducing costs requires the adoption of advanced lithography technologies, which, together with peripheral support facilities such as a defect monitoring system, account for one-third to one-fourth of the total cost of a manufacturing line. Furthermore, bit cost is saturated around 20 nm nodes [35,36] due to unavoidable invisible defect reduction. Unless there is sufficient yield, the total cost will increase even if high-resolution lithography is employed. This is the main reason why multiple, small microprocessor dies (chiplets) are integrated [37,38]. In short, while useful for reducing chip size, scaling is extremely burdensome in terms of capital investment. Large-scale investments to the new fabrication facilities (Fabs) have so far been made considering the technologies that will be available two to three generations ahead without any major technology changes. This is based on the empirical rule in the semiconductor that profits are made several generations after investments for reasons involving the trade-offs between products sales and facility depreciation.

According to this empirical rule, an investment in recently developed 7 nm technology needs to be made in consideration of its applicability to 2–3 nm technologies. In the case of ArF (λ = 193 nm), immersion lithography, double or quad patterning for one layer is needed to meet to those critical pattern dimensions. Extreme ultraviolet (EUV; λ = 13.5 nm) lithography has the potential to allow patterning in a single step, and thus EUV is superior to ArF. However, the price of EUV lithography machines is more than 120 million USD [39], which is more than twice that of ArF immersion (iArF) lithography machines, and their current throughput is less than that of iArF machines. When converted into the processing capacity of current large-scale Fabs (e.g., 50,000 incoming wafers per month), based on this system performance, an investment of about 2 billion USD will be required for EUV technology. Assuming that the lifelong sales for each generation are about 10-times the corresponding business investment, the corresponding market size necessary for this investment is more than 20 billion USD. Although, this estimate is based on the 440 billion USD total worldwide semiconductor sales in 2020, this market size for one product and one manufacturer is not realistic.

In conclusion, this is one of the limits of two-dimensional scaling in light of the economics of the industry, and it is difficult to find a scenario of victory at present, especially beyond nanometer node.

3. Paradigm Shift to Bumpless Build Cube Integration

Extending structures into vertical space (z-direction), for example, by three-dimensional stacking, in combination with conventional two-dimensional integration, is anticipated to overcome the problems noted above. The concept of Bumpless Build Cube (BBCube) is a solution to the problems of next-generation 2.5D (side-by-side arrays) and 3D stack systems, in which device dies and interposers are connected without bumps, described in Section 8.3.

Figure 1 shows a comparison of the bump and bumpless interconnects using TSVs, assuming eight dies for a memory core and one logic controller. Since a chip-level stack formed by Chip-on-Chip (COC) technology using bump connections needs pick-and-place for chip transfer, the die thickness is limited by the mechanical stiffness requirements and warpage, resulting in a chip pitch of around 80–100 µm. The mechanical stiffness decreases with die thickness [40,41].

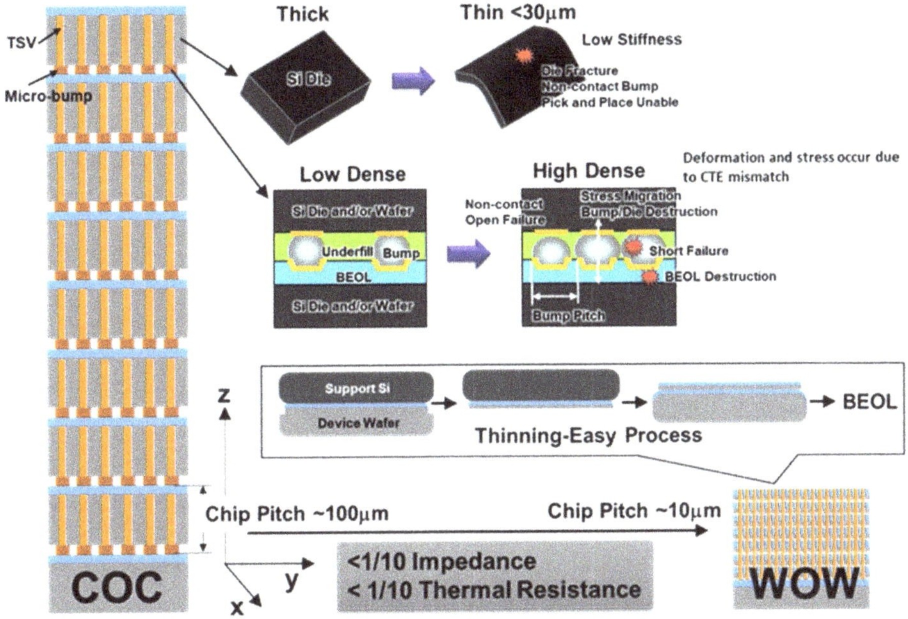

Figure 1. A comparison of bump and bumpless interconnects using TSVs for 3D logic/memory stack structures, assuming eight dies for a memory stack and one logic controller. Since the thickness of the die and density of bumps are limited by mechanical and process difficulties, chip pitch becomes as large as approximately 100 μm, in accordance with the die thickness and bump height. Shortened bumpless interconnects can be formed with higher density (narrower pitch) compared with TSVs and bumps due to the limitations of bump size and pitch. By using wafer thinning and bumpless interconnects, the chip pitch becomes about 1/10, and impedance and thermal resistance become less than 1/10 due to the absence of bumps characterized by high electrical and thermal resistance.

If the bump height varies, some bumps will not come into contact with the electrodes on the chip surface. When high pressure is applied to avoid such contact failures, bonding failures due to plastic deformation are mitigated. However, if excessive pressure is applied, problems such as electrical shorts and destruction will occur between bumps due to the lateral deformation of bumps and the multi-interconnects under bumps due to vertical concentrated stress [42–44]. These problems become more significant when the bump pitch is narrowed. This limits the density of TSVs that can be achieved using a combination of TSVs and bumps.

The WOW process consists of *bonding-first* using a thinned wafer and then the formation of TSV interconnects. Thus, the wafer thickness is determined by whether thinning degrades the device characteristics. There was no damage when a DRAM Si wafer was thinned to 4 μm [45–48]. The wafer (chip) stack pitch was around 10 μm, which is 1/10th as thick as that of COC. The WOW process enabled wafer thinning from 775 μm to 1 μm, as shown in Figure 2. For thinning of a DRAM wafer, the effects of Si thickness, the thinning method, and Cu contamination from the backside on the device characteristics of 20 nm-node DRAMs were evaluated [49]. No obvious degradation of the retention characteristics occurred, even when the Si thickness was reduced to 3 μm, as shown in Figure 3. The refresh time was improved by increasing the thickness of the backside defects layer using grinding. The backside defects act as a trapping site for the Cu diffusion and thus Cu diffusion is prevented when the backside has sufficient defects. From the perspective of reliability, due to the poor gettering ability of the CMP finished surface, is necessary to optimize the gettering ability if there is concern about Cu contamination during the process.

This suggests that it is important to design the diffusion length of defects carefully to prevent defects entering the depletion layer, taking the standby currents and the retention characteristics into account, as shown in Figure 4.

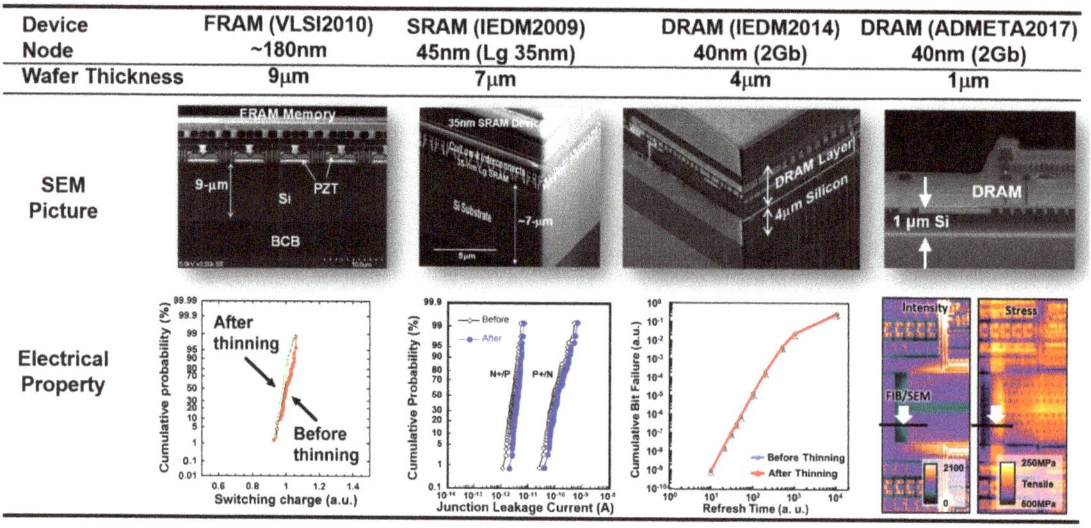

Figure 2. Cross-sectional SEM images and electrical properties after device wafer thinning. Thinning was carried out from 9 to 1 μm for FRAM, SRAM, and DRAMs, respectively. There was no degradation in the electrical characteristics after thinning, and the circuit area at the critical layer could be observed from the back side of the wafer when the silicon thickness became one micrometer.

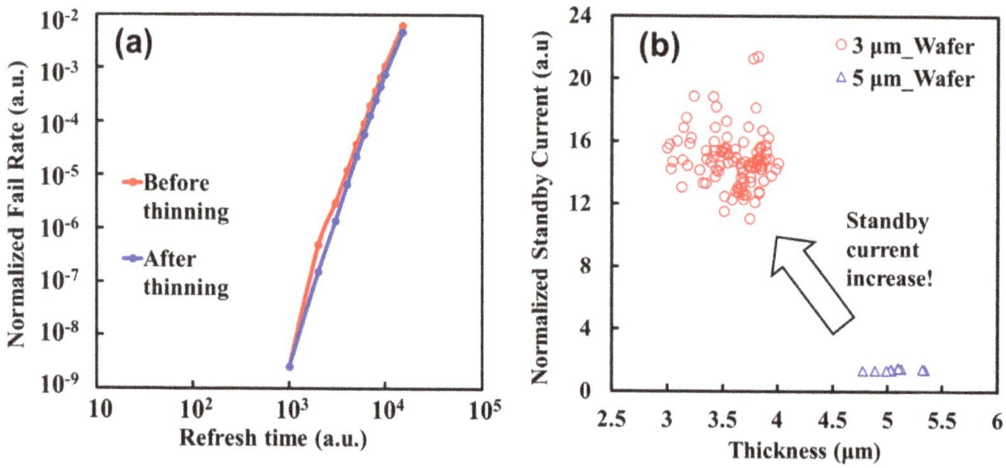

Figure 3. (**a**) Comparison of retention time distributions of same chip before and after fine grinding (#2000 grit abrasive) to 3 μm followed by Cu contamination at ~10^{14}/cm^2, which is 1000-times higher than that of the BEOL process (<10^{11} atoms/cm^2); (**b**) Standby current as a function of Si thickness by fine grinding with and without Cu contamination. Standby current were measured from 3 μm (100 points) and 5 μm (10 points) wafers.

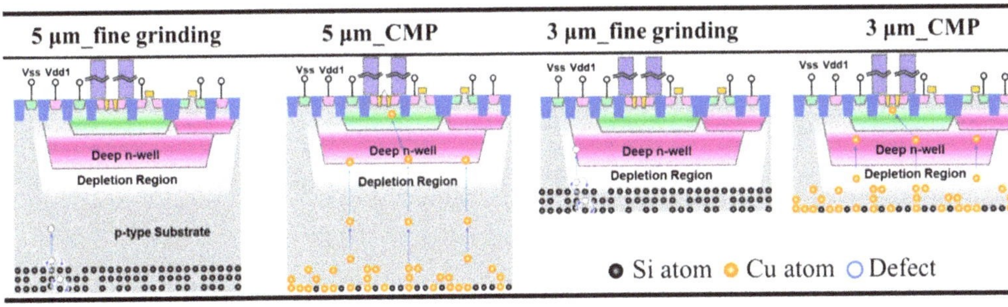

Figure 4. Schematic diagram of cross-sectional image of transistor and the degradation model for DRAMs after thinning. The depth of a deep N-well is about 2.0–3.0 μm. The depth of the depletion region between the deep N-well and the substrate is calculated to be 3.0–4.0 μm. In the case of a Si thickness of 5.0 μm, the defects do not reach the depletion region, and thus both the standby current and retention characteristics do not change. When the Si thickness is being reduced to 3 μm by fine grinding, the defects reach the depletion region. These defects in the depletion region increase the junction leakage current between the substrate (Vss) and the deep N-well (Vdd1). This causes an increase in standby current. On the other hand, since CMP treatment removes the grinding-induced defects and reduces diffused-defects at the depletion region, the standby current is improved.

Since the physical length of TSV interconnects is determined by the wafer thickness, including the device layer and adhesive, the total length in the case of an eight-wafer stack was <80 μm. Trends of TSV interconnects versus the number of stacked chips and/or wafers were estimated as shown in Figure 5 [50–52]. The total height, based on the die-to-die pitch, was less than 0.5 mm, even for a stack of 60 wafers. The TSV density ranged from $10^6/\text{cm}^2$ to $10^7/\text{cm}^2$, which is 10- to 100-times larger than the case of TSVs and bump interconnects.

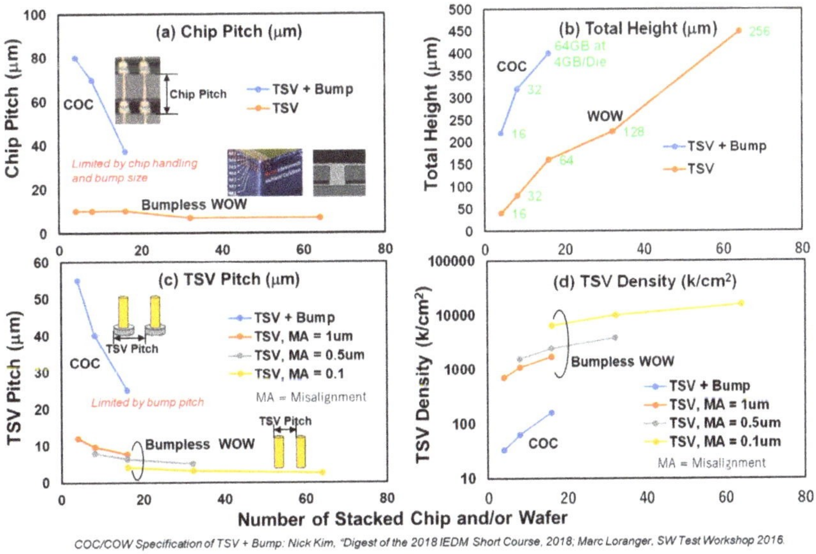

Figure 5. Trends of TSV interconnects as a function of number of stacked layers: (**a**) chip pitch, (**b**) total height, (**c**) TSV pitch, and (**d**) TSV density. For bumpless process, TSV diameter and space between pads were 9 to 1 μm and 1 μm, varied with misalignment (MA), respectively.

It was possible to next make a roadmap to achieve a high-bandwidth system and high-density integration backed up by production costs. Moreover, retaining Cu interconnects technology and the standard 300 mm wafer size for stacking ensures compatibility with existing manufacturing facilities in Front-End processing and helps utilize the mature process technologies that have been developed for wafer processing.

Since the wiring length of the TSVs is determined by the thickness of the wafer, the wiring length becomes shorter when the wafer is thinned down. The conventional wiring length consists of the length of Cu wiring used for TSVs and the bump height, and the total length is about 80–100 μm. The resistance of the bumps is about one order of magnitude higher than that of Cu, e.g., Sn-3.5Ag (12.3 μΩcm) >> Cu (1.68 μΩcm). If only TSVs are used, and there are no bumps with high resistance and the wiring resistance is reduced to <1/10 at a length of 10 μm and a constant diameter. Because of the high density and low resistance of TSVs, high bandwidth and low power consumption can be expected. Details will be discussed in Section 8.

Bumpless Build Cube (BBCube) is a second-generation alternative to the use of TSVs with micro-bumps. The BBCube, bumpless interconnects process involves a *Thinning-First* process before bonding wafers, followed by a *Via-Last* process, meaning that interconnects are formed after bonding the wafers, as shown in Figures 6 and 7. Via-hole etching was carried out, followed by lithography, on a silicon substrate with multilevel interconnects and a device layer after bonding the thinned wafer. Since bumpless Wafer-on-Wafer (WOW) technology uses a back-to-front stack, in principle, any number of thinned 300 mm wafers can be stacked to fabricate large-capacity memory and logic devices. This wafer stacking method is similar to multilevel metallization in the Back-End-of-Line (BEOL), as if replacing dielectric deposition using thinned· wafers and Al and/or Cu metallization with bumpless interconnects using TSVs.

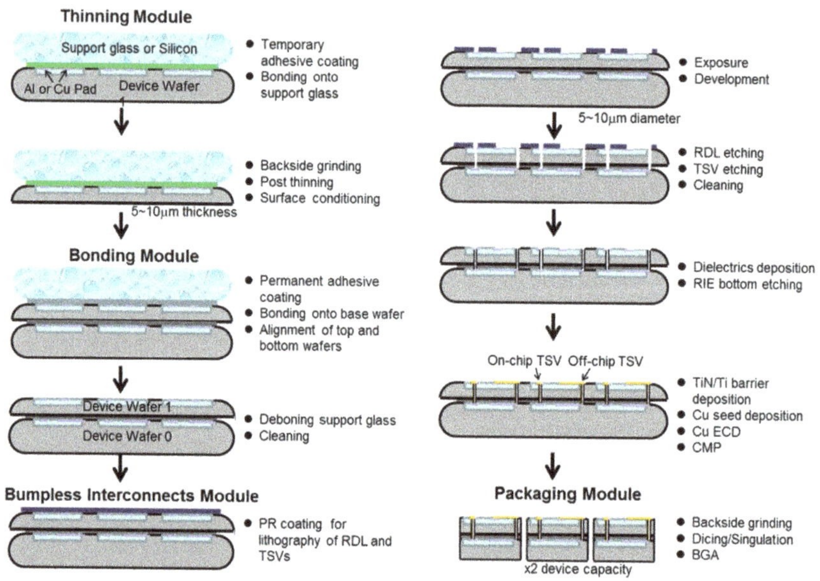

Figure 6. Process flow of bumpless interconnects using TSVs and Wafer-on-Wafer (WOW). Additional wafers can be stacked on top without any limitation on the number of wafers. These modules can also be applied to Chip-on-Wafer (COW) after wafer-level molding. On-chip and off-chip TSV, respectively represent bumpless interconnects formed in the device area and the area around devices, including gap fill (molding) materials in COW.

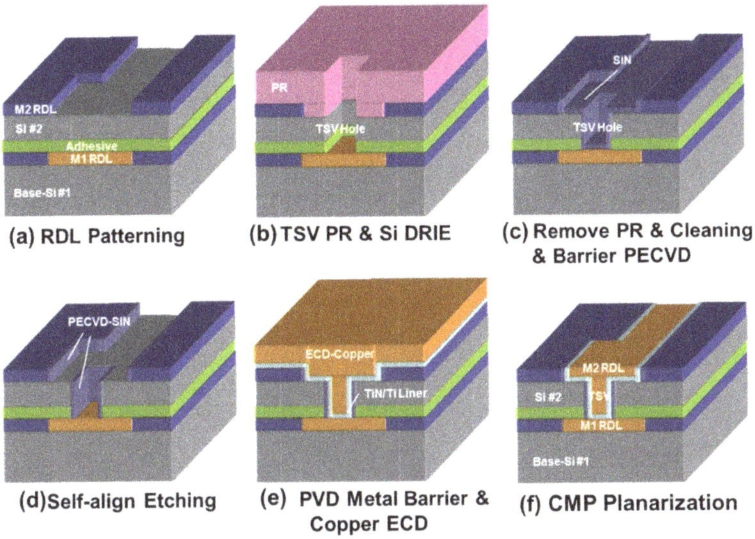

Figure 7. RDL formation and TSV (Cu plug) processes. After bonding of thinned wafer to another wafer surface, (**a**). RDL patterning, (**b**). TSV etching, (**c**). barrier layer formation, (**d**). contact opening, (**e**). Cu plug formation by ECD, and (**f**). planarization by CMP is carried out.

In the case of a chip stack for comparison, a singulation step was needed before stacking. There are several methods for singulating a wafer. For example, in the method of forming a dicing groove on the wafer surface in advance, the wafer surface is attached to a film (DAF: Die-attach Film) and the wafer is thinned with a grinder from the backside. This results in singulation by grinding the back surface to the dicing groove. Each of these singulated chips are picked up by a transfer machine and placed on the surface of a separately prepared wafer. These transfer processes are so called pick-and-place. If the chip thickness is small, the rigidity becomes small, and since the chip transfer method is mechanical, it is easy to break the chip. In addition, stress in the device layer generates chip warpage, causing picking errors. Therefore, in the case of COC and COW, a chip thickness of about 20 μm to 30 μm, which satisfies the requirements of the transfer process and mechanical strength, was used. This is the root cause of the thickness limitation in the chip stacking process. The throughput of pick-and-place was obviously low compared to that of WOW, and there was a trade-off between speed and placing accuracy.

The bumpless WOW process proceeded through the development of four modules, classified along the process flow. The modules included (i) a thinning module for thinning the wafer substrates in which devices are implemented, (ii) a stacking module for bonding and stacking with alignment of the wafers, (iii) a TSV interconnects module for forming Cu interconnects embedded in upper and lower wafers with TSVs, and (iv) a packaging module for singulating the stacked wafers. The TSV interconnects module follows the Dual-Damascene process, forms a so-called redistribution layer (RDL) and vertical interconnects simultaneously, and also serves as a counter electrode for the subsequent stacked wafer.

The thickness of the thinned wafer is a critical dimension for the aspect ratio (depth-to-diameter ratio) of TSVs because the aspect ratio is determined by the diameter and the wafer thickness. Since, in this WOW process, a thinned wafer was bonded on a base wafer, there was no need to take measures for handling ultrathin wafers. The typical Si thickness of a thinned wafer is 4 to 5 μm. When the thicknesses of the device layers in a DRAM and an MPU were assumed to be approximately 5 μm and 10 μm, respectively, the aspect ratio of a TSV was only 5 at maximum for a TSV diameter at 3 μm, whereas conventional TSVs with bumps have aspect ratios more than 10 at a die thickness of 30 μm, including the device layer. With the decreasing aspect ratio, in the TSV processes such as via hole etching, thin-film deposition, and metal filling, the process time decreased to about 1/2 at most, and step coverage significantly improved.

4. Details of BBCube WOW Processes

4.1. Thinning Module

According to the process flow in Figure 6, a wafer with a device layer was bonded to a support substrate (glass or Si wafer) from the device surface with a temporary adhesive in advance. Thinning was performed by mechanical grinding from the back surface of the wafer (Back Grind, or BG) to within several micrometers of the target thickness, followed by polishing until the final thickness was achieved. The final silicon thickness is the thickness at which no degradation of the device characteristics occurs. This was demonstrated with a DRAM device, which is highly sensitive to defects and metal contamination at the diffusion region [53]. The temporary adhesive and the support substrate were removed after thinning the wafer. A permanent adhesive layer with a thickness of 1 to 5 μm was used for wafer stacking [54,55]. The thickness of the permanent adhesive layer can be reduced according to the surface topography of the device wafer, such as the presence of multilevel interconnects and dicing lines.

The reason for using a support substrate is that if the wafer is made thin, it loses its rigidity and bends under its own weight, making it difficult to handle in the wafer process. This can be more intuitively understood by considering that we could not easily handle thin aluminum kitchen foil even at a thickness of ~12 μm. Wafer thinning was carried out from the back side with a grinder using a grinding wheel. In order to grind from the initial thickness of 775 μm to the micrometer level, taking throughput and the wafer flatness into account, two different sizes of abrasive grains, 50 μm and <5 μm were used one after another for high-speed grinding and low-speed grinding with surface conditioning, respectively. When the abrasive grain size is reduced, large defects generated at the wafer surface can be removed [56,57].

The thickness variation of the thinned wafer was determined by the geometric parallelism between the surface of the grinder and the wafer surface. Because of the mechanism of surface grinding, there was a very small angle between the grinding wheel surface and the wafer surface, so the outer thickness of wafer was slightly smaller than the center thickness. On the other hand, the thinned wafer deformed according to the thickness variation and elastic deformation of the temporary adhesive layer as the rigidity of the wafer decreases. Although the Young's modulus of a silicon wafer in the <110> direction remained the same at about 170 GPa down to 10 μm [58], deformation occurred according to the mechanical properties of the adhesive layer and its thickness. Therefore, the total thickness variation (TTV) within the wafer in a micrometer-level thin region is determined by the variation in the thickness of the temporary adhesive layer for the ground surface.

In the optimized grinding method, when the average thickness was 4 μm, which is just 0.5% of the initial thickness, the total thickness variation (TTV) in the 300 mm wafer was about 1 μm, as shown in Figure 8 [47]. Although the thickness of a wafer consists of the thicknesses of the Si substrate and the transistor layer, including multilevel interconnects, the wafer thickness described here is that of the Si substrate. Since the thickness of the transistor layer is 7 to 15 μm for state-of-the-art DRAM and MPU products, the thickness of the transistor layer becomes predominant when the Si substrate reaches the micron level. In the case of such "silicon thin films," the die-level surface geometry of the device layer, steps at dicing lines, IO pads, and particles affect the local thickness variation of the silicon thin film, when the adhesive layer is not thick enough to absorb the surface geometry and particles.

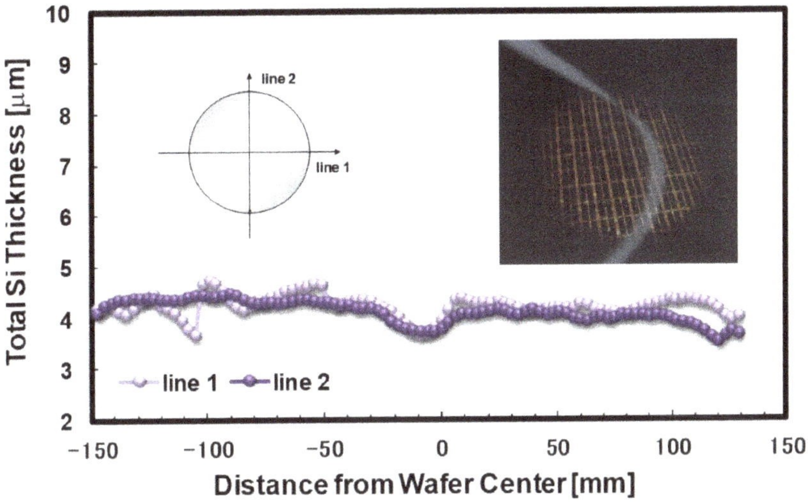

Figure 8. Total thickness of 300 mm DRAM wafer after thinning to 4 μm as a function of wafer position from bottom (notch) to top and from left to right. Light transparency occurs at 4 μm where dark region represents high-density device area.

Since the wafer edge was prepared in a bevel shape, a knife edge shape was unavoidably formed when the thinned wafer surface reached the bevel at the inverse taper angle. It is for this reason that edge fractures and cracking tend to occur during the grinding process. To prevent this, a region of about 0.5 to 2 mm from the edge of the wafer was ground to form a step shape before thinning. This process is called edge trimming [59,60]. However, excess edge trimming over 2 mm of edge exclusion in the Front-End process removes some of the usable device area, which leads to a smaller number of dies within the wafers in a multilevel wafer stack. A novel bevel profile for wafer-level multi stacking technology was therefore proposed by considering the relationship between bevel cracking and bevel angle in wafer thinning, using a grinding process [61]. The bevel angle of the wafer was controlled to 45° to 135°, and bevel cracking after grinding was evaluated with a microscope, as shown in Figure 9. When the bevel angle is smaller than 50°, cracks are noticeably generated during thinning by grinding. According to this result, the bevel profile had a bevel angle of 50° for the area used as the device area after thinning, and a region with a bevel angle of 20° to 30° for the area removed by thinning. This bevel design does not need edge trimming and is expected to reduce wafer area loss without the occurrence of cracking during thinning and wafer transportation.

4.2. Stacking Module

For WOW stacking, the wafers were aligned using alignment marks on the top and bottom wafers just before being attached and permanently bonded. To ensure alignment, infrared light passing through the silicon substrate was used. The wafers bonded to one another in WOW were thin and thus highly transmissive of light. It is necessary to keep a low coefficient of thermal expansion (CTE) mismatch in the stacking to achieve fine pitch alignment and to reduce wafer warpage. When the temperature of one wafer differs from that of another, the two wafer sizes, which are nominally 300 mm, vary due to CTE; for example, at a temperature difference of only 10 °C, the maximum wafer size difference is 11.7 μm, assuming that the CTE of silicon is 3.9×10^{-6}/K. Thus, isothermal heating and warpage-free wafers were needed for submicron-level fine-pitch TSV alignment.

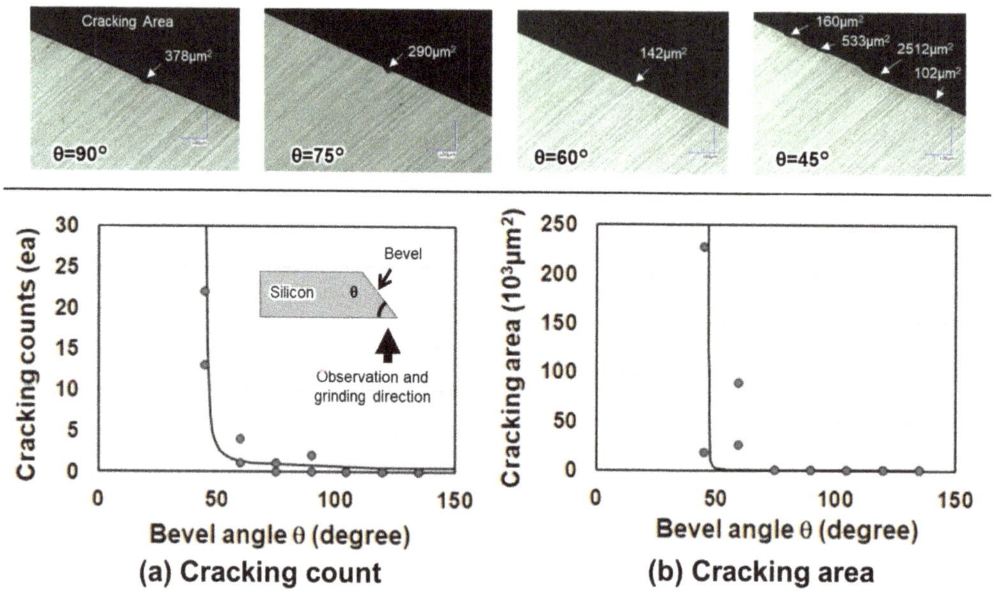

Figure 9. Top-view pictures of the wafer edge after grinding for the bevel angle-controlled sample. (**a**,**b**) show the number of cracks and the cracking area as a function of edge position and bevel angle, respectively. Cracking tends to increase with decreasing bevel angle, significantly below 60 degrees. Since cracking occurs randomly along the wafer edge, it might be caused by the grinding step and conditions of the grindstone.

In general, important issues related to the wafer stack process with alignment are thermal stability of materials during the wafer stacking process and matching of the operating temperatures of the temporary and permanent adhesives. Temporary adhesives are de-bondable by heat, UV light, and/or mechanical force and are useful in conventional stacking methods used to fabricate both COW and WOW. To de-bond a device wafer from a support substrate with low stress, a heat de-bondable adhesive called a hot-melt adhesive, with a wide range of operating temperatures is required. Most of the compounds for permanent adhesives with high thermal stability, such as benzocyclobutene (BCB) resins (curing temperature, 250 °C) [62], require high temperatures for the curing process. On the other hand, low-temperature curable compounds generally have poor thermal stability. One candidate is a reactive hot-melt type temporary adhesive (DTB-TP005, Daicel Co., Tokyo, Japan) and permanent adhesive (DPAS100, Daicel Co.) consisting of an organic-inorganic hybrid structure [63]. Figure 10 shows the experimental setup and the glass transition temperature (Tg) of the temporary adhesive controlled at the bonding and de-bonding temperatures. The device wafer and carrier wafer were bonded from the device surface at a temperature of around 130 °C with a temporary adhesive layer of less than 10 μm in thickness formed by a spin-on technique. The device wafer, which was fixed to the carrier wafer with a temporary adhesive layer, was thinned to about 10 μm by grinding and polishing using DGP8761HC (DISCO Corp., Tokyo, Japan), and then coated with an adhesion promoter containing dual functionality in the molecular structure. The adhesion promoter and permanent adhesive layer were sequentially formed in this order on the surface of another device wafer. In the next process, the coated device wafer was stacked on a thinned wafer coated with an adhesion promoter. The thickness of the permanent adhesive layer was about 2.5 μm. After a curing process, there were no voids between the two stacked wafers. The carrier was de-bonded by mechanical peeling-off at 80 kPa using a differential pressure de-bonder, and then the residual temporary adhesive was able to be removed with an organic solvent.

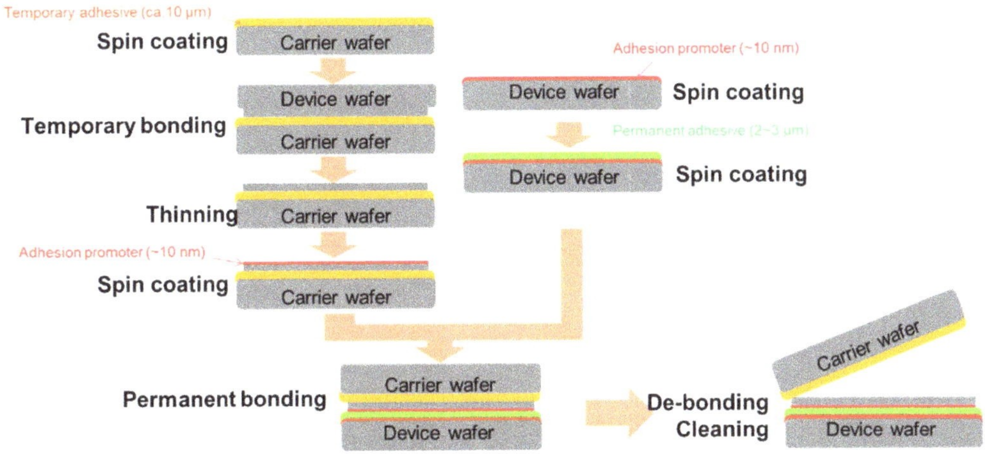

Figure 10. Process flow to evaluate temporary and permanent adhesives for design of experiment (DOE) of wafer stacking.

Permanent adhesives of DPAS100 have high thermal stability and have a maximum operating temperature up to 300 °C suited to the thermal budget in Back-End processes. Because the permanent adhesive needs to be cured within the operating temperature range of the temporary adhesive, its functional chemical groups and curing temperature are optimized. In order to increase the bonding strength between the Si surface and the organic polymer, a different chemical reactivity needs to be created between the two incompatible materials using an adhesion promoter. In this process, a silane coupling agent having an epoxy group was applied as an adhesion promoter. For the permanent adhesive layer, no outgassing by the Gas Chromatography-Mass Spectrometry (GC-MS) spectrum was observed during heating. The weight loss of the permanent adhesive after heating at 300 °C for 30 min was less than 1% by weight. These observations indicated that the adhesive layer had very little residual solvent, unreacted material, and degraded material. In addition, no delamination of the singulated thin wafer and no change in the structure of the permanent adhesive layer was observed after the Temperature Cycle Test (TCT) from −55 °C to 150 °C for 13.5 min hold time of 1000 cycles. The properties and process conditions for both the temporary adhesive and permanent adhesive are shown in Table 1.

Table 1. Properties and process conditions.

	Permanent Adhesive on Adhesion Promoter	Adhesion Promoter	Temporary Adhesive
Thickness	0.5–5 μm	~10 nm	2–20 μm
Softening Temperature	around 50 °C	-	over 100 °C
Solidification	135 ± 5 °C, 30 min and 170–195 °C, 30 min	100–120 °C, 5 min	-
De-bonding	-	-	mechanical peeling at 80 kPa or over 200 °C
Cleaning Solvent	-	-	DTB-Cleaner
Modified Tape Peel Test of Stacked Wafers, After TCT 1000 Cycles	no delamination	-	-

4.3. *Through-Silicon-Via (TSV) Module*

For bumpless TSV interconnects including RDLs, the Damascene method, a mature method based on a Cu/Low-k BEOL process [64] was used to simplify the processes. In the case of TSV processing, dry etching through the dielectrics in BEOL, device layer including shallow trench isolation (STI), Si, and adhesive layer was carried out. Bumpless TSVs with a small aspect ratio, for example <3, have the advantage of shortening the process time for both etching and metal filling compared with conventional deep TSVs. For instance, assuming that the etching rate follows the mass transport limit reaction, the etching times, t and t_1, at different TSV diameters, D and D_1, and depths, d and d_1, followed $t_1/t = (D_1/D)^2 \times (d_1/d)$; that is, $t_1/t = 0.1$ at $D = D_1$, $d = 50$ μm, and $d_1 = 5$ μm, which suggests 1/10th the etching time for the same TSV diameter and 1/10th the depth.

After TSV etching and wet cleaning, a low-temperature Plasma Enhanced Chemical Vapor Deposition (PECVD)-SiN or SiO_2 film was deposited to provide electrical insulation from the Si substrate. The barrier dielectric at the bottom of the TSV was removed by bias sputtering of Ar ions, and Ti/TiN (or Ta/TaN) and Cu were deposited on the barrier metal and the seed layer, respectively, by sputtering. For Cu plug interconnects and RDLs, Electrochemical Deposited Cu (ECD-Cu) was used. ECD-Cu planarization was carried out by chemical mechanical polishing (CMP) to polish-off of the Cu overburden.

Figure 11 shows the leakage current of TSVs varied with annealing temperature as a function of applied voltage, comparing Bosch and direct dry etching methods [65,66]. Since Bosch etching was carried out by repeated alternating isotropic-etching and deposition for sidewall passivation/protection, micro-steps called scalloping were formed in the side walls. The scalloping caused cracks and poor step coverage in the dielectrics and metal layers for thin films deposited by Chemical Vapor Deposition (CVD) and Physical Vapor Deposition (PVD). In contrast, anisotropic dry etching resulted in a smooth surface profile along the side wall and no discontinuous layer was observed. The leakage current in Bosch etching was one order of magnitude higher than that in anisotropic dry etching. The leakage current was caused by Cu diffusion at the side wall of the TSV, which took place at a thinner part of the dielectrics containing cracks. Thus, anisotropic etching is suitable for TSV interconnects and enables the use of low-aspect-ratio vias in the BBCube.

The stress inside the Cu was induced by a mismatch in the CTE between Cu and Si decreases with decreasing aspect ratio of the TSV. Figure 12 shows the results of the Finite Element Method (FEM) analysis of the maximum principal stress for different via heights with 10 μm thick device layers and BEOL interconnects [67]. The stress inside the Cu via with 110-μm thick Si was about twice that of the thinner, with a 30 μm thick via with a constant diameter. The Cu stress at the high-aspect-ratio via exceeded the yield stress (286 MPa) of Cu. The stress distribution showed the following critical points: (a) the BEOL region on the via side, (b) the interior of the via, (c) the bulk region under the via, and (d) the BEOL region under the via. The effect of Si thickness, TSV diameter, adhesive layer thickness and CTE of the adhesive material on the TSV stress was analyzed by sensitivity analysis of the DOE (Design of Experiments) method. The TSV compresses the device surface because the CTE of the adhesive (~50 ppm) was higher than those of Cu (16.6 ppm) and Si (2.6 ppm), and tensile stress was generated due to the strain around the BEOL. Stress at the center of the Cu plug decreased in proportion to the thickness of the Si wafer where the concentrated stresses at thicknesses of 20 μm and 100 μm were 225 MPa and 525 MPa, respectively. Thus, the small aspect ratio provided by an ultrathin wafer had the advantage of reducing stresses generated in the silicon itself, in the bottom and top Cu-TSVs, and in interface regions, even with different CTEs.

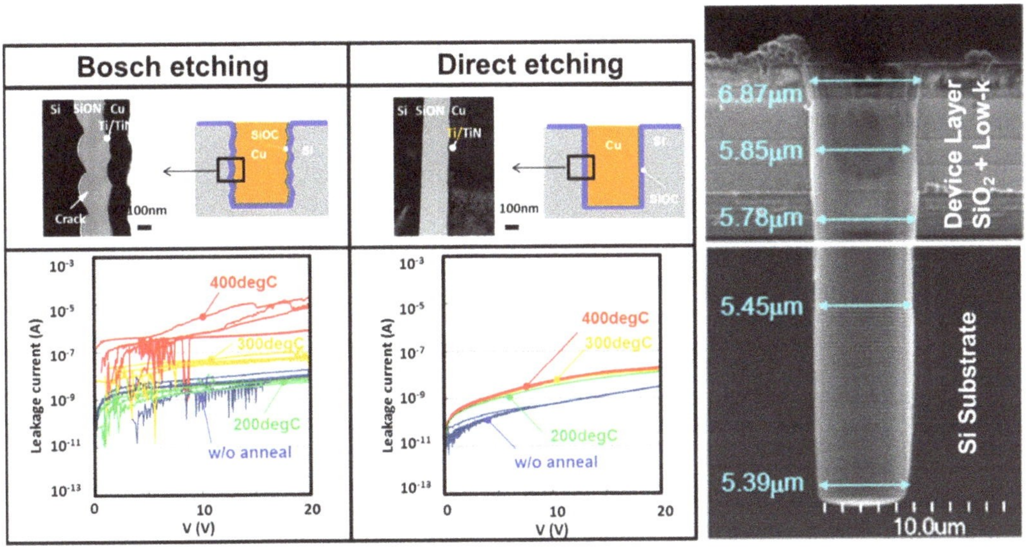

Figure 11. Schematic diagram, cross-sectional TEM images, and leakage current of two types of TSV samples made by Bosch etching and anisotropic dry etching (**left**). Cracks are observed in the Bosch-etched sample, which had a rough interface due to scalloping. The leakage current as a function of applied voltage after annealing at temperatures up to 400 °C was measured. With increasing temperature, the leakage current increased but was two orders of magnitude higher in Bosch etching. SEM images of TSV etched off through Cu/Low-k BEOL layer, device layer, and Si after optimization of scalloping shape (**right**). Fine etching profile through BEOL and Si is achieved.

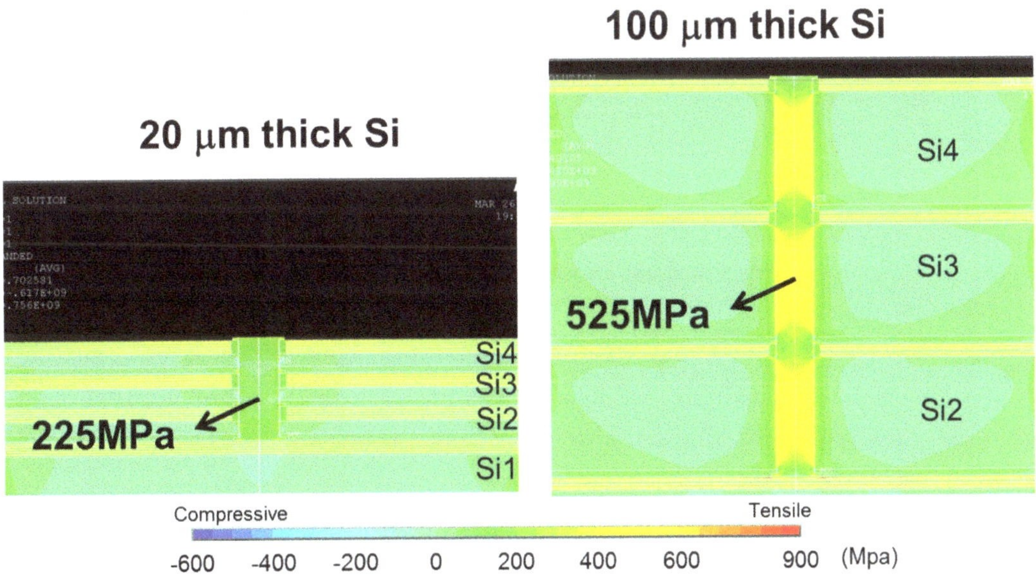

Figure 12. Stress simulation of Cu TSV using the FEM for a Si thicknesses of 100 μm (**right**) and 20 μm (**left**) after three wafers stacking. TSV diameter is 30 μm. A 10 μm Cu/low-k BEOL layer is formed on every wafer surface, and thus the depths of the TSVs are 110 μm and 30 μm, respectively.

4.4. Singulation/Packaging Module and Reliability

After multi-level wafer stacking and TSV interconnects, when the stacked die was applied to the interposer, the same procedure as in conventional packaging (micro-bumps, singulation by dicing, die attach) was followed. After dicing the seven-level wafer stack, the adhesive layer and silicon chips were found to be free of defects or delamination. After the stacked chips were packaged with epoxy resin, they were subjected to heat stress testing at temperatures of −65 °C to 150 °C. Scanning acoustic tomography (SAT) is adopted for internal observation, and after up to 100 repeated heat stress tests, no delamination and voids were found at the interfaces between the molding compound and chips, nor at the chip stack interfaces [68].

A temperature cycling test (TCT) following JESD22A-104 was performed to examine whether the bumpless structure is able to withstand the mechanical stresses caused by extreme temperature variation. The daisy chain (total number of vias n = 216) resistance of the structure with via bottom cleaning showed negligible change after the TCTs, indicating that the structure can tolerate extreme temperature changes despite the presence of a polymer with a high CTE within the structure, as shown in Figure 13a [69]. In general, a high moisture content may decrease the glass transition temperature of the polymer and damage the structure. A high accelerated stress test (HAST) following JESD22A-118 was performed to further investigate the reliability of the structure under specified temperature and moisture conditions. In Figure 13b, the daisy chain (n = 216) resistance of the structure with via bottom cleaning only slightly increased after the HAST, showing that the bumpless structure can successfully protect the polymer from moisture. According to the TCT and HAST tests, the designed structure has good fabrication integrity and is highly reliable.

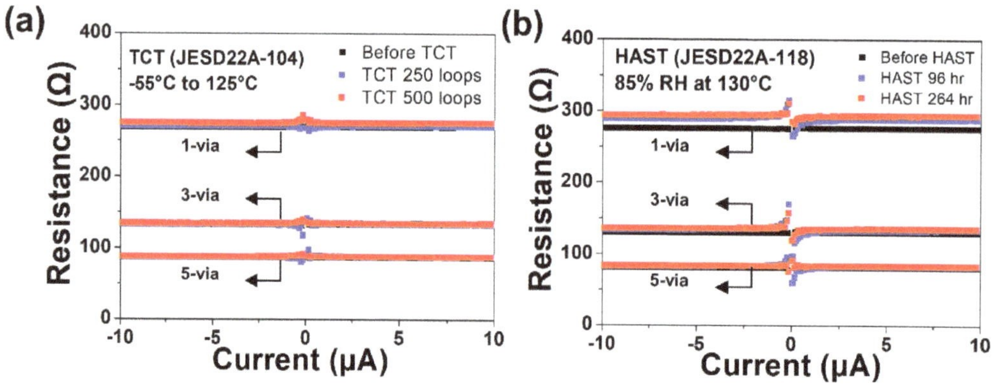

Figure 13. Rsistance change of daisy chain (n = 216) measurement before and after (**a**) thermal cycling test (TCT) and (**b**) highly accelerated stress test (HAST), respectively.

Electromigration (EM) test and SEM analysis was performed to determine the failure site, as shown in Figure 14. The cross-sectional images of bumpless TSVs before and after current stressing indicate that the failure site is located at the bottom RDL close to the corner with the TSV, which is consistent with the simulation results. The failure at this location mainly results from the small thickness of the RDL. Thus, the mean time to failure (MTF) of the bumpless structure can be increased by increasing the RDL thickness [70]. A comparison of the electromigration characteristics between the bumpless TSV structure and the conventional TSV structure with microbumps is shown in the table in Figure 14, indicating that the bumpless TSV technology has better mechanical properties and the ability to withstand longer current stressing time [71].

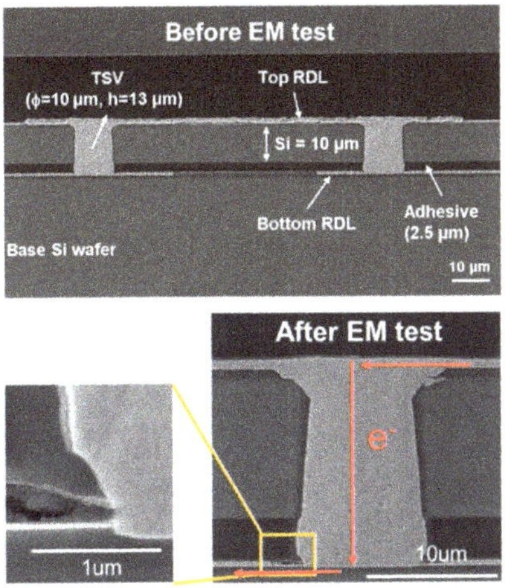

Ref.	Akamatsu et al. ECTC 2016	This work
Structure	Conventional TSV structure with microbump	Bumpless TSV structure
Diameter	TSV (10 μm) IMC joint (30 μm)	TSV (10 μm)
Temperature	100 °C	200 °C
Current density	3.8×10^5 A/cm^2	7×10^5 A/cm^2
Criteria	10%	10%
MTF	4 hr	137 hr

Figure 14. Cross-sectional SEM images of bumpless TSV interconnects before and after electromigration test (EM) (**left**). A comparison of electromigration characteristics between bumpless TSV structure and the conventional TSV with microbump (**right**) [71].

5. BBCube COW Processes

5.1. Heterogenous 3D Integration Process

For heterogeneous integration using chiplet logic devices, memory, and passive devices such capacitors, BBCube COW was developed [72]. Figure 15 shows the bumpless COW process flow used for a 3D functional interposer. First, a permanent adhesive material of Bis-benzo-cyclobutene (DOW, CYCLOTENETM 3022-46) was coated on a 300 mm Si wafer (base wafer) to a thickness of 5 μm. The Si base wafer was patterned to make fiducial marks on the surface to determine the Si capacitor die placement positions. A dummy Si for reducing the volume ratio of dies and mold material was bonded to the Si base wafer. Then. a Si capacitor die was placed on the adhesive from the front side with a surface mounter tool. The Si capacitor was a commercial product (Murata Manufacturing Co., Ltd., Kyoto City, Japan, EMSC series), with dimensions L = 3.07 mm, W = 2.07 mm, T = 100 μm, a capacitance of 1 μF and an equivalent series resistance (ESR) of 100 mΩ. Si capacitors based on deep-trench metal-oxide-semiconductor (MOS) capacitor technology combined with a unique mosaic design and distributed trench capacitors were developed, as shown in Figure 16. As an example, a 100 nF equivalent series inductance (ESL) Si capacitor was made of 200 elementary cells of 470 pF distributed over the chip and combined in parallel with a 10 pF metal insulator metal (MIM) capacitor to lower the impedance at higher frequencies. A Si capacitor is one candidate for overcoming the scaling issue faced by capacitor components.

The permanent adhesive was cured to bond the Si capacitor and dummy Si after die attachment. Epoxy resin with silica-based filler was molded on the die-attached side of the base wafer with a compression molding method. Then, epoxy resin was thinned to a thickness of several tens of micrometers above the Si capacitor. To reduce the wafer warpage in the COW process, a 300 mm Si wafer (carrier wafer) was bonded on the side of the thinned resin mold. The base wafer was thinned down from a thickness of 775 μm to 20 μm with grinding and polishing. The TSV and re-distribution line (RDL) were formed by the Damascene method to make interconnects between the Si capacitor and the RDL.

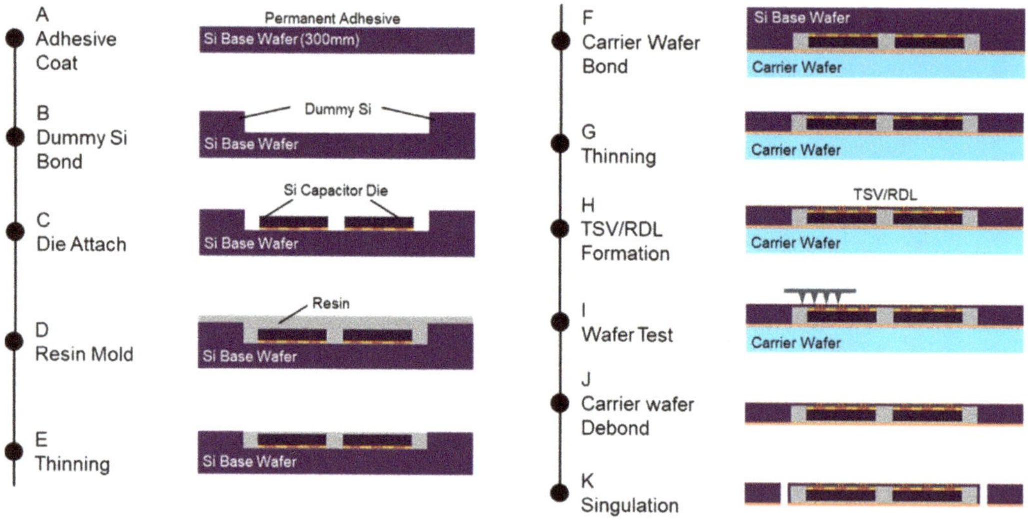

Figure 15. Bumpless COW process flow. Face-down die attachment and Front-side bumpless TSV interconnects are carried out.

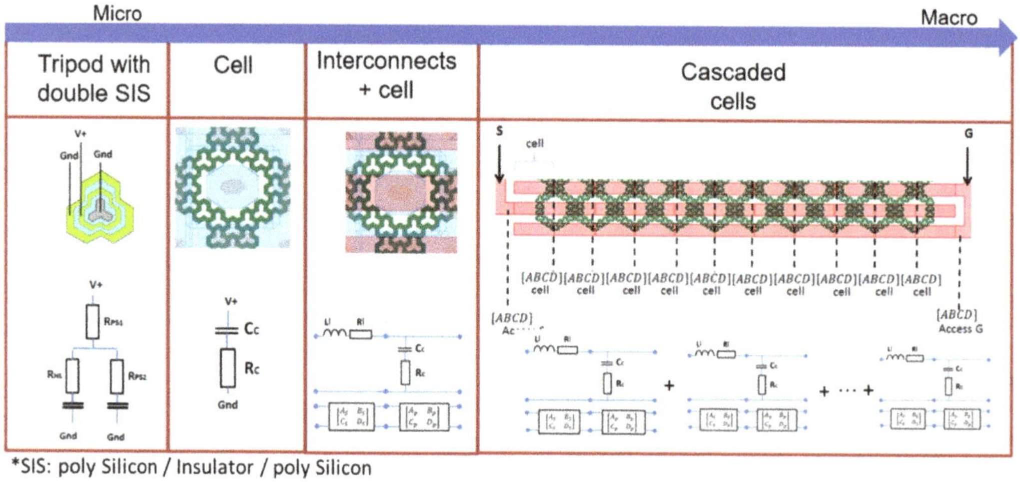

Figure 16. Design of Si capacitor based on deep-trench metal-oxide-semiconductor (MOS) capacitor technology combined with a unique mosaic design and distributed trench capacitors. Die size is L = 3.07 mm, W = 2.07 mm, T = 100 µm. A capacitance and ESR are 1 µF and 100 mΩ, respectively.

5.2. Wafer Warpage Control

Wafer warpage, due to a mismatch in the CTE between organic materials and Si, is a major problem in wafer-level-packaging (WLP) integration. Huge wafer warpage causes wafer cracking and even wafer breakage in the worst case. Even small wafer warpage in the millimeter range causes wafer chucking problems in tools such as the grinder/polisher and wafer bonder, and all vacuum process tools for the TSV/RDL processes. To satisfy the vacuum process and the TSV/RDL formation step, wafer warpage should be less than 300 µm, even though the wafer has a multi-layer structure that includes Si capacitors and molded resin. In particular, the molded resin has a quite different CTE compared to Si, so how to reduce the volume of molded resin is the most important parameter for controlling wafer warpage in the COW process [73].

Figure 17 shows wafer warpage as a function of the molded resin thickness (a), and the definition of the mold cap and its thickness (b), wafer warpage increases linearly as the mold resin thickness increases. Wafer warpage became more than 1.6 mm at the resin thickness of 300 μm, indicating a smile shape. The mold cap was set to 100 μm for bumpless COW process integration. Figure 18 shows the details of wafer warpage in each step of the COW process. In the resin molding step, wafer warpage shows a maximum value of 800 μm. After mold thinning, wafer warpage decreased significantly from 800 μm to 100 μm due to the reduced volume of the molded resin. At the beginning of the TSV/RDL step, wafer warpage was 100 μm, which is low enough to process the TSV/RDL formation step as a result of an appropriate mold cap.

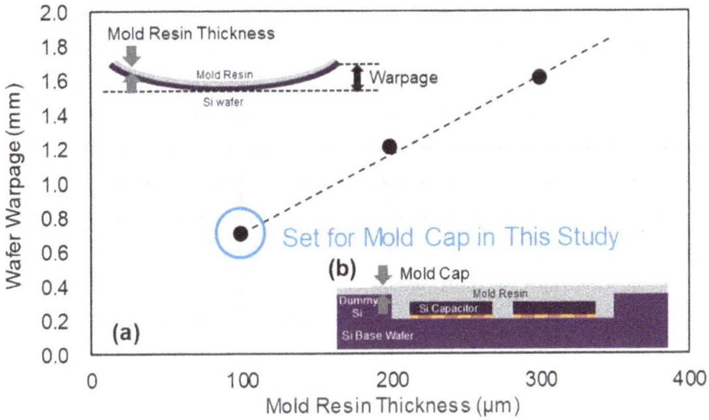

Figure 17. Wafer warpage in resin molding step. (**a**) Wafer warpage as a function of molded resin thickness, and (**b**) cross-sectional image of mold cap in COW process.

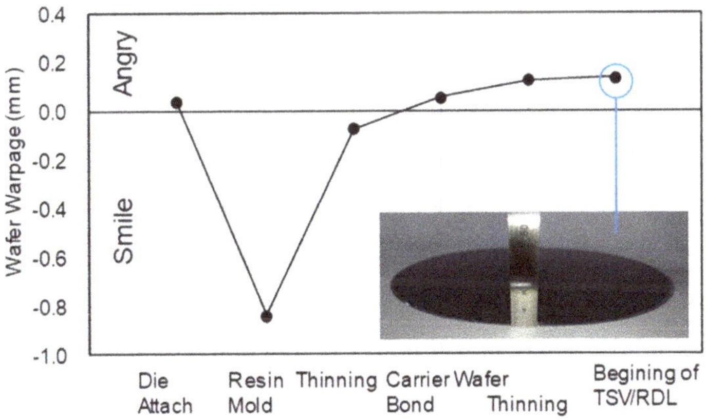

Figure 18. Details of wafer warpage in each step of COW process.

5.3. Accuracy of Die Placement and Reduction of Voids in Adhesive

Die placement is one of the primary technologies in the COW process, whether COW has bumps or not. This is because the die placement accuracy strongly affects the reliability of connectivity and the allocation of high-dense TSVs [74]. In addition, to increase the placement accuracy, there is a trade-off in the throughput. For the bumpless COW with adhesive, misalignment of the Si capacitor in the die placement tool directly causes

TSV interconnect failure. Even if the die placement tool has highly accurate placement positioning of the Si capacitor, die shifting may occur in the adhesive curing step because the adhesive sometimes contains voids that move to the outer region of die. In addition, when the large voids are located at the TSV region, discontinuous barrier layer formation of CVD SiO_2 and PVD metals occurs, which causes void formation in the ECD-Cu metallization and leakage current between TSVs.

To overcome voids and die shifting, a cyclical pressure profile from low-pressure to atmospheric pressure of a pressure oven was used for the adhesive curing step. Figure 19 shows the observation of voids in adhesive and the die placement accuracy before and after the adhesive curing. Figure 20 shows a cross-sectional SEM image of the Si capacitor die after the die attaching step and the adhesive curing. No voids are observed after the cyclical pressure, and die shifting is maintained within +/−25 μm even after the adhesive curing. As a result of the low number of voids and the small die shifting, the thickness of the adhesive between the Si base wafer and the Si capacitor pad was stabilized at 5.2 μm.

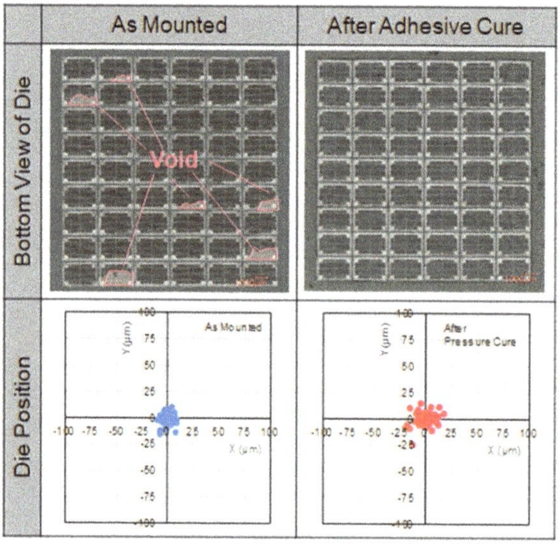

Figure 19. Observation of void in adhesive between the base wafer and Si capacitor (pictures in **top-left** and **right**); and die placement accuracy before and after the adhesive curing step (**bottom-left** and **right**).

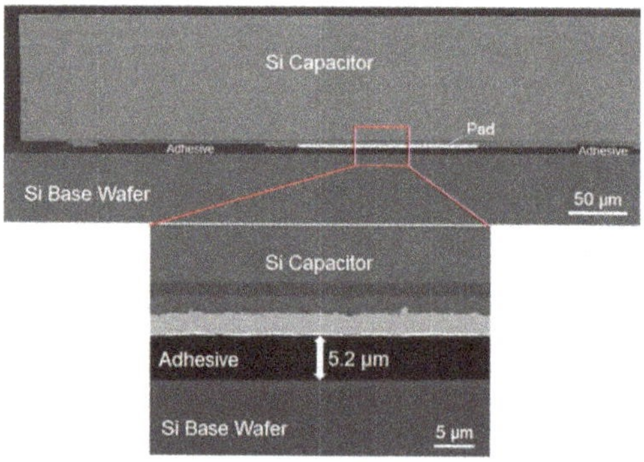

Figure 20. Cross-sectional SEM image of Si capacitor die after die attaching step. No voids at the pad interface are observed.

5.4. Process Sequence of TSV/RDL

The process sequence of TSV/RDL formation is shown in Figure 21. TSV/RDL formation is based on Cu Dual Damascene interconnects. After Si capacitor placement by the COW process as described, the Si base wafer was thinned down to 20 μm with a grinding and polishing tool (DISCO, DGP8761). Wafer thinning is usually carried out in three steps: (a) coarse grinding for a high removal rate of Si, (b) fine grinding for reducing back-side damage, and (c) stress relief such a CMP for removal of damage. According to measurements of the thickness uniformity of the thinned Si, a very low total thickness variation (TTV) of less than 2.0 μm, <10% un-uniformity at 20 μm of remained Si was achieved within the 300 mm wafer, which was low enough for the subsequent steps. After thinning of the Si base wafer, a dielectric layer such a SiO_2 was deposited using a low-temperature plasma-enhanced CVD. Silicon dioxide was patterned for the RDL using the lithography and an etching process. TSV formation is carried out with lithography and etching of the Si and adhesive until the pad of Si capacitor. To protect the Si sidewall of the TSV from Cu, a SiO_2 liner was deposited conformally and then etched to remove the bottom SiO_2. Finally, Cu metallization and planarization were performed. These steps were characterized as a *TSV-last* process from the front side. Via bottom cleaning, such as with O_2 plasma, wet cleaning, and Ar sputtering, removed contaminants, including residues and by-products from the etching process.

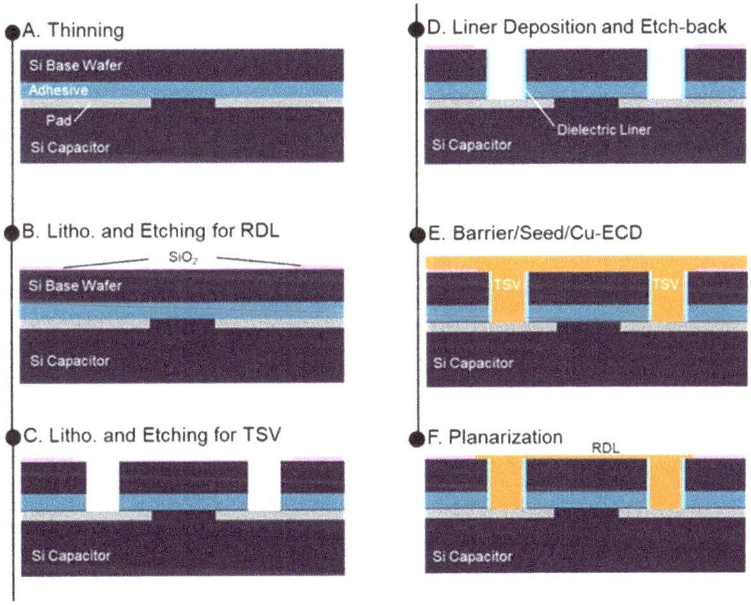

Figure 21. Process sequence of TSV/RDL formation after chip placement and wafer bonding.

The length of the TSV interconnect was 25 μm, which is equal to the total thickness of the thinned Si and the adhesive layer, and the diameter was 10 μm, as shown in Figure 22. A fine plug profile without voids was observed in a cross-sectional SEM image of the TSV/RDL. There was no significant oxide residue in the interface between the TSV and Si capacitor pad. The TSV resistance measured with the Kelvin method was 10 mΩ and had excellent uniformity, which indicated low local warpage of the Si capacitor and low curing shrinkage of the adhesive embedded in the interposer. If the interposer had huge warpage and shrinkage, the resistance of the TSV would have a large deviation and open failure due to unstable contact resistance.

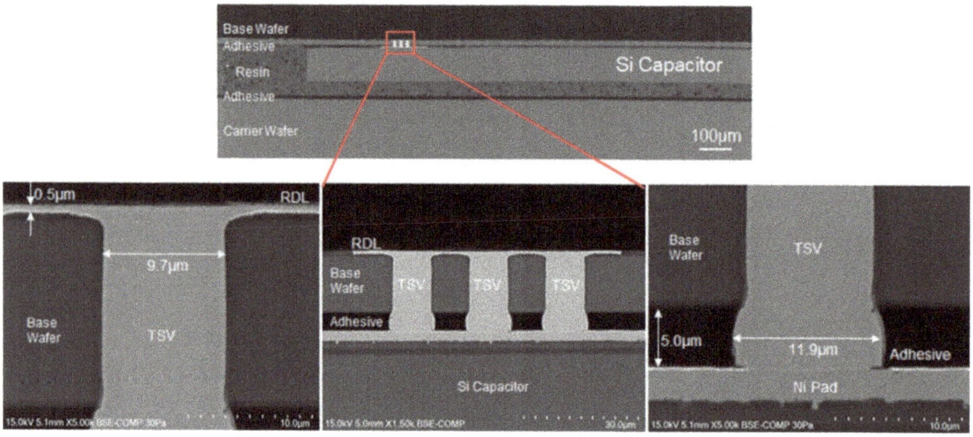

Figure 22. Cross-sectional SEM images of TSV/RDL. Fine plug profile without voids and no significant oxide residue in the interface between the TSV and Si capacitor pad are observed.

5.5. Electrical Characteristics of Embedded-Si Capacitor in COW

The electrical characteristics for high frequency of the Si capacitor embedded in the functional interposer were evaluated by measuring the S-parameter with the shunt through method from 10 kHz to 8.5 GHz. For the measurement, an Al contact pad is formed on the RDL. A vector network analyzer (VNA) was used for the measurement. The impedance, Z, was calculated with Equation (1), shown below. Then, the capacitance, C, and ESR were calculated with Equation (2). Here, Z_0 is a reference impedance of 50 Ω, S_{12} is the measured S-parameter, and ω is angular frequency.

$$Z = \frac{Z_0}{2} \times \frac{S_{12}}{1 - S_{12}} \tag{1}$$

$$Z = ESR + \frac{1}{j\omega C} \tag{2}$$

The measured and calculated RF characteristics are shown in Figure 23, where (a) is the impedance, (b) is the capacitance and (c) is the ESR. The measured capacitance is 0.95 μF at 100 kHz, and ESR is 43.5 mΩ at 100 MHz. These results for the embedded Si capacitor are same as the values measured on the wafer of Si capacitor before the COW process, which means that the bumpless COW process did not have any electrical loss.

5.6. Benefits and Performance of Bumpless Functional Interposer

From the point of view of electrical performance using bumpless COW integration, the principal advantages of a 3D functional interposer are its significantly low energy consumption and low parasitic capacitance, which were expected due to the shortest interconnect length between MPUs and the Si capacitor. However, the TSV formed a metal-oxide-semiconductor (MOS) capacitor at the sidewall of TVS interconnects as well as a Si capacitor, so the parasitic capacitance of the TSV should be considered. Parasitic capacitance of the TSV consists of accumulation capacitance and depletion capacitance along the interconnect line. In the case where the maximum accumulation capacitance is considered, the accumulation capacitance was the same as the liner oxide capacitance, C_{ox}, as expressed by previous studies [75]:

$$C_{ox} = \frac{2\pi\varepsilon_0\varepsilon_{ox} l_{TSV}}{\ln\left(\frac{r_{TSV}+t_{ox}}{r_{TSV}}\right)} \tag{3}$$

where ε_0 is the permittivity of vacuum (8.854×10^{-12} F/m), ε_{ox} is the relative permittivity of the liner oxide (in this study, the liner oxide is SiO_2, hence ε_{ox} is 3.8), l_{TSV} is the length of the TSV, r_{TSV} is the radius of the TSV Cu, and t_{ox} is the thickness of the liner oxide.

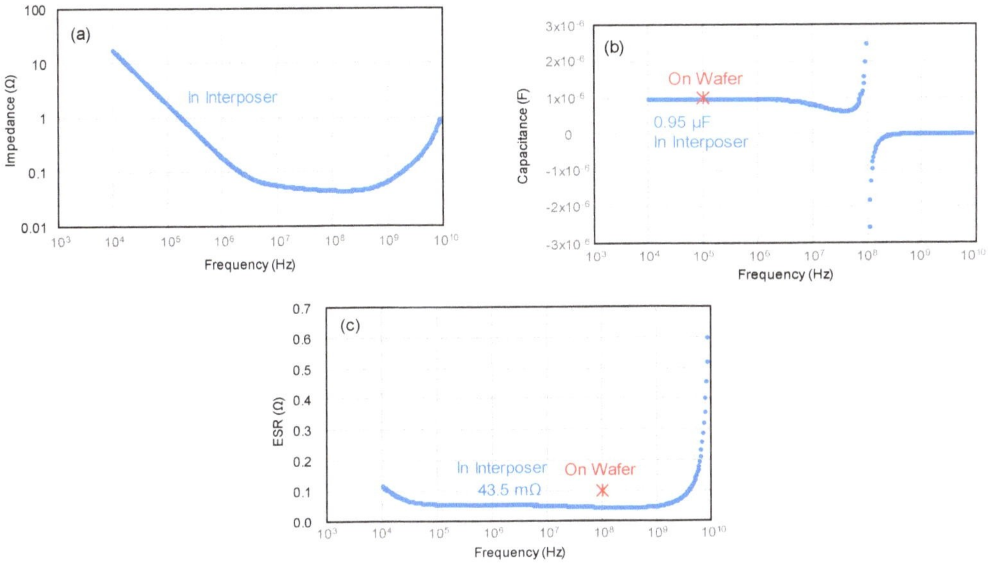

Figure 23. RF characteristics of Si capacitor embedded in functional interposer from 10 kHz to 8.5 GHz: (**a**) impedance, (**b**) capacitance, (**c**) ESR, compared to the Si capacitor before COW process indicated "On Wafter" red symbol.

Figure 24 shows the schematic diagram of TSV structure used to estimate parasitic capacitance, and the calculated parasitic capacitance as a function of the TSV or line length. Parasitic capacitance of the TSV used in our 3D functional interposer is calculated by Equation (3). The parasitic capacitance was significantly reduced to 1/150th compared to that of 2.5D, side-by-side capacitor layout. This extreme reduction in the parasitic capacitance was able to realise lower noise of the power supply for MPUs, and thus a lower applied voltage V_{dd} such a <0.7 V could be used, taking the power distribution network (PDN) into account. As a result of the lower V_{dd}, the power consumption was decreased as the power consumption is proportional to $V_{dd}{}^2$.

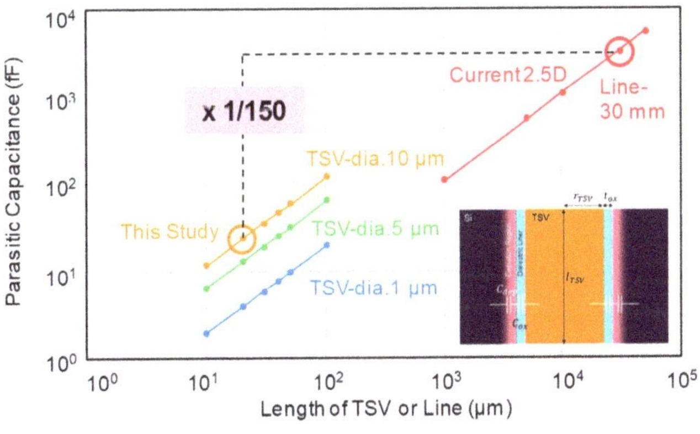

Figure 24. Parasitic capacitance as a function of the length of TSV or wiring line.

6. BBCube Technology Roadmap

Since the bonding process after thinning with a support wafer allowed thinning of silicon wafers down to 4 μm without any degradation of the device characteristics, the total wafer thickness, including the device layer and the adhesive layer was only 10 to 20 μm. This is 1/3rd to 1/5th the thickness of conventional bump interconnects using TSVs. Therefore, even if the number of stacked wafers is 100, assuming that the wafer thickness is 10 μm, the total thickness after stacking is 1 mm. This total height satisfies current packaging standards. Following these multilevel stacking processes, when four, eight, sixteen, etc. of these devices are stacked with a conventional memory device fabricated by a memory density of 30 Gb/cm^2, e.g., 22 nm technology, the total capacity of the 3D memory device can be linearly increased to 120 Gb, 240 Gb, 480 Gb, etc., respectively, as shown in Figure 25.

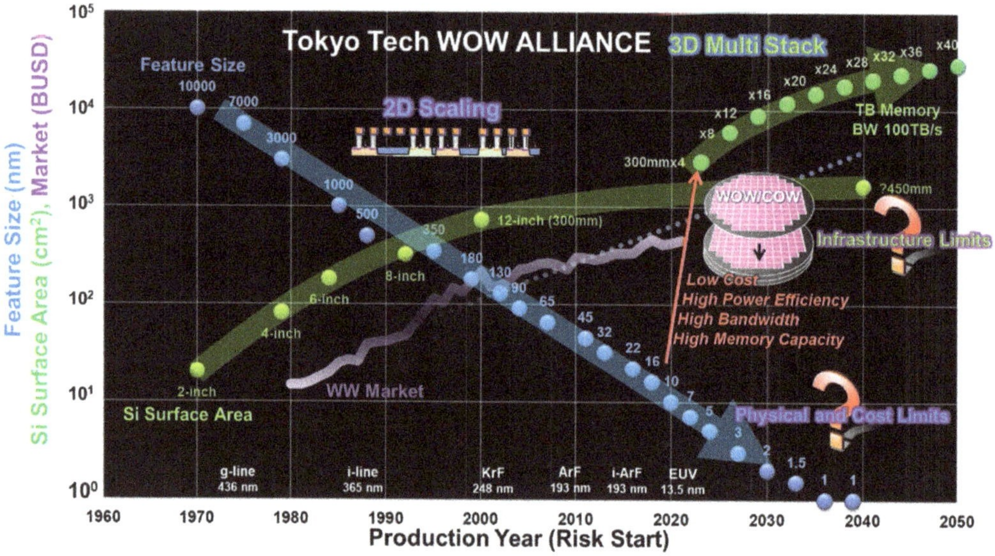

Figure 25. Historical trends and prospect of feature size of transistors, Si wafer area, and semiconductor market. See for example; Available online: http://www.wow.pi.titech.ac.jp/index.html (accessed on 25 December 2021).

Terabit-capacity 3D memory can be realized by stacking 40 layers. In contrast, to achieve equivalent capacity with a single wafer using extreme scaling would require 1 nm node technology, e.g., equivalent dimension about four times of the Si–Si bond length $d_{Si\text{–}Si}$ of 0.23 nm. Consequently, innovative technology not only for 3D transistors but also for 3D chip stacks is needed, as described in Section 2. Considering the technology roadmap, the issues of scaling technology and technology for fabricating 3D structures are often discussed separately. It has been considered that the packaging may take charge of 3D structure. However, these two technologies are not always mutually exclusive. Scaling would be relieved of the stringent requirements by using 3D high-density integration technology combined with mass-production technology. In other words, a sufficiently long learning period would be ensured, and further cost reductions could be expected by concentrating on the control of variations among generations and shortening the process.

Figure 26 shows a schematic diagram of the chip-level configuration for die-to-die connections. The configuration is an evolution from side-by-side to chip-stack in order to reduce signal latency, IR drop, and footprint on the package board. The BBCube is one candidate that satisfies those requirements. The bumpless connections and ultra-thinning enable the shortest wiring and high density TSVs, as well as improved misalignment in the wafer stacking. High-density TSVs are useful because the parallel communication provides high bandwidth. According to the above capabilities, the BBCube architecture provides a solution to the long-standing discussion regarding signal propagation, power distribution, and heat dissipation in the high dense LSIs [76–78], described in the following sections.

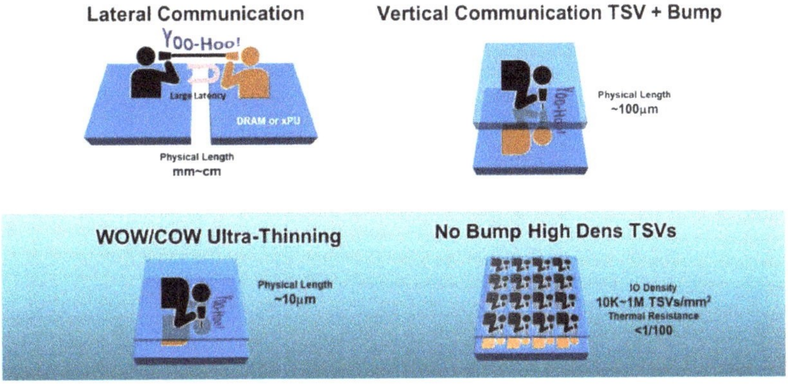

Figure 26. Schematic diagram of chip-level configuration. A physical length of side-by-side (lateral communication) and vertical stack is millimeters–centimeters and about 100 μm, respectively. With no bumps and ultra-thinned wafer, the physical length becomes approximately 10 μm and a high TSV density of more than $10^4/\text{mm}^2$ can be designed.

In fact, the bandwidth of recent high-bandwidth memory (HBM) tended to saturate due to bump pitch constraints, as shown in Figure 27. In the case of BBCube, an order of magnitude higher bandwidth can be realized, since BBCube uses high density TSVs and a novel memory architecture. According to the WOW Alliance, TSV pitch will be narrowed every three years, taking the maturity of bonding alignment into account.

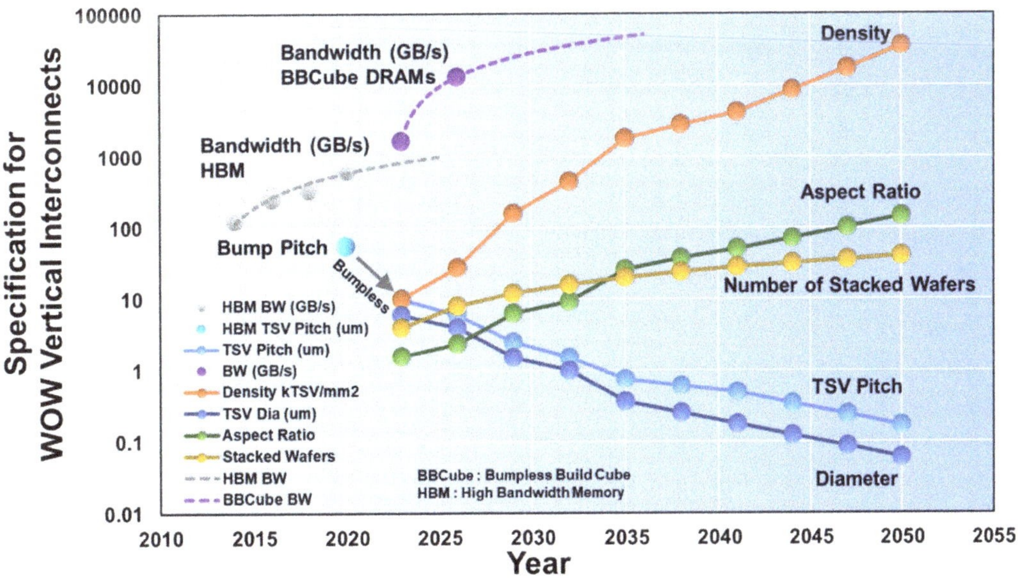

Figure 27. Trends of vertical interconnects of TSVs according to WOW Alliance.

7. BBCube Memory

There are three key challenges in the history of computing systems: (1) size reduction, (2) reduced power, and (3) higher speed. Among these key elements, size reduction is the most imperative challenge, because both low power and high speed can be achieved by size reduction itself. Figure 28 shows the computing system roadmap. In 2035, the target device volume will be 50 mm^3 with a power consumption of 0.5 mW, according to the extrapolating trend. Such a device might be something similar to an AI robotic bee, with CPU/GPU, DRAM [79], NAND flash memory [80,81], and sensors. It would serve human users, so that the AI robotic bee could observe the users' surroundings, protect them, and serve as an administrative secretary.

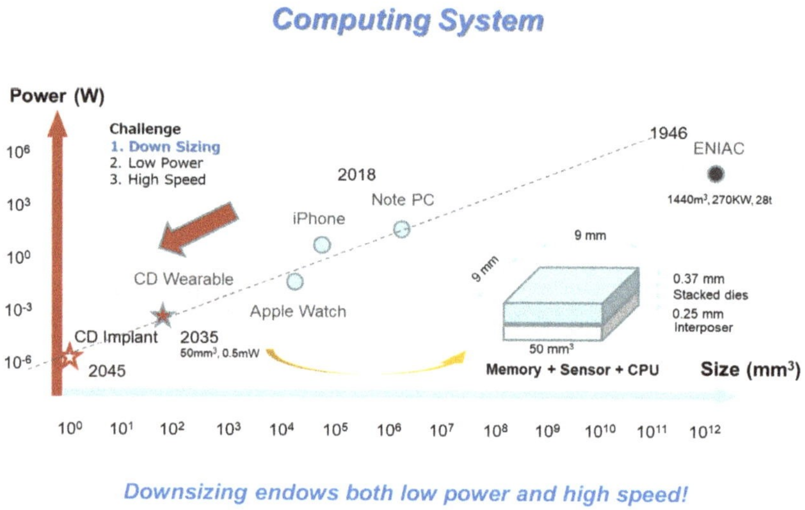

Figure 28. Computing system roadmap for the power consumtion as a function of system volume.

8. BBCube DRAM

TSVs with micro-bumps are conventionally used for the high-bandwidth memory (HBM) [82–87], as shown in Figure 29. However, there are several issues when using micro-bumps. One major problem is that it will be difficult for even HBM to catch up with the increasing speed of GPUs or CPUs. For example, Pascal manufactured by NVIDIA has a processing speed of 1 TB/s, so that four sets of HBMs with 256 GB/s would have to be used. GPU/CPU vendors are constantly striving to increase the speed of their products, for example, to 2 TB/s and 4 TB/s, focusing on AI systems. HBM will have to increase the I/O pin speed by 2.5 times, such as the 5.0 Gb/s/pin [87] from the 2.0 Gb/s/pin [84], and therefore, power and heat will also be increased.

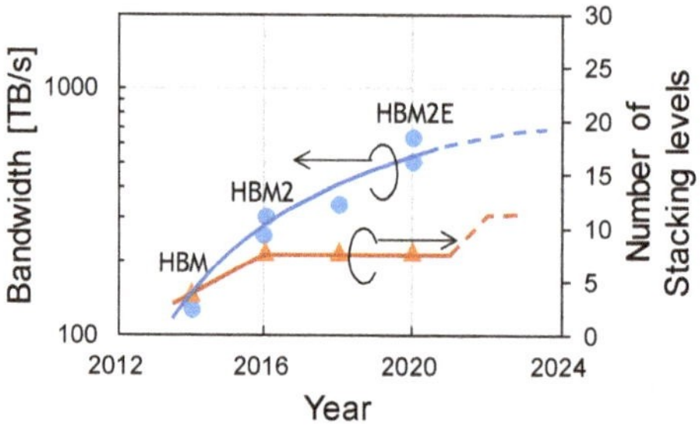

Figure 29. HBM road map with micro-bumps, where HBM [82], HBM2 [83–85], HBM2E [86,87], respectively.

8.1. Electrical Characteristics of BBCube

The electrical characteristics of the BBCube structure were calculated by 3D EM field analysis and compared to conventional 3D integration (3DI) with micro-bumps [34]. The TSV model used for conventional 3DI with micro-bumps, such as HBM, is shown in Figure 30a, and the TSV model used for BBCube is shown in Figure 30b. In conventional 3DI, on top of the TSVs, bumps consisting of copper pillars and solder were formed [88]. We assumed dimensions equivalent to those of HBM. The stacking pitch was 87 μm, and the TSV pitch was 55 μm. In comparison, BBCube's Si was thinned to 4 μm, and the stacking pitch became 10 μm. The TSV pitch was 12 μm. The physical dimensions and material properties are shown Tables 2 and 3 respectively.

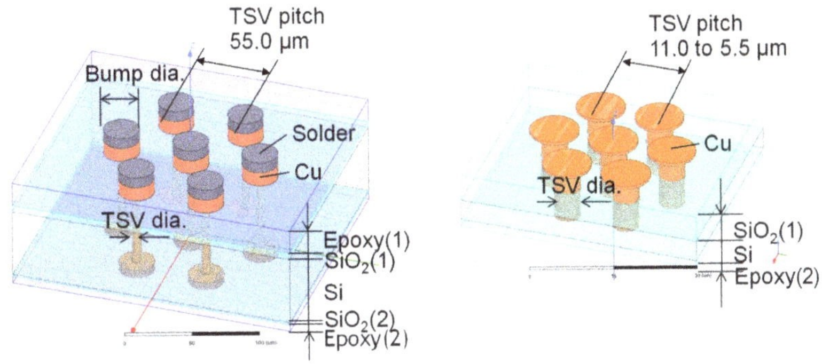

Figure 30. TSV model for 3D EM field analysis; (**a**) 3DI with micro-bumps and (**b**) BBCube.

Table 2. Physical dimension.

		3DI w/Micro-Bumps	BBCube WOW	
			Gen1	Gen2
Thick. (μm)	Epoxy (1)	19.5	-	
	Epoxy (2)	6.0	1.5	
	SiO_2 (1)	5.0	4.5	
	SiO_2 (2)	1.5	-	
	Si	55.0	4.0	
	Total	87.0	10.0	
TSV dia. (μm)		8.0	4.0	2.0
TSV pitch (μm)		55.0	12.0	5.5

Table 3. Material properties.

	Relative Permittivity	Bulk Conductivity (Siemens/m)
Copper	1.0	5.8×10^7
Si	11.9	10
SiO_2	4.0	0
Solder	1.0	7×10^6
Epoxy	3.6	0

Figure 31a shows the frequency characteristic of the TSV capacitance. Due to a slow-wave mode [89], it increased below 3 GHz. As shown in Figure 31b, the liner thickness determines the TSV capacitance below 3 GHz. The TSV diameter and Si thickness also determines the TSV capacitance, which can be reduced by employing BBCube. As shown in Figure 31c, the TSV pitch did not affect the TSV capacitance. Therefore, when the TSV diameter was 5 μm, BBCube was able to shorten the TSV pitch to 11 μm without increasing the capacitance. Moreover, BBCube was able to shorten the TSV pitch to 5.5 μm when the TSV diameter was 2 μm. Compared to the conventional 3DI, the TSV capacitance in the case of BBCube became 1/20th. The frequency dependence of the TSV resistance as shown in Figure 31d increased over 5 GHz due to skin effect, but this was higher than the operating frequency of BBCube, so it did not have any influence. The TSV inductance as shown in Figure 31d was flat to frequency. Due to the shorter TSV, the resistance and inductance in the BBCube case were much smaller than conventional 3DI. In the case of inductance, it reduced 1/10th to 1/15th than that of the conventional one.

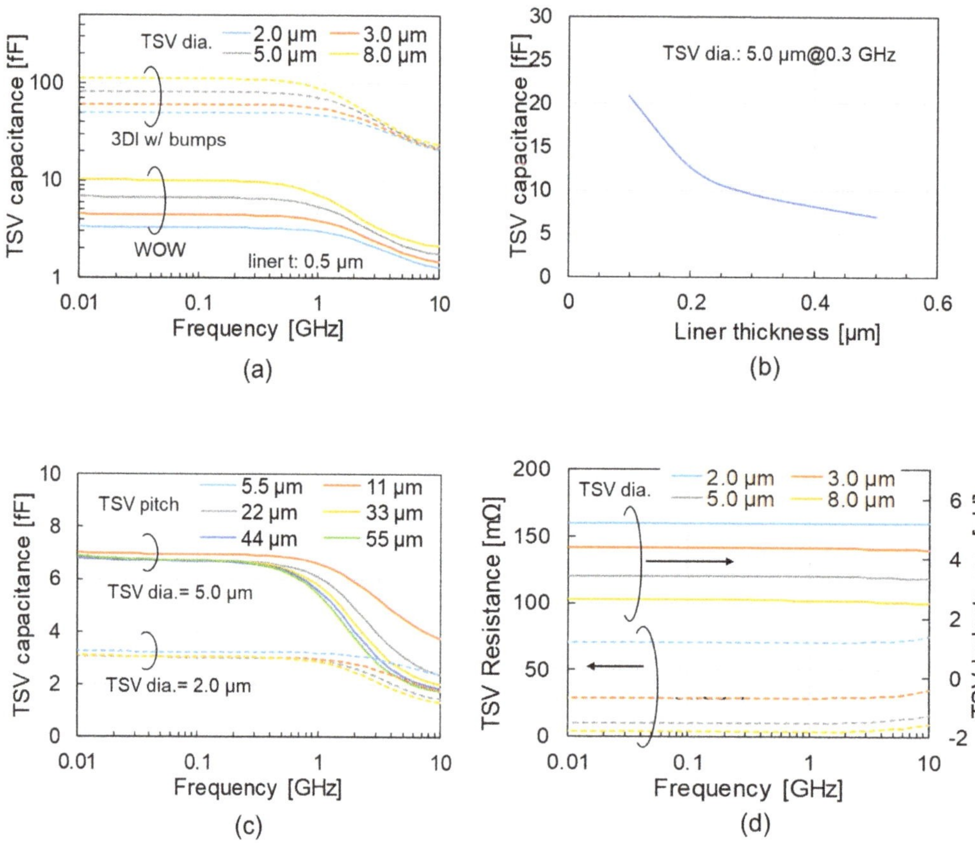

Figure 31. Calculation results of TSV capacitance: (**a**) frequency characteristics of capacitance, (**b**) impact of liner thickness, (**c**) impact of TSV pitch and (**d**) frequency characteristics of resistance and inductance.

Circuit simulation was used to estimate the power consumption of the I/O circuit. An eye diagram was calculated, and the I/O current that satisfies the eye mask was determined [34]. The structure of BBCube is presented in Figure 32 and a block diagram of the simulation is shown in Figure 33a. By utilizing the capabilities of the dense TSVs, the data rate was set at only 800 Mb/s, which is lower than the HBM2E of 3.2 Gb/s. Nine DRAM die was stacked in case of BBCube. An additional die was used to implement novel 3D-based redundancy. The I/O current was assumed to be proportional to the output resistance and capacitance of the I/O buffer circuit. The circuit parasitic capacitances C_o (driver output capacitance), C_d (capacitance at dropping point of TSV) and C_{in} (receiver input capacitance), assumed to be proportional to the driver output resistance R_o. T_1 was an s-parameter model of the TSV, calculated by 3D EM analysis. The HBM and BBCube results are compared in Figure 33b. BBCube achieved 30-times higher I/O power efficiency. In the stacked memory as a whole, BBCube could realize a four-times higher bandwidth with only 13% of the I/O power consumption compered to HBM, as shown in Figure 34. The TSV area in the DRAM die became 64% using 12 µm pitch, and the I/O circuit area becomes about 1/30th. Furthermore, as the process technology matures, the next generation (Gen2) should be able to achieve 32-times higher bandwidth with the same I/O power consumption as HBM.

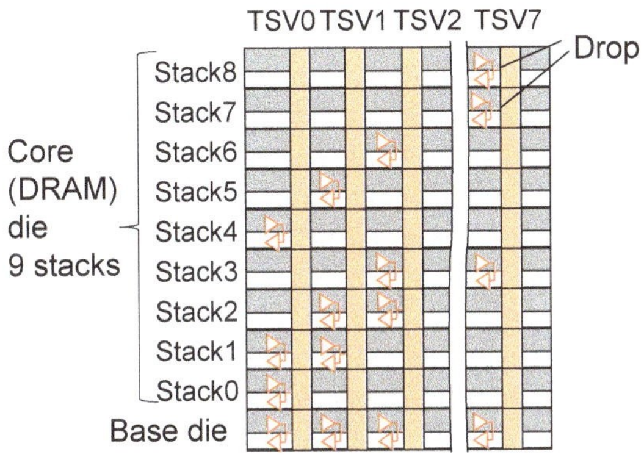

Figure 32. Configuration of BBCube.

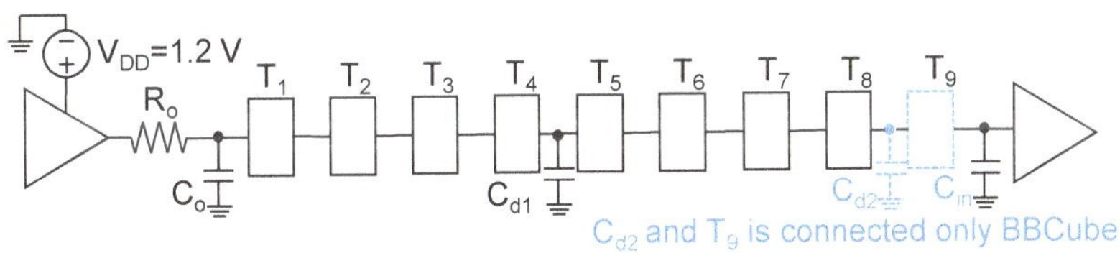

(**a**)

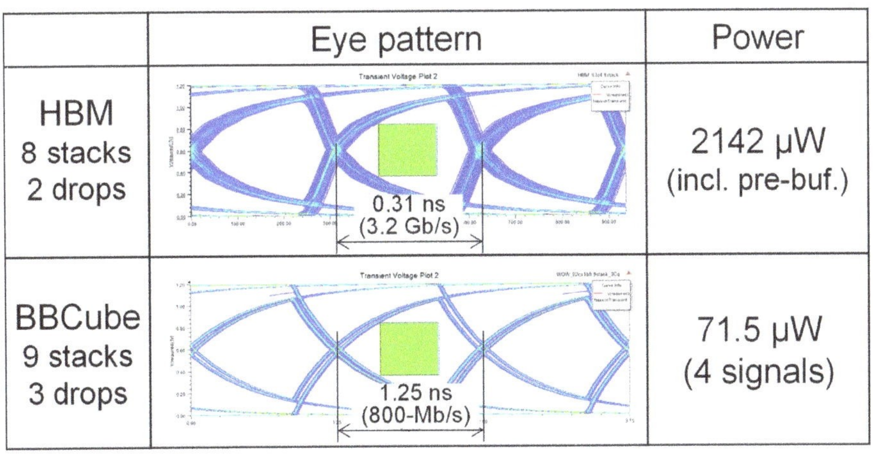

(**b**)

Figure 33. Eye diagram and data transfer power within a TSV: (**a**) block diagram of simulation and (**b**) comparison between HBM and BBCube.

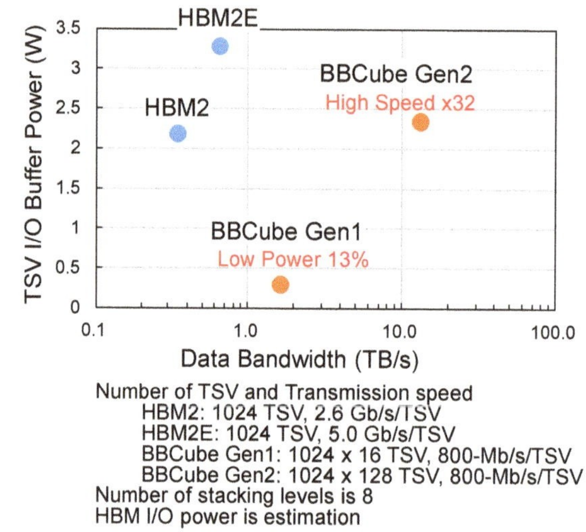

Figure 34. A comparison of the data bandwidth and TSV I/O power consumption between HBM and BBCube, where HBM2 [85], HBM2E [86], repectively.

8.2. Thermal Characteristics of BBCube

The temperature of a DRAM cell influences its retention time and limits the number of stacks [90]. Therefore, a thermal analysis of stacked DRAMs was performed. The TSVs in the BBCube were connected directly to the bottom die, whereas with conventional 3DI, it is necessary to put a solder and a BEOL layer between the TSVs, which increases the thermal resistance. The thermal resistance in the BBCube case was 1/4th that of conventional 3DI [91]. Figure 35 shows the temperature difference between the top of the stacked DRAM, which was at room temperature, and the highest temperature part of the DRAM cell. Under the stacked DRAM, a base die with the same power consumption was placed in both the HBM and BBCube. For the BBCube with 9 stacks, the difference of the DRAM cell temperature was 8.3 °C due to the low thermal resistance. Even when 34 dies were stacked, the difference in temperature in BBCube was 16 °C, which was about two-thirds that of HBM with 8 stacks. BBCube allowed stacking of 4-times more dies than HBM. This allowed the memory capacity to reach 64 GB using 16 Gb DRAM dies.

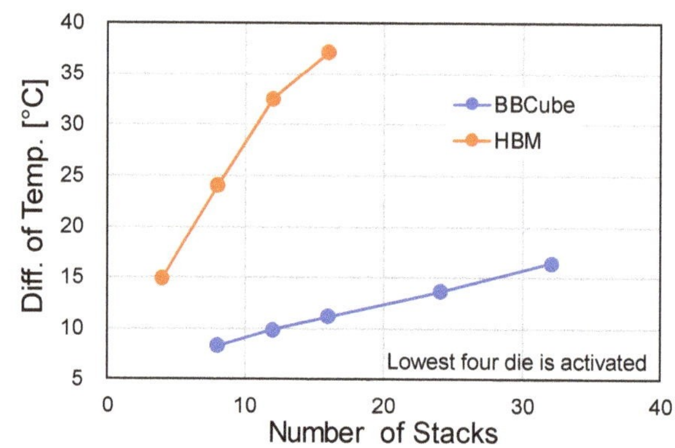

(a)

Figure 35. *Cont.*

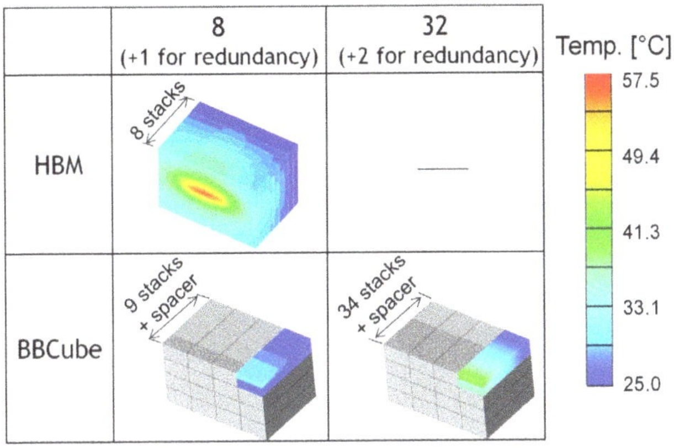

(b)

Figure 35. Thermal simulation results: (**a**) the temperature difference between the top of the stacked dies, which is set to room temperature, and the highest temperature part of the DRAM cell and (**b**) heat distribution.

8.3. Competitive BBCube DRAM Structure

The competitive BBCube DRAM structure is one that enables 8-die stacking with bumpless TSVs. By increasing the number of channels and lowering the TSV impedance, ultra-high bandwidths of 1, 4, and 8 TB/s should be achievable, as illustrated in Figure 36.

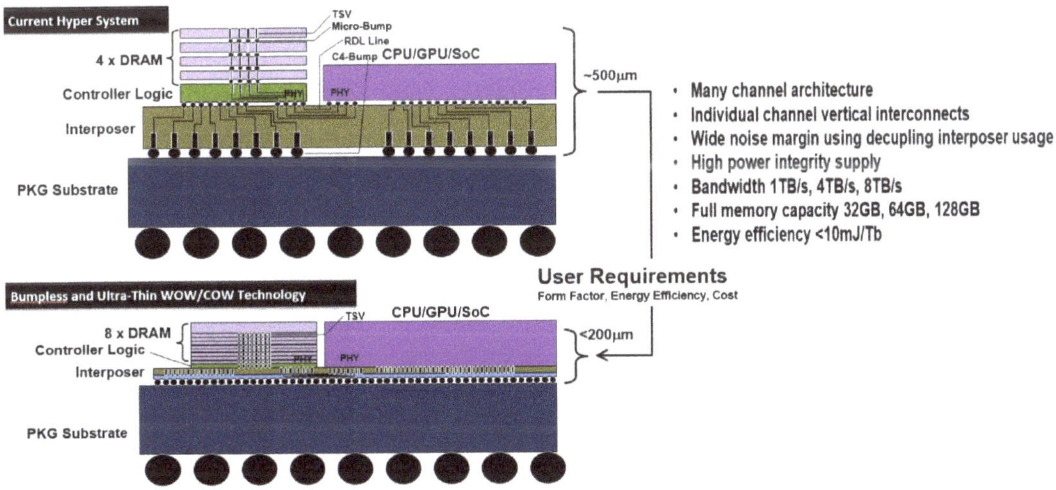

Figure 36. A comparison of conventional DRAM stack such as HBM and BBCube memory for 2.5D system. Competitive 3D structure accompanied with lowering system height and large memory capacity is anticipated.

Figure 37 shows the HBM data bandwidth roadmap. By realizing the parallelism enhancement due to the increase in the number of I/O's, the bandwidth of the HBM, which had no bumps, was expected to be ever-increasing. As for the I/O power consumption, the first target of the bumpless HBM was one-thirtieth that of the current HBM2 [28], as shown in Figure 38.

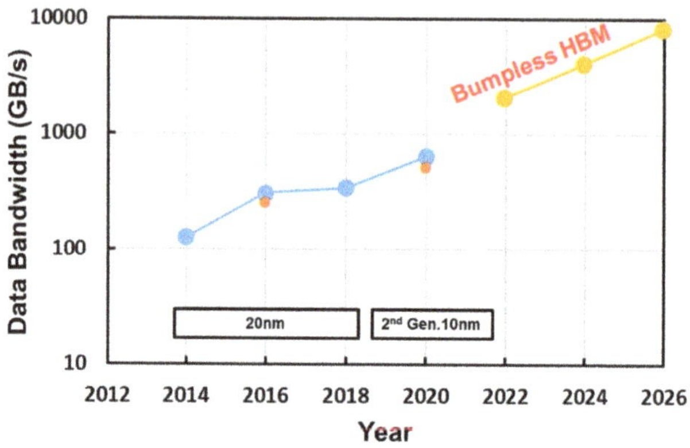

Figure 37. Data bandwidth roadmap [82–87].

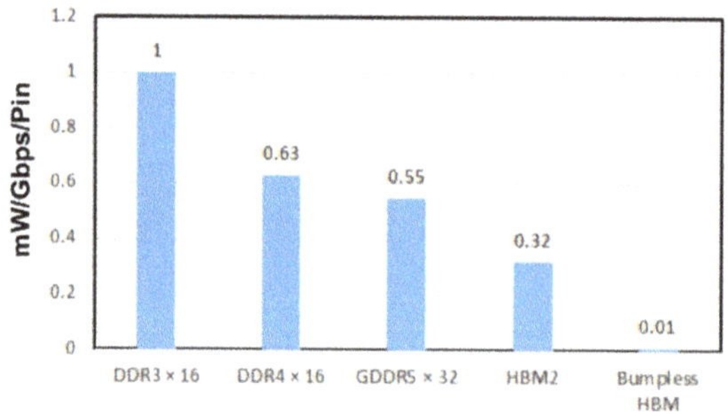

Figure 38. I/O power efficiency.

Figure 39 compares HBM2 with bumps [31–33] and bumpless HBM with respect to their data bandwidth and I/O buffer power, according as the number of I/O's [28,29,34,92,93].

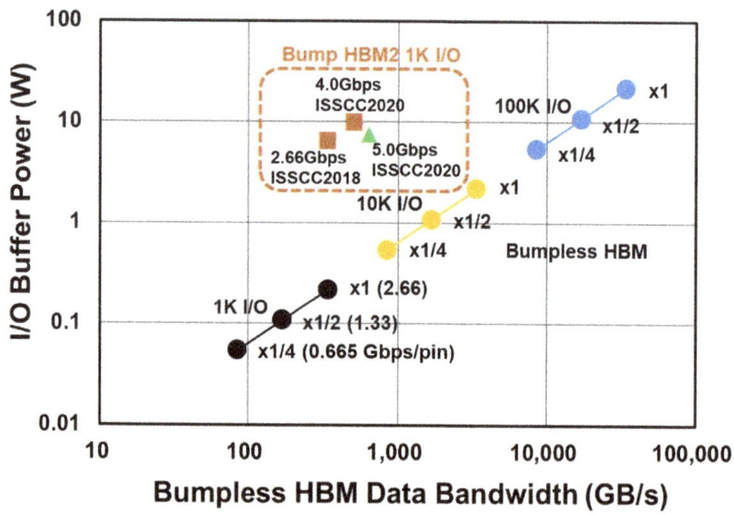

Figure 39. Data bandwidth and I/O buffer power comparison [85–87].

Bumpless HBM can achieve an ultra-high data bandwidth by increasing the I/O number to 1 K, 10 K and 100 K, and can lower the I/O buffer power to 1/2 or 1/4 by reducing the I/O pin frequency with a four-phase shielded I/O scheme, as illustrated in Figure 40. The great advantage of this scheme is validated in Figure 33.

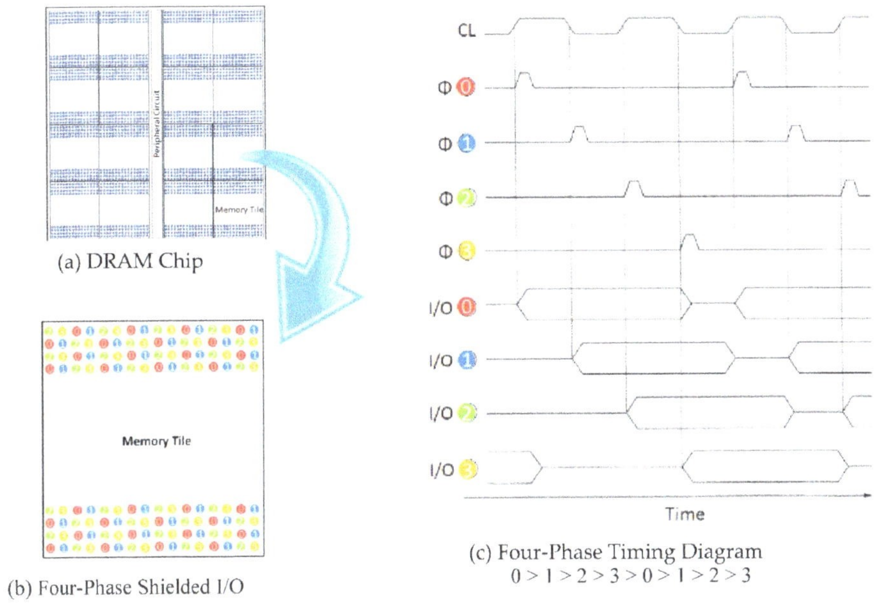

Figure 40. Four-Phase Shielded I/O Scheme.

9. BBCube NAND

9.1. Limitations of Stacked WL Tiers in 3D NAND Chip

i. If 64 tiers by one etching shot is limited by the highest aspect ratio, 512 tiers would require 8 times cell process, such as 64×8. Therefore, there is a large heat budget to enhance the source/drain diffusion of the transistors for both a CMOS Under Array (CUA) and a CMOS Next Array (CNA). As a result, the peripheral transistors would be very large, and their performance would be degraded.
ii. If the number of cells per string should increase to 128, 256, and beyond, the cell current would be very small, so that random page access would become slower.
iii. In the case of (2), both the page count and block size must be large, which would be user unfriendly for reprogramming, such as data copying and moving.
iv. A solution to the issues in (2) and (3) would be a multiple vertical bitline architecture, but this would make routing and wiring difficult within the tight XY bitline pitch.
v. The number of high-voltage transistors for NAND string drivers must also increase according as the number of stacked WL tiers, which would occupy a huge Si area in spite of the CUA structure.

Thereby, the number of the stacked WL tiers would be limited at the 256 tiers, which is produced by 128×2 of the twice cell process.

9.2. Vertical Bitline Architecture

Figure 41 shows the stacked 3D NAND chip in the case of one 3D NAND chip. Figure 42 illustrates four stacked 3D NAND chips which are connected with a vertical bitline (BL), as well as bumpless TSVs [92,93].

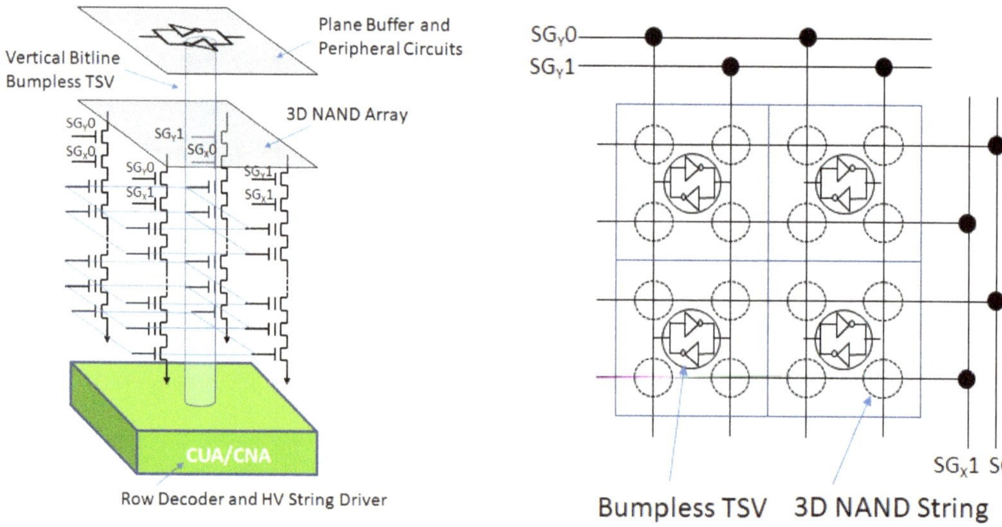

(a) (b)

Figure 41. Vertical bitline architecture. (**a**) bird's eye view; (**b**) top view.

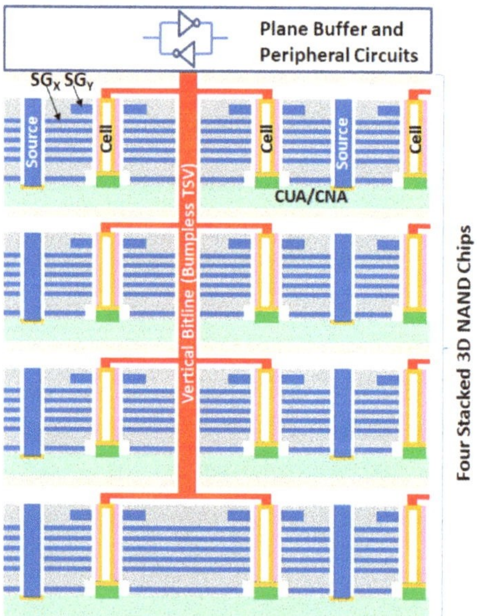

Figure 42. Four 3D NAND chips connected through Vertical BL.

9.3. Word Plate Access NAND

Toshiba proposed an original 3D NAND device for the first time [94]. Multiple CGs and Lower SGs were merged in each plate to reduce the number of HV-driver transistors, as well as to tighten the pillar pitch. In this study, the polysilicon word plate was replaced by a damascene tungsten metal gate, but the basic 3D NAND structure of the merged multiple CGs and Lower SG was the same as the original one. Figure 43 presents a future 3D stacked memory design [95]. By increasing the number of TSVs, a part of a peripheral circuit of the first memory chip can be located in the second memory chip. A stacked DRAM combined with a NAND flash memory is illustrated below as an example.

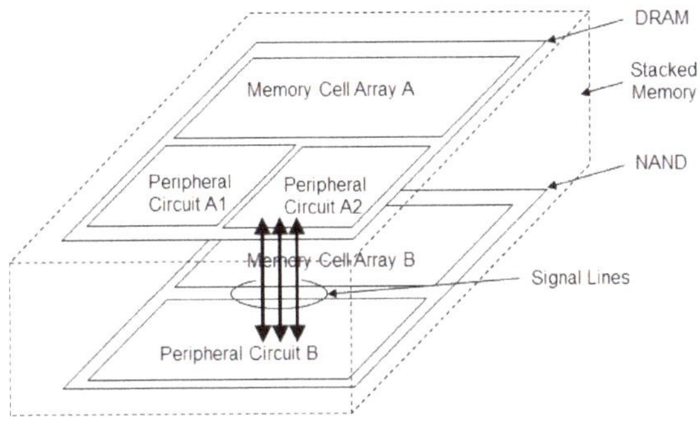

Figure 43. Future 3D stacked memory system [95].

Figure 44 shows a proposed BBCube NAND with multiple BL layers. Thanks to the original architecture of 3D NAND, a word line is expanded to a word plate, so that the 3D NAND can be read and programmed *by plane* instead of *by line*.

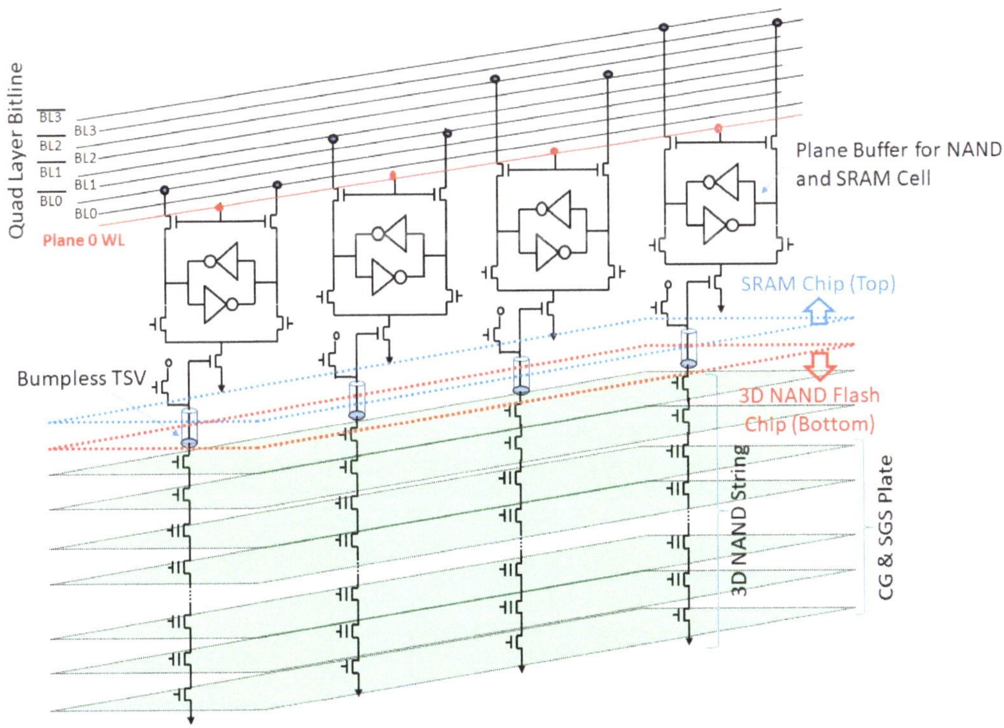

Figure 44. Circuit diagram of BBCube NAND composed of two chips.

10. BBCube Memory Application

WOW technology with bumpless interconnects using TSVs for three-dimensional stacking in wafer form has been described. An optimized thinned wafer thickness of 4 μm can increase the number of TSVs per chip with the fine pitch of the TSVs and can reduce the impedance of the TSV interconnects with no bumps. Therefore, an even higher-speed and higher-density HBM, namely the BBCube DRAM, can be realized with the four-phase shielded I/O scheme. Additionally, the BBCube NAND with the vertical BL architecture, which can be read and programmed *by plane* instead of *by line* by using the bumpless TSV, has been proposed. The BBCube DRAM for RAM and the BBCube NAND for ROM are sister memories with the high bandwidth.

As the number of the stacked memory chips is increased, the total memory density should be huge, similar to an enterprise. Therefore, an AI robotic bee, as an example, that can be used as a human assistant, which has a CPU, ultra-small enterprise, BBCube DRAM, BBCube NAND, and sensors, should be eventually realized in 50 mm^3 with 0.5 mW power consumption, as proposed in Figure 45.

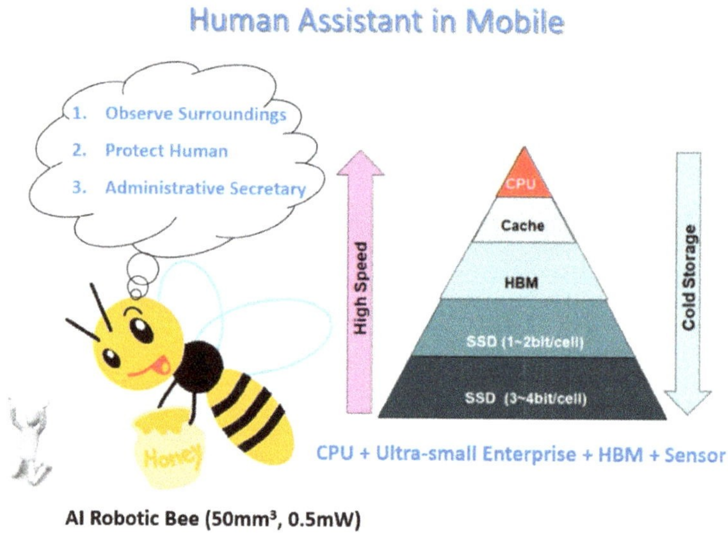

Figure 45. AI Robotic Bee (50 mm^3, 0.5 mW) as a human assistant.

11. Introduction to 3D Redundancy

In this section, our research motivation is to realize wafer-level fabrication, by which we can provide higher density and lower impedance TSVs with excellent heat conductivity, as discussed in this paper.

Figure 46 shows the configuration of the stacked DRAM system in BBCube generation one. It consists of 8 stacked dies, with one extra die for 3D redundancy, which will be discussed later. Each die is equipped with 16 tiles, and each tile has 4 or more banks. These tiles are memory arrays with 1024 I/Os vertically connected by TSVs. Therefore, massively high parallelism of 16 k I/Os was realized. Within each bank, sub-arrays of DRAM cells with extra sub-arrays are provided to perform intra die redundancy of a layer-by-layer scheme. We called this two-dimensional (2D) redundancy.

The superior properties of TSVs, such as lower impedance and higher heat conductance compared with existing micro-bump structures, originate from a technique for ultra-thinning Si substrates [34]. The quality and reliability of TSVs are supported by a copper dual damascene (DD) technique, which is very common for front end of line (FEOL) processes in device manufacturing. Since devices are already placed on wafers with the placement accuracy of lithography tools, wafer form fabrication is potentially capable of achieving layer-to-layer alignment with nm-level precision. These techniques have been completely proven in the manufacturing processes of devices and materials, such as CMOS image sensors (CIS), silicon on insulator (SOI) substrates, and microprocessors

(MPUs) [96–98]. Wafer form fabrication is the key to utilizing these techniques, and to enjoying the benefits of the maturity of manufacturing equipment.

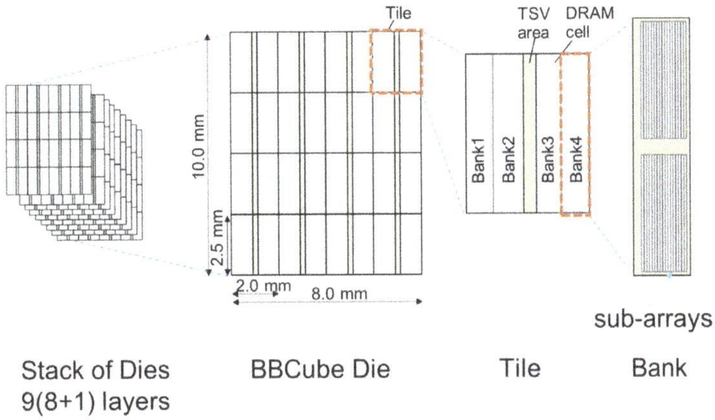

Figure 46. Configuration and structural hierarchy of the stacked DRAM system, BBCube. Banks consisted of sub-arrays for 2D redundancy, and stacked dies of 9 layers for 3D redundancy are illustrated.

To realize wafer form fabrication, it is essential to investigate defect management design, especially for random defects, since the probability of randomly defective portions being included in the module stack cannot be eliminated, as illustrated in Figure 47a. On the other hand, conventional KGD processes performed wafer testing, so that it is possible to stack defect free dies, as shown in Figure 47b. By simple arithmetic, the stacked device yield of the KGD process, Y_{3D}^{KGD}, was equal to the wafer test yield, Y_{device}, as defined in Equation (4). Besides, the yield of the wafer stacking case, Y_{3D}^{WoW}, was calculated from the wafer test yield to the power of the number of stacked layers, k, without any remedy, as expressed in Equation (5).

$$Y_{3D}^{KGD} \equiv Y_{device} \tag{4}$$

$$Y_{3D}^{WoW} = (Y_{device})^k \tag{5}$$

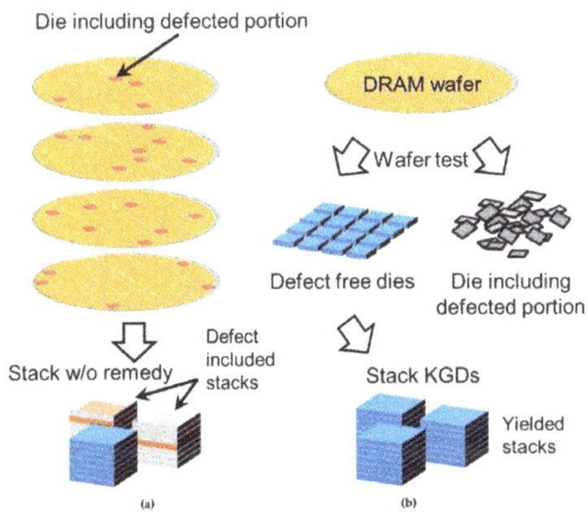

Figure 47. (**a**) For the wafer form fabrication, the probability of randomly defective portions being included in the module stack cannot be eliminated. (**b**) On the other hand, conventional KGD process includes wafer testing, so that it is possible to stack defect-free dies.

Here, we describe 3D redundancy [99,100] as applied to the configuration of the first generation BBCube. Using a general stacked DRAM system configuration, we illustrated the techniques constituting 3D redundancy in detail. It is apparent that these techniques are more practical and rational.

11.1. Method of 3D Redundancy

11.1.1. Typical Configuration of Stacked Synchronous DRAM Systems

Figure 48 shows a schematic diagram of a typical configuration of stacked DRAM devices [101], in which the circuit design hierarchy is the same as that of BBCube. In general, each die consists of banks [85]. These banks include sub-arrays, with redundant sub-array(s) to replace defective sub-array(s) while performing 2D redundancy. This 2D redundancy was carried out on a die-by-die basis within each layer of the stack. Recent DRAM devices were provided with extra cell arrays occupying 10% to 20% of the total area to reduce bit error rate (BER), in both cases of block sparing type redundancy and error check and correction (ECC) [86]. In this paper, we used a simple sub-array sparing model to be discussed later, to evaluate the area overhead.

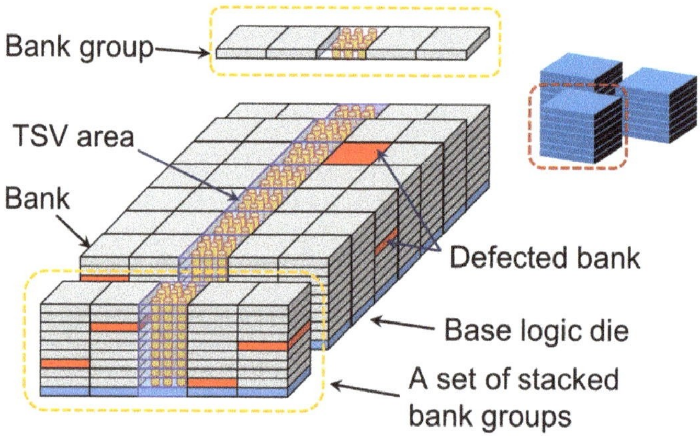

Figure 48. Schematic diagram of typical configuration of stacked DRAM devices. The circuit design hierarchy of BBCube is common. Each die consists of banks. These banks include not only sub-arrays for memory capacity, but also redundant sub-arrays to replace defective sub-array(s) to perform 2D redundancy.

When the 2D redundancy fail, the defective banks remain in the dies. In the case of the KGD process on the other hand, the dies were disposed of, which was in vain [101]. The TSV area connected neighboring banks to other layers, to avoid longer intra-die wiring [85–87]. The neighboring banks associated with the same TSV "digits" were grouped into bank groups, which corresponded to tiles in the case of BBCube. With WOW technology, the calculated delay basis distance in the z-direction between neighboring layers was approximately 30 μm. Therefore, in a set of stacked bank groups, banks were mutually compatible and replaceable with each other. It is possible to prefetch data from banks in different layers through the bypassing of vertical global wiring (Copper TSVs) in front of the buffer circuit block.

A base logic die includes a test circuit, a high-speed interface (HSIF), and a DRAM control circuit.

11.1.2. Techniques for 3D Redundancy

Based upon the typical configuration described above, we introduced a 3D redundancy, which consisted of three techniques, and a derivative extension at the sub-array level. The target was to provide the maximum number of stacked devices from the total fabricated

DRAM silicon area. Note that 3D redundancy is combined with 2D redundancy to reduce defect density, so as to be applicable to a vertically replaceable memory block architecture.

11.1.3. Layer Addition to Cover Circuit Resources

As illustrated so far, if defective banks result from the defect rate exceeding the 2D redundancy capability, the total number of non-defective banks is not enough for stack device operation. To enable repair of such devices, we added an extra layer to supplement the required number of non-defective banks, which is illustrated in Figure 49a. Note that this is not adding redundant cell arrays. The supplementally stacked die(s) were completely compatible with other dies inside the stacked layers below.

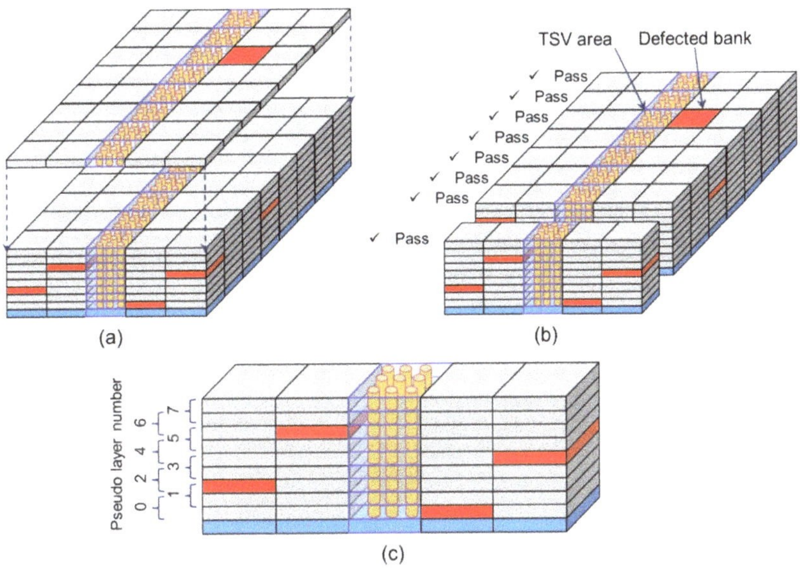

Figure 49. 3D redundancy techniques for: (**a**) layer addition to cover circuit resources, (**b**) bank replacement among a set of stacked bank groups, and (**c**) pseudo layer-by-layer operation.

11.1.4. Bank Replacement within a Set of Stacked Bank Groups

The next technique was to replace banks within a set of stacked bank groups. The banks, which belong to TSV "digits" were grouped together as a bank group. These bank groups were stacked as a set of stacked bank groups. Basically, banks supported a closely located set of TSV "digits", to avoid an increase in internal wiring capacitance, as illustrated in the previous section. Therefore, we were able to replace defective banks within individual sets of stacked bank groups to provide the overall functionality of individual TSV "digits". Note that, in different TSV "digits" (set of bank groups), the combination of selected functionable banks can be different from those of others. This scheme gives us another degree of freedom for memory repair optimization. Eventually, the entire functionality of all TSV "digits" will be individually accomplished, as indicated in Figure 49b.

11.1.5. Quasi Layer-by-Layer Operation

So far, in this discussion, it has been assumed that all stacked-die layers are connected to TSV equally. However, connected I/O transistors behave as load capacitance, even if they are not in use. TSVs should be connected to a minimum number of stacked-die layers. The third technique for achieving 3D redundancy is quasi layer-by-layer operation. For a functional stacked device, in a set of bank groups, the maximum possible number of defective banks must be equal to the number of banks included in the added extra layer or layers. This means that if we allocate twice as many layers as extra added layer(s) to a pseudo layer, we can obtain quasi layer-by-layer operation, as illustrated in Figure 49c. With this technique, logical layer can be defined as same as conventional stacked DRAM devices. In the case of Figure 49c, it is possible to share total required I/Os of data bus in 9 layers instead of 8 layers. Therefore, it is possible to reduce required silicon area of I/O transistors for a layer [102].

11.1.6. Derivative Extension Case: 3D Redundancy at Sub-Array Level

In the discussion so far, we have assumed the typical configuration of stacked DRAM systems such as HBM. In the BBCube case, each bank provided 1024 bit-wide I/Os in a tile, so that mutual compatibility within the set of the bank groups (tiles) was guaranteed. However, the tiles of BBCube, which were already highly fine-grained and partitioned into narrower pseudo banks from a 1024 bit-wide bank should be considered for better energy efficiency [103]. In such cases, each bank group (tile) may involve only one bank, or one pseudo bank, as illustrated in Figure 50a. Such a case makes 3D redundancy much less effective. We investigated whether we could use the sub-array level, which is the next level in the hierarchy below the bank level in typical DRAM and BBCube configurations.

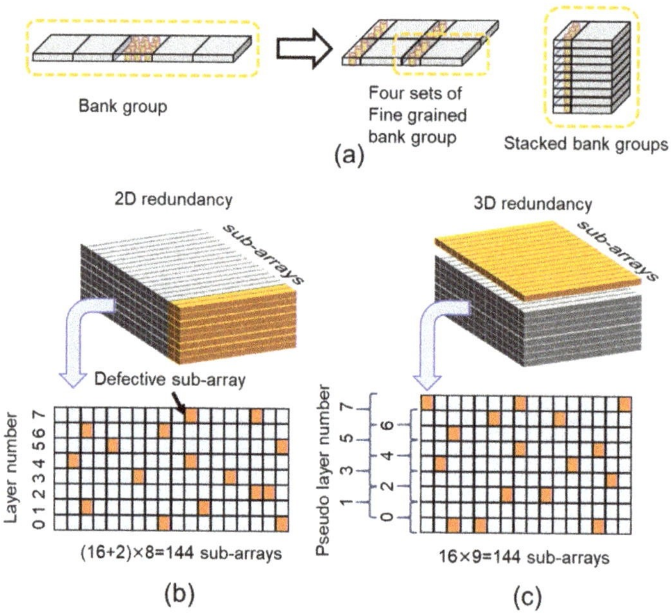

Figure 50. (**a**) Cases where each bank group (tile) may involve only one bank, or one pseudo bank. (**b**) In the case where 2 more redundant sub-arrays give near 100% yield, more than 127 non-defective sub-arrays are included in 144 sub-arrays. (**c**) In case of the same defective rate as (**b**), stack of single bank tiles with 1 extra layer tile includes 16 defective sub-arrays at most.

The sub-array level is used for 2D redundancy, to achieve excellent yield for the KGD process. With a certain amount of area penalty, near 100% yield can be obtained, as illustrated in Figure 50b.

In the case where two more redundant sub-arrays give near 100% yield, the maximum number of defective sub-arrays in the stack of tiles must be 16. Thus, $(16 + 2) \times 8 = 144$ sub-arrays must include 128 fine sub-arrays, as shown in Figure 50b. With a different configuration, $16 \times 9 = 144$ sub-arrays also must include 128 fine sub-arrays. This means that two vertically neighboring physical layers should include 16 or fewer defective sub-arrays. As a result, assignment of 8 pseudo layers out of 9 physical layers is possible, as illustrated in Figure 50c.

This "two more sub-arrays are enough" situation is realized by yield improvement activities and more nested 2D redundancy. The data transferred from replaced sub-arrays needs to be bypassed across physical layers, before the data multiplexers that provide bank data to the I/O buffers.

Accordingly, sub-array level 3D redundancy is feasible, and the bank configuration in a tile should be flexible to achieve energy efficiency optimization.

11.1.7. Parameter Definition

The definitions of parameters for the yield calculation are illustrated in Figure 51. In the case of BBCube, the number of tiles corresponded to the number of bank groups of typical DRAM systems in the figure. The scheme in which stacked bank groups support TSV "digits" becomes clearer when considered on a tile-by-tile basis.

m	Number of banks in a tile
n	Number of tiles in a layer
k	Number of required stacked layers
ℓ	Number of redundantly added layer(s)
λ	Defect density
S	Die size

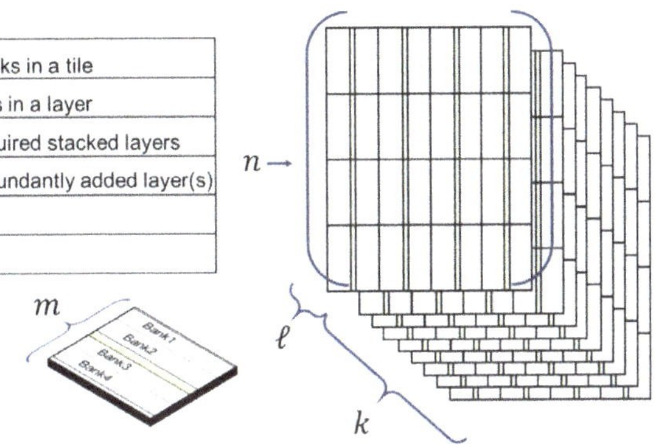

Figure 51. BBCube stack configuration and symbols for calculation.

When performing 2D redundancy, it is assumed that each bank includes 16 sub-arrays with redundant sub-arrays. Therefore, one extra sub-array incurs a 6.25% (i.e., 1/16) area penalty. The calculation was carried out with a simple sub-array replacement model.

11.1.8. Yield Calculation

In this paper, a Poisson distribution model was assumed as for the random defect yield model, which is consistent with the discussion below on intrinsic random defects, and does not result in a loss of generality of the whole discussion. Device yield, Y_{device}, is expressed by:

$$Y_{device} = Y_s \times Y_r \rightarrow Y_{device} \equiv Y_r = \exp(-\lambda S) \tag{6}$$

$$Y_{Bank} = \exp\left(-\frac{\lambda S}{n \times m}\right) = Y_r^{\left(\frac{1}{n \times m}\right)},\ Y \equiv Y_{Bank} \tag{7}$$

where Y_s is systematic yield, Y_r is random defect yield, S is the die size, and λ represents the area density of random defects [104].

The systematic yield Y_s is linked to various root causes from its definition as "systematic". They are identified and dealt with in the scope of process development, yield improvement activities, design for manufacturing (DFM) techniques, and big data analytics from manufacturing processes [105]. These efforts also target randomly distributed defects, which extrinsically cause random defect yield loss.

On the other hand, there are random defects that cannot be sufficiently reduced, even with intensely run yield improvement activities. For example, the variation of DRAM cell retention time due to impurity profile fluctuations can be minimized, by engineering efforts, in controllable portions of the processes, but there are remaining portions where intrinsic fluctuations exist. Fabrication engineers are struggling to achieve yield improvements, but sometimes encounter non-visible defects. This type of randomness is an essential barrier to yield improvement for both WOW devices and leading-edge devices, because we have already entered an era in which the number of atoms should be considered as an index of pattern pitch [106].

Problems that have root causes can be solved by eliminating them. Therefore, we assume that Y_s is close enough to "1" and concentrate on intrinsic Y_r.

In this analysis, we used the term "KGD case", which means a process that involves testing, and screened die stacking. It may be possible to apply 3D redundancy for stacked diced devices with a micro-bump structure. In that case, the calculation result will be the same between "WOW" and "stacked diced device" cases. An aim to identify something that could act as a benchmark led us to a comparison of the fabrication process differences. In this paper, we focus on the differences in redundancy procedures, and we do not go into the differences in the processes. We assumed the same 3D integration process yield of 100% for both WOW and KGD cases. For performance evaluation in other sections, we were able to deal with the differences in device structures.

The following Equation (8) presents the model for the yield of BBCube, Y_{3D}, with the 3D redundancy scheme illustrated in this paper. The parameters are described in Figure 51. When the single layer die yield (wafer test yield), Y_{device}, is given by Equation (6) and the bank yield is calculated from Equation (7), the BBCube yield, Y_{3D}, can be expressed as:

$$Y_{3D} = \left[\sum_{i=k\times m}^{(k+\ell)\times m} \{ ({}_{(k+\ell)\times m}C_i) \cdot Y^i \cdot (1-Y)^{(k+\ell)\times m - i} \} \right]^n \quad (8)$$

$$Y_{3D}^{KGD} \equiv Y_{device} \quad (9)$$

In the formula, the total yield of a tile is calculated by summing all products of bank yield and defect rate weighted by number of its combination. The random defect yield of the targeted tile is calculated as the term in brackets in Equation (8), so that the yield of the whole BBCube system can be obtained as the tile yield to the power of the number of tiles. To compare with BBCube yield, Y_{3D}^{KGD} by the KGD process is equal to Y_{device} itself, because it is possible to select a functional die by testing, which is expressed as in the definition in Equation (9), which is the same as Equation (4).

11.2. Results and Discussion

11.2.1. Results of BBCube Yield

Our goal was to yield the maximum number of stacked devices (BBCube) from the total fabricated silicon wafer area of the device.

As indicated in Figure 52a, BBCube fabricated by the WOW process showed better yield than that of the KGD stacking case, for all single layer die yields Y_{device}. The reason is as follows. For the KGD case, if the required number of banks is not available in a silicon die, the die area is wasted. On the contrary, for the WOW case, when there is an error in banks, the necessary number of banks can be substituted from other layers in the stack of bank groups. Therefore, when the single layer die yield Y_{device} is greater than or equal to 50%, the BBCube yield becomes more than 99% with 3D redundancy. If we want to achieve such excellent yield with only 2D redundancy, we must prepare more redundant sub-arrays. The area penalty evaluation of 2D redundancy is shown in Figure 52b.

Figure 52b illustrates, when a die yield without redundancy is given, how much area penalty is required to achieve the target yield by 2D redundancy. The die yield without redundancy is sometimes called the "perfect yield". The target yield cases evaluated are greater than 50% (">50%"), ">99.5%" and ">99.99%".

For productivity comparison, we needed to consider the area penalty of redundancy schemes. In the case of 3D redundancy, 9 wafers were consumed to realize the device function of 8 layers. Therefore, the area penalty of 3D redundancy was 12.5%, without a consideration of the area penalty of 2D redundancy, so that, in the case where the single layer die yield Y_{device} is greater than 87.5%, the KGD process seemed to be more productive.

However, to realize such an excellent single layer die yield Y_{device}, an area penalty of 12.5% or more was necessary for 2D redundancy only. Moreover, for almost the entire practical range of die yield without redundancy, to achieve a target yield of ">99.5%," 2D redundancy needed 12.5% or more die area than the case of a target yield of ">50%," as illustrated in Figure 52b. For 3D redundancy, a target yield of ">50%" was enough to achieve a single layer die yield Y_{device} of more than 99% of the BBCube yield. Therefore, 3D redundancy realizes better productivity, even under such conditions.

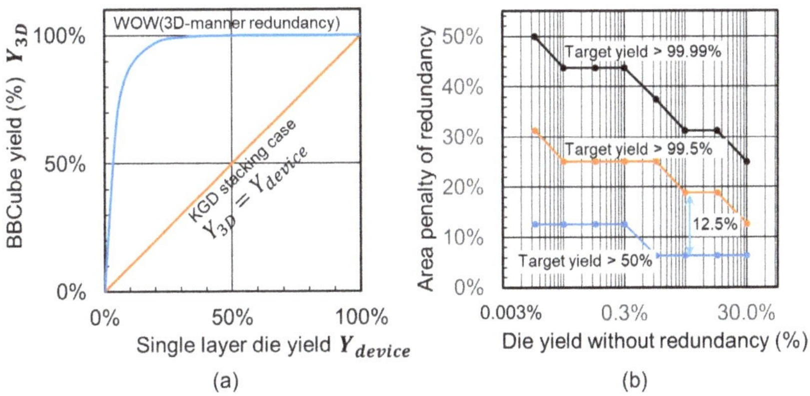

Figure 52. (**a**) BBCube yield comparison between WOW and KGD cases. (**b**) Area penalty of 2D redundancy required to achieve target yields of ">50%", ">99.5%" and ">99.99%".

Figure 53 shows a yield comparison of the cases for BBCube when more layers are aggressively stacked. Cases in which 9 (8 + 1), 17 (16 + 1), and 33 (32 + 1) layer are stacked are shown. Even in the case of 33 layers, with a single layer die yield Y_{device} of more than 80%, more than 99% BBCube yield was achieved, which indicates that 3D redundancy can support wafer form fabrication up to such an aggressive number of stacked layers. The portion for the added layer overhead for 3D redundancy is lowered to 3.125% (1/32) in this case.

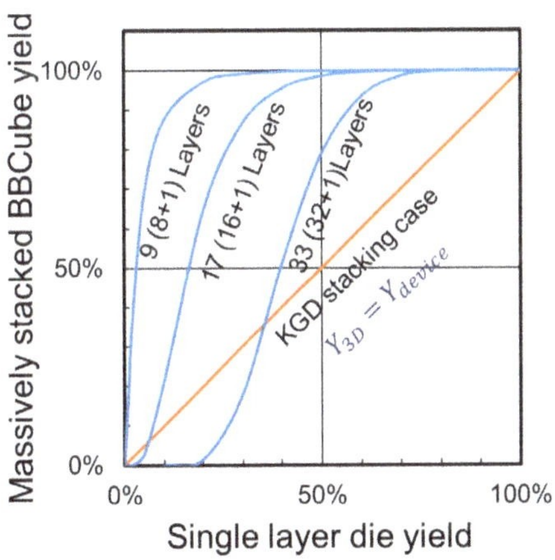

Figure 53. Yield comparison of the cases for BBCube, when more layers are aggressively stacked. Results for 9 (8 + 1), 17 (16 + 1), and 33 (32 + 1) stacked layers are illustrated. Even in the case of 33 layers, BBCube with 3D redundancy indicates superior yield compared with conventional technology at higher yield region.

This result shows that the WOW process with 3D redundancy provides better productivity than the KGD stacking case, even for future applications.

11.2.2. Discussion

In both cases of 2D redundancy and 3D redundancy, excellent yield can be realized if we prepare a certain memory cell area. In 3D redundancy, freedom of circuit block replacement, which is orthogonal to 2D redundancy, is provided. Therefore, it is possible to define redundancy success rate "digits" by the "digits" of TSVs, which leads to a higher total number of combinations. Superficially, it looks less productive to add extra wafers for redundancy purposes. By replacing banks within a set of stacked bank groups and introducing sophisticated vertical bank group allocation to realize quasi layer-by-layer operation, the orthogonality of the "digit" by "digit" basis become clear. These 3D redundancy techniques appear to be more practical, and rather rational.

11.3. Conclusion of 3D Redundancy

The excellent performance of BBCube due to the WOW technology and the application of 3D redundancy to utilize wafer form manufacturing have been presented in this chapter. Wafer-on-wafer fabrication was realized with the support of a 3D redundancy scheme, which led us to conclude that BBCube could be the next system scaling enabler.

12. Thermal Resistance Comparison of BBCube and Micro-Bumps

In 3D stacking technology, thermal management problems become more difficult due to the vertical thermal resistance of interconnection layers and back end of line (BEOL) [107–110]. Therefore, the temperature of stacked dies increase when they contain more IC chips [111]. Recently, a bumpless 3D multi-stack process using ultra-thin technology was proposed [45–48]. This approach is expected to decrease the vertical thermal resistance. Hence, the total thermal resistance of 3D stacked ICs with and without solder bumps was estimated.

12.1. Thermal Resistance Calculation Method

Figure 54 shows an example of 3D stacked ICs. This structure consists of a Si substrate, Si with TSVs, back end of line (BEOL), vertical interconnections (micro-bumps), and direct contact by TSVs (bumpless). The thermal conductivity of the vertical interconnection was calculated using the Finite Element Method (FEM), and then thermal resistance was calculated using that thermal conductivity. Additionally, the total thermal resistance was calculated using a thermal network method.

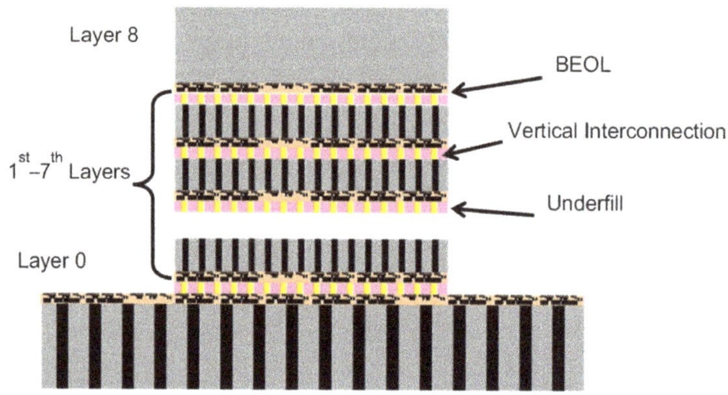

Figure 54. A typical 3D stacked memory structure consists of base layer (layer 0) and memory dies (layer 1st to layer 8th). Each die has BEOL layer and vertical interconnect layer.

The temperature rise calculation had four primary steps:

i. Make assumptions about the IC stack structure,
ii. Estimate the effective thermal conductivity of each layer, and
iii. Calculate thermal resistance of each layer, and
iv. Calculate the temperature rise using the thermal network method.

Figure 55 provides a structural comparison of the micro-bump and bumpless types for IC stacks with 8 layers.

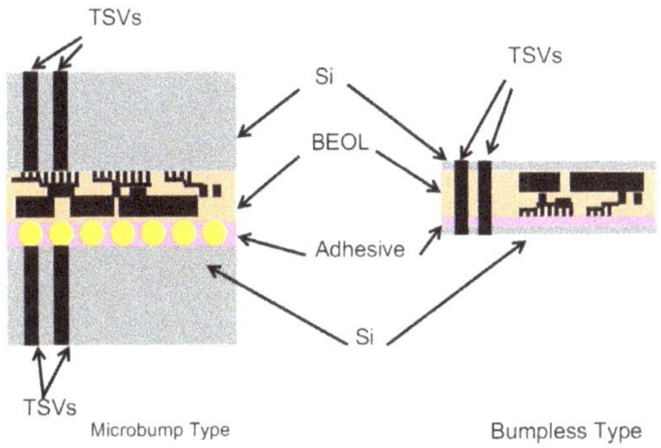

Figure 55. A comparison of bump and bumpless interconnects using TSVs for 3D logic/memory stack structures. Each structure consists of a Si layer, a BEOL layer, an interconnection layer, and TSVs. As for the bumpless structure, TSVs are fully transfixed from the top layer to bottom layer.

12.2. Thermal Resistance of Micro-Bump Vertical Interconnection

When thermal conductivity of each layer is calculated, the total thermal resistance of the micro-bump 3D stacked ICs can be calculated. The thermal conductivities of micro-bump interconnect were reported by Matsumoto et al. [111]. Additionally, 148, 160.5, and 1.44 W/m/K for Si, Si with TSVs, and BEOL were used, respectively. The thermal conductivity of the vertical interconnection using micro-bumps depends on the bump size, bump pitch, and underfill. Additionally, FEM was used to calculate the thermal performance of micro-bumps. Figure 56 shows the FEM models, and Figure 57 shows the calculation result using the bump occupancy definition shown in Figure 58. From the result, we can see that the use of underfill material is advantageous only when the TSV occupancy is less than 0.05. Using these material thermal conductivities, the thermal resistance of each layer was calculated, and then the total thermal resistance was calculated. In this calculation, the size of the micro-bump was 25 µm and the micro-bump pitch was 50 µm. The total thermal resistance was 1.54 Kcm^2/W, which confirms that the thermal resistance of the BEOL and interconnection is too large to reduce the temperature rise.

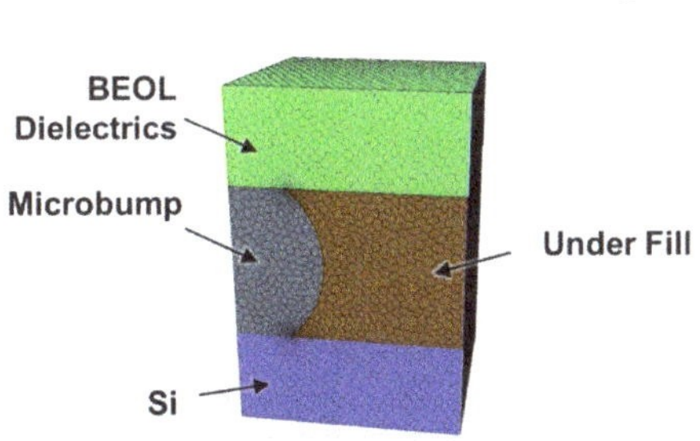

Figure 56. Schematic diagram of thermal conductivity of microbump type calculated using a Finite Elements Method Model (FEM model). In this FEM model, the BEOL layer, Si layer, Microbumps, and underfill are modeled.

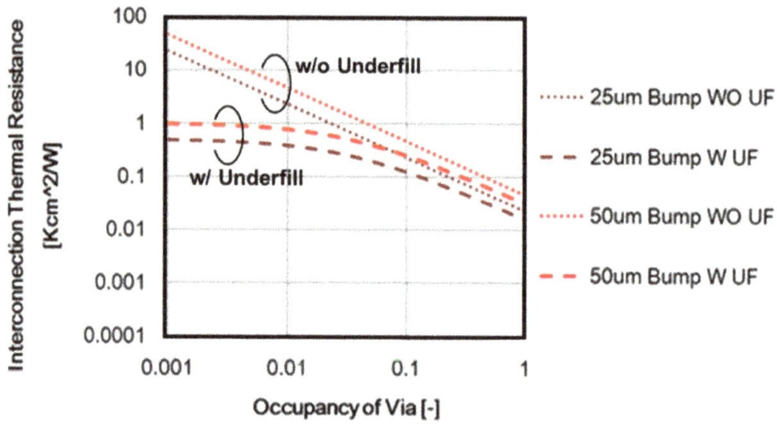

Figure 57. Interconnection thermal resistance as a function of TSV occupancy. The definition of TSV occupancy is shown in Figure 58. Two different microbump sizes (25 µm and 50 µm) and the impact of underfill are calculated, where W UF = with underfill material and WO UF = without underfill material. In the case of a 25 µm microbump with and without underfill, thermal resistance is smaller than that of 50 µm. As for the underfill, the thermal resistance with underfill is small compared to no underfill, especially at low occupancy of <0.01.

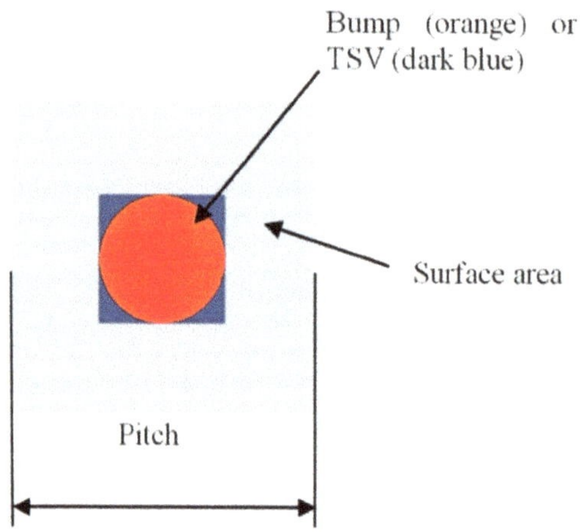

Figure 58. The definition of TSV occupancy. TSV occupancy is defined by the area of a bump of TSV divided by the surface area.

12.3. Thermal Resistance of BBC

The thermal conductivity of the BEOL with Cu TSV interconnections was calculated using the FEM. Figure 59a shows the FEM model for the BEOL and interconnection in bumpless IC stacks. Figure 59b compares the interconnection thermal resistance for the micro-bump and bumpless types. The thermal resistance for the bumpless type was two orders of magnitude lower than that for the conventional structure. This suggests that only 1% of the total metal area of bumpless TSVs is required to achieve the same thermal resistance, in comparison with the conventional structure. Table 4 shows the thermal resistance results for both types of IC stacks.

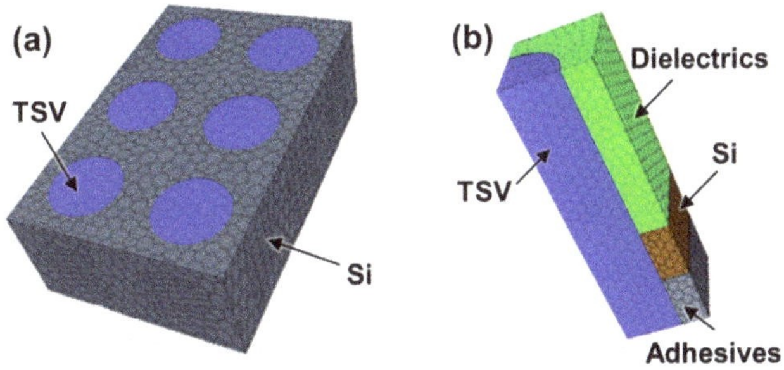

Figure 59. Schematic diagram of thermal conductivity of (**a**) bumpless structure calculated using FEM model, where Si layer and TSVs are molded; (**b**) TSV interconnects with microbumps, where the dielectrics (BEOL) layer, Si layer, adhesive layer, and TSV are modeled.

Table 4. Total thermal resistance of TSV with microbump and bumpless TSV.

				TSV with Micro Bump 51.4 µm Pitch 25 µm Bump		**Bumpless TSV 512 × 16 TSV 5 µm Gap**	
		Components	Equivalent Thermal Conductivity (W/mK)	Thickness (µm)	Thermal Resistance (Kcm^2/W)	Thickness (µm)	Thermal Resistance (Kcm^2/W)
DRAM	Top Chip Rth1	Si	148	150	0.049	5	0.00034
		BEOL	1.44	15	0.104	-	-
			3.99	-	-	15	0.038
		Interconnection Micro Bump	2.54	20	0.079	-	-
		Interconnection Bumpless	2.56	-	-	5	0.02
	2–8 Chip 7 Layers Rth2–8	Si with TSV	160.5	50	0.003	5	0.00031
		BEOL	1.44	15	0.104	-	-
			3.99	-	-	15	0.038
		Interconnection Micro Bump	2.54	20	0.079	-	-
		Interconnection Bumpless	2.56	-	-	5	0.02
Logic RthL		Si with TSV	160.5	50	0.003	5	0.0003
Total Thermal Resistance		Rth1 + 7 × Rth2–8 + RthL	-	-	**1.54**	-	**0.46**

As shown in Figure 59, the bumpless process is a kind of via last process, and the TSVs fully go through from bottom to top. In this case, TSVs are formed by copper and their thermal conductivity is around 400 (W/m/K). This value is around 280 times larger than BEOL thermal conductivity and around 160 times larger than microbump interconnect thermal conductivity. Thus, only a 1% volume fraction is effective for effective thermal conductivity improvement. In addition, the bumpless interconnection thickness is 4 times thinner than microbump one, hence the thermal resistance of interconnection is more than 4 times smaller. The interconnection and BEOL thermal resistances for the bumpless type were almost 4 and 3 times smaller, respectively, than those for the conventional structure.

12.4. Temperature Rise Calculation Result

The temperature rise for each layer was calculated using:

$$T_M = \sum_{k=1}^{M} \left(R_k * \sum_{l=k}^{M} Q_{N-l+1} \right) \tag{10}$$

where,

$$T_M \;:\; \text{Temperature rise of layer M } \left({}^{\circ}\mathrm{C}\right)$$

$$R_k \ : \text{Thermal Resistance of layer k} \left(\frac{\text{Kcm}^2}{\text{W}}\right)$$

$$Q_k \ : \text{Heat Generation of layer k (W)}$$

The temperature rise was caused by the product of its own thermal resistance and the heat generated by the layer below it.

Figure 60 shows that the temperature increased as a function of the number of DRAM dies, and a comparison for the micro-bump and bumpless types. "Layer x" represents the DRAM dies. The maximum temperature rise ΔT for the no microbump case (BBCube) was around 5.8 °C, which is almost one-fourth that of the microbump case.

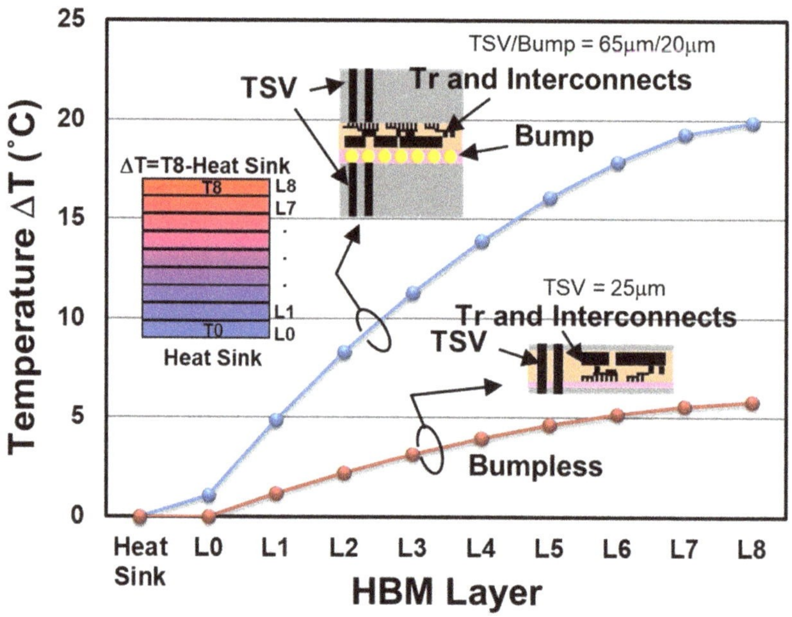

Figure 60. Temperature rises with and without microbumps as a function of the number of stacked dies. "Layer x" represents the DRAM dies. The maximum temperature rise for the no microbump case (BBCube) is around 5.8 °C, which is lower than that of the microbump case, i.e., 20 °C.

12.5. Thermal Resistance Comparison Conclusion

We established a calculation method for evaluating the thermal resistance of 3D stacked ICs following the method in Matsumoto et al. [111]. Using this method, we calculated the temperature rise for each layer in 3D IC stacks with both microbump and bumpless vertical interconnections. For the ICs modeled in this study, the total thermal resistance of organic layers 20 and 5 μm thick were 1.54 and 0.46 Kcm^2/W, respectively.

13. Summary and Conclusions

Due to the demand for post-scaling in device structures, three-dimensional integration technologies are expected to be increasingly employed. By doing so, when wafers with micrometer thickness are stacked, the total thickness is reduced, and the transistor capacity increases in proportion to the number of wafers. Increasing the TSV interconnects density enables terabyte-level bandwidth without sacrificing energy efficiency. Power consumption and heat dissipation are especially important for high-density modules, such as 2.5D and 3D systems. 2.5D, which is not a physical term, refers to a high-speed, high-bandwidth system that incorporates and integrates a three-dimensional memory such as HBM, GPU

(Graphic Processing Unit), and MPU on an interposer, and is a general term for the back-end processes. In recent years, it has become a product differential technology to combine multiple chips and passive components with different functions into one system module. The authors' research organization, the "WOW Alliance", has proposed the BBCube architecture using WOW and COW processes for 2.5D and 3D systems including passive devices, as described in this paper.

As the number of stacked wafers increases, the number of incoming wafers in manufacturing increases proportionally [112]. Recently, volume production with 80,000 wafers per month has been used. To maintain the same throughput with stacks of 8 DRAM wafers, the number of incoming wafers will be 640,000 per month. Without considering facility costs and running costs, it would be possible to increase the size of fabrication plants. However, a production line with eight-times larger footprint may not balance the production costs. Thus, in the future, enlarged wafer size or an alternative approach such as a combination of reducing total process steps with very high throughput may be reconsidered in this situation.

If the alignment accuracy of wafer stacking is improved, about 1 to 10 million TSVs can be formed per square centimeter. Such large-scale I/O is too high for DRAM stacking, but if scaling down of TSVs and layout flexibility evolve, it will be possible to stack MPU logic and SRAM cache memory individually. If the power distribution and ground can be located directly beneath of SRAM cell, stable current and low applied voltage <0.7 V with low noise can be realized because they can be connected with equivalent lengths and high parallelity by micrometer-level short interconnects. Such high-density TSV interconnects in conjunction with BBCube (low power consumption) will help to reduce the excess heat of 3D systems.

In summary, it is possible to achieve the next step in the semiconductor roadmap by employing three-dimensional integration technology, as discussed. Although it is necessary to develop 3DI technology with high productivity, such as front-end wafer technology, many of those mature processes can be applied. Thus, the new technology for 3DI is only the thinning and stacking processes. These technologies can also be improved as there are well-known technologies from the front-end and novel material candidates, which are expected to become mature by applying the know-how gained over many years in the semiconductor industry.

Author Contributions: Conceptualization, T.O.; BBCube WOW and COW processes, T.O.; BBCube memory design, K.S. and N.C.; BBCube memory redundancy, S.S.; BBCube thermal analysis, H.R.; writing—review and editing, Sections 1–6 and 13, T.O., Sections 7–10, K.S. and N.C.; Section 11, S.S., Section 12, H.R.; supervision, T.O. All authors have read and agreed to the published version of the manuscript.

Funding: This research was funded by the WOW Alliance of Tokyo Institution of Technology (Tokyo, Japan).

Acknowledgments: This study was carried out based on the three-dimensional integration development program of the WOW Alliance at the Tokyo Institute of Technology, and the authors thank the alliance members of more than 30 companies for their cooperation.

Conflicts of Interest: The authors declare no conflict of interest.

References

1. Kilby, J. Miniaturized Electronic Circuits. U.S. Patent 3,138,743, 6 February 1959.
2. Noyce, R. Semiconductor Device-and-Lead Structure. U.S. Patent 2,981,877, 30 July 1959.
3. Moore, G. Cramming More Components Onto Integrated Circuits. *Electronics* **1965**, *38*, 114–117. [CrossRef]
4. Dennard, R.H.; Gaensslen, F.H.; Rideout, V.L.; Bassous, E.; LeBlanc, A.R. Design of ion-implanted MOSFET's with very small physical dimensions. *IEEE J. Solid-State Circuit* **1974**, *9*, 256–268. [CrossRef]
5. Furukawa, S. Three-dimensional Device. *J. Inst. Telev. Eng. Jpn.* **1982**, *36*, 1060–1067.
6. Kurokawa, K.; Aiso, H. Polynomial Transformer. In Proceedings of the 7th Symposium on Computer Arithmetic, Urbana, IL, USA, 4–6 June 1985; IEEE Computer Society Press: Piscataway, NJ, USA, 1985; Volume 7, pp. 153–158.

7. Nakano, M. Oyo Buturi. *Jpn. Soc. Appl. Phys.* **1985**, *54*, 652–659.
8. Akasaka, Y. Three-Dimensional IC Trends. *Proc. IEEE* **1986**, *74*, 1703–1714. [CrossRef]
9. Spiesshoefer, S.; Schaper, L. IC Stacking Technology Using Fine Pitch Nanoscale Through Silicon Vias. In Proceedings of the 53rd Electronic Components and Technology Conference (ECTC), New Orleans, LA, USA, 27–30 May 2003; pp. 631–633.
10. Yasumoto, M.; Hayama, H.; Enomoto, T. Promissing New Fabrication Process Developed for Stacked LSI's. In Proceedings of the 1984 International Electron Devices Meeting, San Francisco, CA, USA, 9–14 December 1984; pp. 816–819.
11. Hayashi, Y.; Wada, S.; Kajiyana, K.; Oyama, K.; Koh, R.; Takahashi, S.; Kunio, T. Fabrication of Three-Dimensional IC Using 'CUmulatively Bonded IC' (CUBIC) Technology. In Proceedings of the Digest of Technical Papers. 1990 Symposium on VLSI Technology, Honolulu, HI, USA, 4–7 June 1990; pp. 95–96.
12. Ramm, P.; Buchner, R. Method of Making a Three-Dimensional Integrated Circuit. U.S. Patent 5,563,084, 8 October 1996.
13. Ramm, P.; Bollmann, D.; Braun, R.; Buchner, R.; Cao-Minh, U.; Engelhardt, M.; Enmann, G.; Graβ, T.; Hieber, K.; Hübner, H.; et al. Three dimensional metallization for vertically integrated circuits. *Microelectron. Eng.* **1997**, *37*, 39–47. [CrossRef]
14. Matsumoto, T.; Satoh, M.; Sakuma, K.; Kurino, H.; Miyakawa, N.; Itani, H.; Koyanagi, M. New Three-Dimensional Wafer Bonding Technology Using the Adhesive Injection Method. *Jpn. J. Appl. Phys.* **1998**, *1*, 1217–1221. [CrossRef]
15. Tummala, R.; Madisetti, V.K. System on chip or system on package. *IEEE Des. Test Comput.* **1999**, *16*, 48–56. [CrossRef]
16. Lu, J.-Q.; Kumar, A.; Kwon, Y.; Eisenbraun, E.T.; Kraft, R.P.; McDonald, J.F.; Gutmann, R.J.; Cale, T.S.; Belemjain, P.; Erdogan, O.; et al. 3-D Integration Using Wafer Bonding. *MRS Proc. Vol.* **2001**, *16*, 515–521.
17. Shigetou, A.; Itoh, T.; Suga, T. Bumpless Interconnect of Cu Electrodes in Millions-Pins Level. In Proceedings of the IEEE 56th Electronic Components and Technology Conference (ECTC), San Diego, CA, USA, 30 May–2 June 2006; pp. 1003–1008.
18. Takahashi, K.; Terao, H.; Tomita, Y.; Yamaji, Y.; Hoshino, M.; Sato, T.; Morifuji, T.; Sunohara, M.; Bonkohara, M. Current Status of Research and Development for Three-dimensional Chip Stack Technology. *Jpn. J. Appl. Phys.* **2001**, *40*, 3032–3037. [CrossRef]
19. Klumpp, A.; Merkel, R.; Wieland, R.; Ramm, P. Chip-to-wafer stacking technology for 3D system integration. In Proceedings of the IEEE 53rd Electronic Components and Technology Conference (ECTC), New Orleans, LA, USA, 27–30 May 2003; pp. 1080–1083.
20. Umemoto, M.; Tanida, K.; Nemoto, Y.; Hoshino, M.; Kojima, M.; Shirai, Y.; Takahashi, K. High-Performance Vertical Interconnection for High-Density 3D Chip Stacking Package. In Proceedings of the IEEE 54rd Electronic Components and Technology Conference (ECTC), Las Vegas, NV, USA, 4 June 2004; pp. 616–623.
21. Knickerbocker, J.U.; Andry, P.S.; Buchwalter, L.P.; Deutsch, A.; Horton, R.R.; Jenkins, K.A.; Kwark, Y.H.; McVicker, G.; Patel, C.S.; Polastre, R.J.; et al. Development of next-generation system-on-package (SOP) technology based on silicon carriers with fine-pitch chip interconnection. *IBM J. Res. Dev.* **2005**, *49*, 725–753. [CrossRef]
22. Fukushima, T.; Yamada, Y.; Kikuchi, H.; Koyanagi, M. New Three-Dimensional Integration Technology Using Self-Assembly Technique. In Proceedings of the IEEE International Electron Devices Meeting, Washington, DC, USA, 5 December 2005; pp. 359–362.
23. Tanaka, N.; Yoshimura, Y.; Naito, T.; Miyazaki, C.; Uematsu, T.; Hanada, K.; Toma, N.; Akazawa, T. Low-Cost Through-hole Electrode Interconnection for. 3D-SiP Using Room-temperature Bonding. In Proceedings of the IEEE 56th Electronic Components and Technology Conference (ECTC), San Diego, CA, USA, 30 May–2 June 2006; pp. 814–818.
24. Brunnbauer, M.; Fürgut, E.; Beer, G.; Meyer, T.; Hedler, H.; Belonio, J.; Nomura, E.; Kiuchi, K.; Kobayashi, K. An Embedded Device Technology Based on a Molded Reconfigured Wafer. In Proceedings of the IEEE 56th Electronic Components and Technology Conference (ECTC), San Diego, CA, USA, 30 May–2 June 2006; pp. 547–551.
25. Topol, A.W.; la Tulipe, D.C.; Shi, L.; Frank, D.J.; Bernstein, K.; Steen, S.E.; Kumar, A.; Singco, G.U.; Young, A.M.; Guarini, K.W.; et al. Three-Dimensional Integrated Circuits. *IBM J. Res. Dev.* **2006**, *50*, 491–506. [CrossRef]
26. Burns, J.A.; Aull, B.F.; Chen, C.K.; Chen, C.-L.; Keast, C.L.; Knecht, J.M.; Suntharalingam, V.; Warner, K.; Wyatt, P.W.; Yost, D.-R.W. A Wafer-Scale 3-D Circuit Integration Technology. *IEEE Trans. Elect. Dev.* **2006**, *53*, 2507–2516. [CrossRef]
27. Miyakawa, N.; Hashimoto, E.; Maebashi, T.; Nakamura, N.; Sacho, Y.; Nakayama, S.; Toyoda, S. Multilayer stacking technology using wafer-to-wafer stacked method. *ACM J. Emerg. Technol. Comput. Syst.* **2008**, *4*, 1–15. [CrossRef]
28. Sakui, K.; Ohba, T. Three-dimensional Integration (3DI) with Bumpless Interconnects for Tera-scale Generation—High Speed, Low Power, and Ultra-small Operating Platform. In Proceedings of the 2019 IEEE Custom Integrated Circuits Conference (CICC), Austin, TX, USA, 14–17 April 2019; pp. 22–26.
29. Sakui, K.; Ohba, T. High Speed, Low Power, and Ultra-small Operating Platform with Three-dimensional Integration (3DI) by Bumpless Interconnects. In Proceedings of the IEEE 11th International Memory Workshop (IMW), Monterey, CA, USA, 12–15 May 2019; pp. 60–63.
30. Kim, J.-S.; Oh, C.S.; Lee, H.; Lee, D.; Hwang, H.-R.; Hwang, S.; Na, B.; Moon, J.; Kim, J.-G.; Park, H.; et al. A 1.2V 12.8GB/s 2Gb Mobile Wide-I/O DRAM with 4 × 128 I/Os Using TSV-Based Stacking. *IEEE J. Solid-State Circuits* **2012**, *47*, 107–116. [CrossRef]
31. Ohba, T.; Maeda, N.; Kitada, H.; Fujimoto, K.; Suzuki, K.; Nakamura, T.; Kawai, A.; Arai, K. Thinned Wafer Multi-stack 3DI Technology. *Microelectron. Eng.* **2010**, *87*, 485–490. [CrossRef]
32. Ohba, T. Three-Dimensional (3D) Integration Technology. *Electrochem. Soc. Trans.* **2011**, *34*, 1011–1016. [CrossRef]
33. Ohba, T.; Kim, Y.S.; Mizushima, Y.; Maeda, N.; Fujimoto, K.; Kodama, S. Review of Wafer-Level Three-Dimensional Integration (3DI) using Bumpless Interconnects for Tera-Scale Generation. *IEICE Electron. Express* **2015**, *12*, 1–14. [CrossRef]

34. Chujo, N.; Sakui, K.; Ryoson, H.; Sugatani, S.; Nakamura, T.; Ohba, T. Bumpless Build Cube (BBCube): High-Parallelism, High-Heat-Dissipation and Low-Power Stacked Memory Using Wafer-Level 3D Integration Process. In Proceedings of the 2020 IEEE Symposium on VLSI Technology, Honolulu, HI, USA, 16–19 June 2020.
35. Helms, C.R. Semiconductor Technology & Manufacturing Status, Challenges, and Solutions—A New Paradigm in the Making. *AIP Conf. Proc.* **2003**, *683*, 63–73.
36. Available online: https://www.semi.org/en/semiconductor-industry-2015-2025 (accessed on 25 December 2021).
37. Green, D.S. DARPA's CHIPS Program, and Making Heterogeneous Integration Common. In Proceedings of the 14th Annual Conference on 3D Architectures for Semiconductor Integration and Packaging (3D-ASIP), San Francisco, CA, USA, 5–7 December 2017; Available online: https://www.darpa.mil/news-events/2016-07-19 (accessed on 25 December 2021).
38. Su, L.T.; Naffziger, S.; Papermaster, M. Multi-Chip Technologies to Unleash Computing Performance Gains over the Next Decade. In Proceedings of the 2017 IEEE International Electron Devices Meeting (IEDM), San Francisco, CA, USA, 2–6 December 2017; pp. 1–8.
39. Available online: https://www.brookings.edu/techstream/the-chip-making-machine-at-the-center-of-chinese-dual-use-concerns/ (accessed on 25 December 2021).
40. Schoenfelder, S.; Ebert, M.; Landesberger, C.; Bock, K.; Bagdahn, J. Investigations of The Influence of Dicing Techniques on the Strength Properties of Thin Silicon. *Microelectron. Reliab.* **2007**, *47*, 168–178. [CrossRef]
41. Huang, P.S.; Tsai, M.Y. Nonlinearities in Thin-Silicon Die Strength Tests. In Proceedings of the International Microsystems, Packaging, Assembly and Circuits Technology Conference (IMPACT), Taipei, Taiwan, 19–21 October 2011; pp. 91–95.
42. Ohba, T. Multilevel Interconnect Technologies in SoC and SiP for 100-nm Node and Beyond. In Proceedings of the IEEE 6th International Conference on Solid-State Integrated Circuit Technology (ICSICT), Shanghai, China, 22–25 October 2001; pp. 46–51.
43. Wang, G.; Merrill, C.; Zhao, J.H.; Groothuis, S.K.; Ho, P.S. Packaging Effects on Reliability of Cu/Low-k Interconnects. *IEEE Trans. Device Mater. Reliab.* **2003**, *3*, 119–128. [CrossRef]
44. Bang, W.H.; Kim, C.-U.; Kang, S.H.; Oh, K.H. Fracture Mechanics of Solder Bumps During Ball Shear Testing: Effect of Bump Size. *J. Electron. Mater.* **2009**, *38*, 1896–1905. [CrossRef]
45. Maeda, N.; Kim, Y.S.; Hikosaka, Y.; Eshita, T.; Kitada, H.; Fujimoto, K.; Mizushima, Y.; Suzuki, K.; Nakamura, T.; Kawai, A.; et al. Development of Sub 10-μm Ultra-Thinning Technology Using Device Wafers for 3D Manufacturing of Terabit Memory. In Proceedings of the IEEE Symposium on VLSI Technology, Honolulu, HI, USA, 15–17 June 2010; pp. 105–106.
46. Kim, Y.S.; Tsukune, A.; Maeda, N.; Kitada, H.; Kawai, A.; Arai, K.; Fujimoto, K.; Suzuki, K.; Mizushima, Y.; Nakamura, T.; et al. Ultra Thinning 300-mm Wafer down to 7-μm for 3D Wafer Integration on 45-nm Node CMOS using Strained Silicon and Cu/Low-k Interconnects. In Proceedings of the IEEE International Electron Devices Meeting (IEDM), Baltimore, MD, USA, 7–9 December 2009; pp. 365–366.
47. Kim, Y.S.; Kodama, S.; Mizushima, Y.; Maeda, N.; Kitada, H.; Fujimoto, K.; Nakamura, T.; Suzuki, D.; Kawai, A.; Arai, K.; et al. Ultra Thinning down to 4-μm using 300-mm Wafer proven by 40-nm Node 2Gb DRAM for 3D Multi-stack WOW Applications. In Proceedings of the IEEE Symposium on VLSI Technology, Honolulu, HI, USA, 9–12 June 2014; pp. 26–27.
48. Mizushima, Y.; Kim, Y.S.; Kodama, S.; Nakamura, T.; Ohba, T. Plan view stress distribution at 1 μm underneath of DRAM device using WOW ultra-thinning technology. In Proceedings of the Proceedings Advanced Metallization Conference (AMC), Austin, TX, USA, 13–14 September 2017.
49. Chen, Z.; Kim, Y.S.; Fukuda, T.; Sakui, K.; Kobayashi, T.; Obara, T.; Ohba, T. Reliability of Wafer-Level Ultra-Thinning down to 3 μm using 20 nm-Node DRAMs. In Proceedings of the IEEE International Reliability Physics Symposium (IRPS), Monterey, CA, USA, 21–25 March 2021. [CrossRef]
50. Ohba, T. Size-Reduction of HBW System using WOW Bumpless TSV Interconnects. In Proceedings of the International Conference on Solid State Devices and Materials (SSDM), Nagoya, Japan, 2–5 September 2019; pp. 417–418.
51. Loranger, M.; Moon, S.-W. *Verification of High-Bandwidth-Memory (HBM) through Direct Probing on MicroBumps*; Semiconductor Wafer Test Workshop: San Diego, CA, USA, 2016. Available online: https://www.formfactor.com/wp-content/uploads/S01_02_Loranger_SWTW2016-2.pdf (accessed on 25 December 2021).
52. Kim, N. Future of the Packaging Technologies for HBM. In Proceedings of the IEEE International Electron Devices Meeting (IEDM) Short Course 2, San Fransisco, CA, USA, 1–5 December 2018.
53. Chen, Z.; Araki, N.; Kim, Y.S.; Fukuda, T.; Sakui, K.; Nakamura, T.; Kobayashi, T.; Obara, T.; Ohba, T. Ultra-Thinning of 20 nm-Node DRAMs down to 3 μm for Wafer-on-Wafer (WOW) Applications. In Proceedings of the IEEE Electronic Components and Technology Conference (ECTC), San Diego, CA, USA, 1 June–4 July 2021; pp. 1131–1137.
54. Araki, N.; Kim, Y.S.; Kodama, S.; Hsiao, C.; Chang, H.; Lin, C.; Ohba, T. Development of Micrometer-Thick Bonding Material for Wafer-On-Wafer (WOW) Applications. In Proceedings of the 14th International Microsystems, Packaging, Assembly and Circuits Technology Conference (IMPACT), Taipei, Taiwan, 24–26 October 2018; p. 470.
55. Araki, N.; Maetani, S.; Kim, Y.S.; Kodama, S.; Ohba, T. Development of Resins for Bumpless Interconnects and Wafer-On-Wafer (WOW) Integration. In Proceedings of the IEEE 69th Electronic Components and Technology Conference (ECTC), Las Vegas, NV, USA, 28–31 May 2019; pp. 1002–1007.
56. Nakamura, T.; Mizushima, Y.; Kitada, H.; Kim, Y.S.; Maeda, N.; Kodama, S.; Sugie, R.; Hashimoto, H.; Kawai, A.; Arai, K.; et al. Influence of Wafer Thinning Process on Backside Damage in 3D Integration. In Proceedings of the IEEE International 3D Systems Integration Conference (3DIC), San Francisco, CA, USA, 2–4 October 2013. [CrossRef]

57. Mizushima, Y.; Kim, Y.; Nakamura, T.; Sugie, R.; Hashimoto, H.; Uedono, A.; Ohba, T. Impact of back-grinding-induced damage on Si wafer thinning for three-dimensional integration. *Jpn. J. Appl. Phys.* **2014**, *53*, 05GE04. [CrossRef]
58. Lee, S.; Kim, J.-H.; Kim, Y.S.; Ohba, T.; Kim, T.-S. Effects of Thickness and Crystallographic Orientation on Tensile Properties of Thinned Silicon Wafers. *IEEE Trans. Compon. Packag. Manuf. Technol.* **2020**, *10*, 296–303. [CrossRef]
59. Engineering R&D Division, Operation V. The effects of edge trimming. *DISCO Tech. Rev.* **2016**. Available online: https://www.disco.co.jp/eg/solution/technical_review/pdf/TR16-09_The%20effects%20of%20edge%20trimming_20160610.pdf (accessed on 25 December 2021).
60. Inoue, F.; Visker, J.; Jourdain, A.; Moeller, B.; Yokoyama, K.; Peng, L.; Kosemura, D.; Wolf, I.D.; Rebibis, K.J.; Miller, A.; et al. Edge Trimming for Wafer-to-Wafer 3D Integration. In Proceedings of the Materials for Advanced Metallization (MAM), Brussels, Belgium, 13–20 March 2016; pp. 83–84.
61. Aoki, T.; Hirasawa, M.; Izunome, K.; Ohba, T. Development of Novel Bevel Profile for Wafer-level Stacking Technology. In Proceedings of the International Conference on Electronics Packaging (ICEP), Tokyo, Japan, 12–14 May 2021; pp. 123–124.
62. Mills, M.E.; Townsend, P.; Castillo, D.; Martin, S.; Achen, A. Benzocyclobutene (DVS-BCB) polymer as an interlayer dielectric (ILD) material. *Microelectron. Eng.* **1997**, *33*, 327–334. [CrossRef]
63. Araki, N.; Maetani, S.; Kim, Y.S.; Hirota, T.; Nakamura, T.; Ohba, T. Material Optimization of Permanent and Temporary Adhesives for Wafer-level Three-dimensional Integration. In Proceedings of the IEEE 69th Electronic Components and Technology Conference (ECTC), Orlando, FL, USA, 3–30 June 2020; pp. 56–61.
64. Edelstein, D.; Heidenreich, J.; Goldblatt, R.; Cote, W.; Uzoh, C.; Lustig, N.; Roper, P.; McDevittt, T.; Motsifft, W.; Simon, A.; et al. Full Copper Wiring in a Sub-0.25 μm CMOS ULSI Technology. In Proceedings of the IEEE International Electron Devices Meeting (IEDM), Washington, DC, USA, 7–10 December 1997; pp. 773–776.
65. Kitada, H.; Maeda, N.; Fujimoto, K.; Mizushima, Y.; Nakata, Y.; Nakamura, T.; Ohba, T. Diffusion Resistance of Low Temperature Chemical Vapor Deposition Dielectrics for Multiple Through Silicon Via on Bumpless Wafer-on-Wafer Technology. *Jpn. J. Appl. Phys.* **2011**, *50*, 05ED02. [CrossRef]
66. Diehl, D.; Kitada, H.; Maeda, N.; Fujimoto, K.; Ramaswami, S.; Sirajuddin, K.; Yalamanchili, R.; Eaton, B.; Rajagopalan, N.; Ding, R.; et al. Formation of TSV for the Stacking of Advanced Logic Devices Utilizing Bumpless Wafer-on-Wafer Technology. *Microelectron. Eng.* **2011**, *92*, 3–8. [CrossRef]
67. Kitada, H.; Maeda, N.; Fujimoto, K.; Suzuki, K.; Kawai, A.; Arai, K.; Suzuki, T.; Nakamura, T.; Ohba, T. Stress Sensitivity Analysis on TSV Structure of Wafer-on-a-Wafer (WOW) by the Finite Element Method (FEM). In Proceedings of the IEEE Proceedings of International Interconnect Technology Conference (IITC), Sapporo, Japan, 1–3 June 2009; pp. 107–109.
68. Ohba, T. Wafer-Level Three-Dimensional Integration Using Bumpless Interconnects and Ultra-thinning. In *3D Integration in VLSI Circuits*; Sakuma, K., Iniewski, K., Eds.; CRC Press: Boca Raton, FL, USA, 2018; pp. 86–210.
69. Tsai, Y.-C.; Lee, C.-H.; Chang, H.-C.; Liu, J.-H.; Hu, H.-W.; Ito, H.; Kim, Y.S.; Ohba, T.; Chen, K.-N. Electrical Characteristics and Reliability of Wafer-on-Wafer (WOW) Bumpless Through-Silicon Via. *IEEE Trans. Electron Device* **2021**, *68*, 3520–3525. [CrossRef]
70. Frank, T.; Moreau, S.; Chappaz, C.; Leduc, P.; Arnaud, L.; Thuaire, A.; Chery, E.; Lorut, F.; Anghel, L.; Poupon, G. Reliability of TSV interconnects: Electromigration, thermal cycling, and impact on above metal level dielectric. *Microelectron. Rel.* **2013**, *53*, 17–29. [CrossRef]
71. Akamatsu, T.; Tadaki, S.; Yamazaki, K.; Kitada, H.; Sakuyama, S. Study of chip stacking process and electrical characteristic evaluation of Cu pillar joint between chips including TSV. In Proceedings of the IEEE Electronic Components and Technology Conference (ECTC), Las Vegas, NV, USA, 31 May–3 June 2016; pp. 1827–1833.
72. Funaki, T.; Satake, Y.; Kobinata, K.; Hsiao, C.-C.; Matsuno, H.; Abe, S.; Kim, Y.S.; Ohba, T. Miniaturized 3D Functional Interposer using Bumpless Chip-on-Wafer (COW) Integration with Capacitors. In Proceedings of the IEEE Electronic Components and Technology Conference (ECTC), San Diego, CA, USA, 1 June–4 July 2021; pp. 185–190.
73. Mizushima, Y.; Kitada, H.; Uchibori1, C.J.; Maeda, N.; Kodama, S.; Kim, Y.S.; Fujimoto, K.; Yoshimi, S.; Nakamura, T.; Ohba, T. Impacts of Thermo-Mechanical Stresses on Bumpless Chip in Stacked Wafer Structure. *Jpn. J. Appl. Phys.* **2013**, *52*, 05FE01. [CrossRef]
74. Satake, Y.; Funaki, T.; Tabata, K.; Kobinata, K.; Kim, Y.S.; Ohba, T. Development of Functional Interposer Using Bumpless Chip-on-Wafer. In Proceedings of the International Conference on Solid State Devices and Materials (SSDM), Virtual Conference, 27–30 September 2020; pp. 119–120.
75. Kim, Y.S.; Kodama, S.; Mizushima, Y.; Araki, N.; Hsiao, C.; Chang, H.; Lin, C.; Ohba, T. Optimization of Via Bottom Cleaning for Bumpless Interconnects and Wafer-On-Wafer (WOW) Integration. In Proceedings of the IEEE Electronic Components and Technology Conference (ECTC), San Diego, CA, USA, 29 May–1 June 2018; pp. 1962–1963.
76. Mead, C.; Rem, M. Minimum Propagation Delays in VLSI. *IEEE J. Solid-State Circuits* **1982**, *17*, 773–775. [CrossRef]
77. Kang, S.-M.; Leblebici, Y. *CMOS Digital Integrated Circuits*, 2nd ed.; McGraw-Hill: New York, NY, USA, 1999.
78. Martin, K. *Digital Integrated Circuit Design*; Oxford University Press: Oxford, UK, 2000.
79. Dennard, R.H. Field-Effect Transistor Memory. U.S. Patent 3,387,286, 4 June 1986.
80. Masuoka, F. Semiconductor Memory Device. U.S. Patent 4,437,174, 13 March 1984.
81. Masuoka, F.; Momodomi, M.; Iwata, Y.; Shirota, R. New Ultra High Density EPROM and Flash EEPROM with NAND Structure Cell. In Proceedings of the IEEE International Electron Devices Meeting (IEDM), Washington, DC, USA, 6–9 December 1987; pp. 552–555.

82. Lee, D.U.; Kim, K.W.; Kim, H.; Kim, J.Y.; Park, Y.J.; Kim, J.H.; Kim, D.S.; Park, H.B.; Shin, J.W.; Cho, J.H.; et al. A 1.2V 8Gb 8-Channel 128GB/s High-Bandwidth Memory (HBM) Stacked DRAM with Effective Microbump I/O Test Methods Using 29nm Process and TSV. In Proceedings of the IEEE International Solid-State Circuits Conference (ISSCC), San Fransisco, CA, USA, 9–13 February 2014; pp. 432–433.
83. Lee, J.C.; Kim, J.; Kim, K.W.; Ku, Y.J.; Kim, D.S.; Jeong, C.; Yun, T.S.; Kim, H.; Cho, H.S.; Kim, Y.O.; et al. A 1.2V 64Gb 8-Channel 256GB/s HBM DRAM with Peripheral-Base-Die Architecture and Small-Swing Technique on Heavy Load Interface. In Proceedings of the IEEE International Solid-State Circuits Conference (ISSCC), San Fransisco, CA, USA, 31 January–4 February 2016; pp. 318–319.
84. Sohn, K.; Yun, W.-J.; Oh, R.; Oh, C.-S.; Seo, S.-Y.; Park, M.-S.; Shin, D.-H.; Jung, W.-C.; Shin, S.-H.; Ryu, J.-M.; et al. A 1.2V 20nm 307GB/s HBM DRAM with At-Speed Wafer-Level I/O Test Scheme and Adoptive Refresh Considering Temperature Distribution. In Proceedings of the IEEE International Solid-State Circuits Conference (ISSCC), San Fransisco, CA, USA, 31 January–4 February 2016; pp. 316–317.
85. Cho, J.H.; Kim, J.; Lee, W.Y.; Lee, D.U.; Kim, T.K.; Park, H.B.; Jeong, C.; Park, M.-J.; Baek, S.G.; Choi, S.; et al. A 1.2V 64Gb 341GB/s HBM2 Stacked DRAM with Spiral Point-to-Point TSV Structure and Improved Bank Group Data Control. In Proceedings of the IEEE International Solid-State Circuits Conference (ISSCC), San Fransisco, CA, USA, 11–15 February 2018; pp. 208–209.
86. Oh, C.-S.; Chun, K.C.; Byun, Y.-Y.; Kim, Y.-K.; Kim, S.-Y.; Ryu, Y.; Park, J.; Kim, S.; Cha, S.; Shin, D.; et al. A 1.1V 16GB 640GB/s HBM2E DRAM with a Data-Bus Window-Extension Technique and a Synergetic On-Die ECC Scheme. In Proceedings of the IEEE International Solid-State Circuits Conference (ISSCC), San Francisco, CA, USA, 16–20 February 2020; pp. 330–331.
87. Lee, D.U.; Cho, H.S.; Kim, J.; Ku, Y.J.; Oh, S.; Kim, C.D.; Kim, H.W.; Lee, W.Y.; Kim, T.K.; Yun, T.S.; et al. A 128Gb 8-High 512GB/s HBM2E DRAM with a Pseudo Quarter Bank Structure, Power Dispersion and an Instruction-Based At-Speed PMBIST. In Proceedings of the IEEE International Solid-State Circuits Conference (ISSCC), San Francisco, CA, USA, 16–30 February 2020; pp. 334–335.
88. Jun, H.; Cho, J.; Lee, K.; Son, H.-Y.; Kim, K.; Jin, H.; Kim, K. HBM (High Bnadwidth Memory) DRAM Technology and Architecture. In Proceedings of the 2017 IEEE International Memory Workshop (IMW), Monterey, CA, USA, 14–17 May 2017.
89. Ndip, I.; Curran, B.; Löbbicke, K.; Guttowski, S.; Reichl, H.; Lang, K.-D.; Henke, H. High-Frequency Modeling of TSVs for 3-D Chip Integration and Silicon Interposers Considering Skin-Effect, Dielectric Quasi-TEM and Slow-Wave Modes. *IEEE Trans. Compon. Packag. Manufacuturing Technol.* **2011**, *1*, 1627–1642. [CrossRef]
90. Weis, C.; Jung, M.; Naji, O.; When, N.; Santos, C.; Vivet, P.; Hansson, A. Thermal Aspects and high-Level Explorations of 3D stacked DRAMs. In Proceedings of the 2015 IEEE Computer Society Annual Symposium on VLSI, Montpellier, France, 8–10 July 2015; pp. 609–614.
91. Ryoson, H.; Fujimoto, K.; Ohba, T. A Design Guide of Thermal Resistance down to 30% for 3D multi-stack devices. In Proceedings of the 2017 International Conference on Electronics Packaging (ICEP), Yamagata, Japan, 19–22 April 2017; pp. 522–525.
92. Sakui, K.; Ohba, T. High Bandwidth Memory (HBM) and High Bandwidth NAND (HBN) with the Bumpless TSV Technology. In Proceedings of the IEEE 2019 International 3D Systems Integration Conference, Sendai, Japan, 8–10 October 2019. 3DIC2019.4005.
93. Sakui, K.; Ohba, T. Sophisticated Architecture for High Bandwidth Memory (HBM) and High Bandwidth NAND (HBN) with the Bumpless TSV Technology. In Proceedings of the 29th Materials for Advanced Metallization Conference, Grenoble, France, 16–18 November 2020.
94. Tanaka, H.; Kido, M.; Yahashi, K.; Oomura, M.; Katsumata, R.; Kito, M.; Fukuzumi, Y.; Sato, M.; Nagata, Y.; Matsuoka, Y.; et al. Bit Cost Scalable Technology with Punch and Plug Process for Ultra High Density Flash Memory. In Proceedings of the 2007 Symposium on VLSI Technology, Kyoto, Japan, 12–14 June 2007; pp. 14–15.
95. Sakui, K. Semiconductor Memory Device and Memory System. U.S. Patent 6,594,169B2, 15 July 2003.
96. Maleville, C.; Asper, B.; Poumeyrol, T.; Moricean, H.; Bruel, M.; Auberton-Herve, A.J.; Barge, T.; Metral, F. *Silicon-on Insulator and Devices VII*; Hemment, P.L.F., Cristoloveanu, S., Izumi, K., Houston, T., Wilson, S., Eds.; Electrochem. Soc.: Pennington, NJ, USA, 1996; p. 34.
97. Kagawa, Y.; Fujii, N.; Aoyagi, K.; Kobayashi, Y.; Nishi, S.; Todaka, N.; Takeshita, S.; Taura, J.; Takahashi, H.; Nishimura, Y.; et al. Novel stacked CMOS image sensor with advanced Cu2Cu hybrid bonding. In Proceedings of the IEEE International Electron Devices Meeting (IEDM), San Francisco, CA, USA, 3–7 December 2016; pp. 208–2011. [CrossRef]
98. Natarajan, S. A 32 nm logic technology featuring 2nd-generation high-k + metal-gate transistors, enhanced channel strain and 0.171 μm^2 SRAM cell size in a 291Mb array. In Proceedings of the IEEE International Electron Devices Meeting (IEDM), San Francisco, CA, USA, 15–17 December 2008; pp. 941–943.
99. Sugatani, S.; Chujo, N.; Sakui, K.; Ryoson, H.; Nakamura, T.; Ohba, T. Vertically Replaceable Memory Block Architecture for Stacked DRAM Systems by Wafer-on-Wafer (WOW) Technology. *IEEE Trans. Electron Devices* **2020**, *67*, 4606–4610. [CrossRef]
100. Sugatani, S.; Chujo, N.; Sakui, K.; Ryoson, H.; Nakamura, T.; Ohba, T. Bumpless Build Cube (BBCube) using Wafer-on-Wafer (WOW) Technology with 3D-manner Redundancy Scheme. In Proceedings of the International Conference on Solid State Devices and Materials (SSDM), Virtual Conference, 27–30 September 2020.
101. Jun, H.; Nam, S.; Jin, H.; Lee, J.-C.; Park, Y.J.; Lee, J.J. High-Bandwidth Memory (HBM) Test Challenges and Solutions. *IEEE Des. Test* **2017**, *34*, 16–25. [CrossRef]
102. Toroflux Paradox: Making Things (Dis)appear with Math. Available online: https://www.youtube.com/watch?v=VK7XR-wlpAk (accessed on 29 December 2021).

103. O'Connor, M.; Chatterjee, N.; Lee, D.; Wilson, J.; Agrawal, A.; Keckler, S.W.; Dally, W.J. Fine-Grained DRAM: Energy-Efficient DRAM for Extreme Bandwidth Systems. In Proceedings of the 50th Annual IEEE/ACM International Symposium on Microarchitecture (MICRO), Cambridge, MA, USA, 14–18 October 2017; pp. 41–54.
104. Available online: https://www.youtube.com/watch?v=1r_tSjZCNzg (accessed on 29 December 2021). (In Japanese)
105. Price, D.W.; Robinson, J.; Bhatti, N.; VonDenHoff, M.; Lim, A.; Rathert, R.J.; Sherman, K.; Sutherland, D.; Cappel, R.; Meenakshisundaram, G. Application of Inline Defect Part Average Testing (I-PAT) to Reduce Latent Reliability Defect Escapes. Available online: https://www.kla-tencor.com/documents/KLA_I-PAT_AEC_October_2018.pdf (accessed on 6 February 2020).
106. Clark, R. Advanced Process Technologies Required for Future Scaling and Devices, Short Course 1. In Proceedings of the IEEE Symposium on VLSI Technology, Kyoto, Japan, 9–14 June 2019.
107. Fatemeh, T.; Siavash, E.; Shujuan, W.; Kambiz, V. Analysis of Critical Thermal Issues in 3D Integrated Circuits. *Int. J. Heat Mass Transf.* **2016**, *97*, 337–352.
108. Ankur, J.; Robert, J. Analytical and Numerical Modeling of Thermal Performance of Three-Dimensional Integrated Circuits. *IEEE Trans. Compon. Packag. Technol.* **2010**, *33*, 56–63.
109. Gian, L.; Banit, A.; Navin, S.; Sheng-Chin, L.; Timothy, S.; Kaustav, B. A thermaly-Aware Performance Analysis of Vertically Integrated (3-D) Processor-Memory Hierarchy. In Proceedings of the Design Automation Conference (DAC), San Francisco, CA, USA, 24–28 July 2006. Available online: https://www.cs.ucsb.edu/~{}sherwood/pubs/DAC-3dmodel.pdf (accessed on 25 December 2021).
110. Chatterjee, S.; Cho, M.; Rao, R.; Mukhopadhyay, S. Impact of Die-to-Die Thermal Coupling on the Electrical Characteristics of 3D Stacked SRAM Cache. In Proceedings of the 28th IEEE Semiconductor Thermal Measurement and Management Symposium (SEMI-THERM), San Jose, CA, USA, 18–22 March 2012.
111. Matsumoto, K.; Ibaraki, S.; Sueoka, K.; Sakuma, K.; Kikuchi, H.; Orii, Y.; Yamada, F.; Fujihira, K.; Takamatsu, J.; Kondo, K. Thermal Design Guidelines for a Three-dimensional (3D) Chip Stack, Including Cooling Solutions. In Proceedings of the 29th IEEE Semiconductor Thermal Measurement and Management Symposium (SemiTherm 2013), San Jose, CA, USA, 17–21 March 2013.
112. Ohba, T.; Nakamura, T. 3D stacked integration technologies: Overview and future prospects, Oyo Buturi. *Jpn. Soc. Appl. Phys.* **2020**, *89*, 75–81. (In Japanese)

electronics

MDPI

Article

Assessing the Role of Program Suspend Operation in 3D NAND Flash Based Solid State Drives

Cristian Zambelli [1,*,†], Lorenzo Zuolo [2,†], Antonio Aldarese [2,†], Salvatrice Scommegna [2,†], Rino Micheloni [2,†,‡] and Piero Olivo [1,†]

1 Dipartimento di Ingegneria, Università degli Studi di Ferrara, Via G. Saragat 1, 44122 Ferrara, Italy; piero.olivo@unife.it
2 Flash Signal Processing Labs, Microchip Corp., Via Torri Bianche 1, 20871 Vimercate, Italy; lorenzo.zuolo@microchip.com (L.Z.); antonio.aldarese@microchip.com (A.A.); salvatrice.scommegna@microchip.com (S.S.); rino.micheloni@microchip.com (R.M.)
* Correspondence: cristian.zambelli@unife.it; Tel.: +39-0532-974-993
† These authors contributed equally to this work.
‡ Current address: Freelance Consultant, Via Roma 23, 22010 Moltrasio, Italy.

Abstract: 3D NAND Flash is the preferred storage medium for dense mass storage applications, including Solid State Drives and multimedia cards. Improving the latency of these systems is a mandatory task to narrow the gap between computing elements, such as CPUs and GPUs, and the storage environment. To this extent, relatively time-consuming operations in the storage media, such as data programming and data erasing, need to be prioritized and be potentially suspendable by shorter operations, like data reading, in order to improve the overall system quality of service. However, such benefits are strongly dependent on the storage characteristics and on the timing of the single operations. In this work, we investigate, through an extensive characterization, the impacts of suspending the data programming operation in a 3D NAND Flash device. System-level simulations proved that such operations must be carefully characterized before exercising them on Solid State Drives to eventually understand the performance benefits introduced and to disclose all the potential shortcomings.

Keywords: program suspend; 3D NAND Flash; Solid State Drives

Citation: Zambelli, C.; Zuolo, L.; Aldarese, A.; Scommegna, S.; Micheloni, R.; Olivo, P. Assessing the Role of Program Suspend Operation in 3D NAND Flash Based Solid State Drives. *Electronics* **2021**, *10*, 1394. https://doi.org/10.3390/electronics10121394

Academic Editor: Changhwan Shin

Received: 10 May 2021
Accepted: 7 June 2021
Published: 10 June 2021

1. Introduction

The data storage in enterprise scenarios, including High Performance Computing (HPC) and cloud-based services, requires the low latency and the high throughput that only Solid State Drives (SSD) architectures can deliver [1]. 3D NAND Flash-based SSDs are the preferred solution due to the offered large storage density, the lower total cost of ownership (TCO), and the inherent higher reliability with respect to Hard Disk Drives (HDDs) [2].

Among the many parameters of the drive that exercise the reliability/performance trade-off [3], one of the key aspects that is perceived by the users of the SSD platforms concerns their Quality of Service (QoS) [4–6]. In a generic computing architecture, the host system (i.e., the CPU) issues data requests to SSDs. Every drive includes a dedicated processor with a specific host interface (i.e., the controller) and a set of 3D NAND Flash chips organized in communication channels (see Figure 1). The time taken by the SSD to service the host falls under the QoS umbrella, which is defined following specific criteria that also encompass the error rate, the transmission delay, the jitter, and the network availability [7].

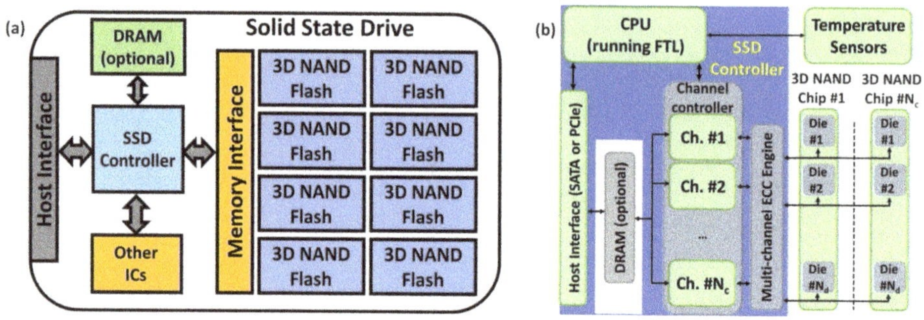

Figure 1. Layout of the components in an SSD (**a**) and an overview of its internal architecture (**b**). Reprinted with permission from [8] under Creative Commons License 4.0 (CC-BY).

From the host standpoint, there are two kinds of data requests, namely write and read. The former gives a freedom degree to the SSD controller in choosing where to physically store the data inside the pool of available 3D NAND Flash chips integrated in the drive. A memory programming operation (i.e., write) features longer latency compared to a read one (1–2 ms versus 75–100 µs [9]); however, since the data to be written can be cached inside DRAM buffers on the SSD the service time (the write QoS) can be reduced [1].

The latter is translated by the SSD controller in a physical 3D NAND Flash address that is exploited in a memory read operation that returns the content of that location the to the host. However, if that address belongs to a device that is already committed in a program operation, the SSD will suffer from prolonged service time and expose large read latency tails [10].

To further complicate the picture, we must remember the 3D NAND Flash characteristics: read and write operation granularity is at the page level (from 4 kB to 16 kB in state-of-the-art devices [11]); however, since the in-place data update operation is prohibited, there is the need of a block erase operation (sized several pages and lasting from 10 ms to 15 ms [9]) that deletes all the invalid data to make room for the updated ones. This free space-reclaim operation, also known as garbage collection (GC) [12,13], can be very long if it encompasses many blocks (several hundreds of milliseconds) and is managed by the SSD controller during the idle times. Read requests can occur during GC, with evident impairment of the QoS in enterprise scenarios where applications rarely provide idle periods. Currently, one of the greatest challenges in the SSD context that must be faced for implementing interactive online services generating thousands of drive accesses [14] is in the experienced read QoS improvement. Either caching techniques or GC preemption cannot provide significant help. A method to handle the QoS-boost shortcoming materializes in the suspension of an ongoing program or an erase operation and the resumption at a later point in time when all prioritized read requests on the memory are serviced [10]. Starting from NAND Flash memory products developed with planar technology [15], vendors provide dedicated commands to suspend the erase operation [14] since this was the one with the highest latency. However, the transition to the 3D counterpart and the increased number of bits stored per cell evidenced that program latency requires suspension as well [9,16].

In this work, we focus on the role of the program operation suspend since we found many unexplored points especially in the design space of 3D NAND Flash-based SSDs. Among them, the operation timing detail and reliability are key features. The former dictates the SSD performance metrics, like the throughput and latency. The latter impacts the error management strategies since the programming operation suspension might alter the memory Raw Bit Error Rate (RBER) metric [17] in non-suspended portions of the memory. Furthermore, since in SSD architectures, there are many 3D NAND Flash devices that are operated simultaneously, the controller must be aware of the potential power consumption surge.

The contributions of this paper are twofold:

1. We perform an electrical characterization with respect to the operation timing and reliability on a commercial Triple Level Cell (TLC) 3D NAND Flash technology for enterprise scenarios that includes the program suspend in its command set.
2. We evaluate, through a co-simulation framework that accounts for the measured 3D NAND Flash program suspend characteristics, the program suspend's impact on the figures of merit of an SSD, including the throughput, latency, QoS, and power consumption. Simulations are performed with synthetic benchmarks using different read/write ratios and different micro-architectural drive parameters for design space exploration.

2. Related Works

The interest in developing solutions for preempting long latency operations with shorter ones has been widely explored in the storage scenario. In Reference [18], a new scheduler was proposed to achieve load balancing on the NAND Flash resources of an SSD, thus, granting prioritization mechanisms that may reduce the latency tails. A similar concept was considered in [19] to discuss how read latency fluctuations in Flash-based storage can be reduced using preemptible programs and erases.

Storage technologies different from SSDs also have uses for this topic. In Reference [20], the authors developed a technique for HDDs in which the host input/output transactions are split in small requests to the drive to guarantee the servicing of high priority requests. In Reference [21], the authors studied write cancellation techniques on a non-volatile memory technology with similar NAND Flash characteristics in terms of the read and write latency disparity, namely the Phase Change Memory (PCM) [22].

In addition to those seminal works, specific studies were conducted in the context of the program and erase suspension dynamics either integrating new peripheral circuits in planar NAND Flash devices or at the SSD firmware level. In Reference [14], the authors proposed two practical erase suspension mechanisms and analyzed the trade-offs between the two mechanisms introducing a timeout-based switching policy between the two. A significant reduction in read tail latency is shown on production-grade SSDs. In Reference [10], the first contribution on the program operation suspension appeared; however, its role in isolation with respect to erase was considered only for a few specific cases.

Two strategies for suspension, namely Inter Phase Suspension and Intra Phase Cancellation were proposed showing a direct outcome on SSD read latency. In Reference [23], the authors investigated finer granular read operations for 3D NAND Flash, starting from the circuit-level implementation to memory architecture. Using a novel Single-Operation-Multiple-Location read operation, they can perform several smaller read operations to different locations simultaneously, so that multiple requests can be serviced in parallel. Several patents also dealt with the program suspension topic to be implemented in NAND Flash products [24,25].

All the studies presented so far were focused primary on the steps to implement the operation suspend on Flash storage paradigms that were mostly Single Level Cell (SLC) or Multi Level Cell (MLC). As far as the SSD QoS is concerned, there is an assessment based on the sole read latency without providing additional details concerning the impact of the suspension on the memory reliability and on the SSD's power consumption. Further, there are no characterization studies concerning 3D NAND Flash products in which the suspension is part of the command set [26] exposed by the memory to the drive.

To the best of our knowledge, this is the first work to investigate the effect of the program operation suspend by characterizing a commercial TLC 3D NAND Flash product integrated in enterprise-class SSD by evaluating the implications on the SSD's reliability and performance. In this work, we propose general memory-driven design methodologies and algorithms that are functional in the firmware design and can be followed as a general guide for other existing storage products based on 3D NAND Flash.

3. Background and Methods Applied

3.1. 3D NAND Flash Program Suspend

The program operation was implemented in 3D NAND Flash memories by following two consecutive steps: first, the data to be programmed in a specific memory location (i.e., a page) are transferred from the host (in SSDs, the channel controller is responsible for managing this task) and loaded in an on-memory structure called *page-buffer* [27]; second, an iterative algorithm based on the Incremental Step Pulse Program (ISPP) and verify concept [28] is applied on the target Flash page. In a TLC device, like the one whose structure is depicted in Figure 2a, the data load phase associated to a single wordline of a memory block actually comprises three different page loads in sequence (lower, center, and upper pages) and separated by a setup time (i.e., the page buffer setup time). The ISPP algorithm is then started and ends when all the memory cells in each TLC page reach a desired amount of charge or a threshold voltage (see Figure 2b). The entire program operation lasts several milliseconds as indicated in [9] for state-of-the-art technologies.

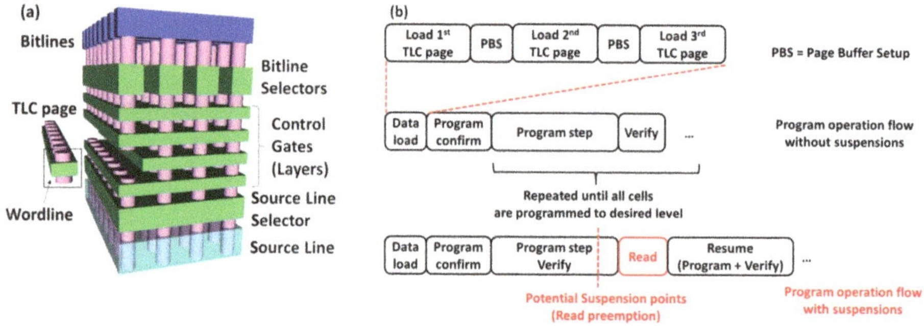

Figure 2. (**a**) The TLC 3D NAND Flash architecture. Reprinted with permission from [29] under Creative Commons License 4.0 (CC-BY). (**b**) An overview of the program operation phase without and with suspension.

Enabling program suspension implies that the ISPP algorithm can be interrupted at any time. The 3D NAND Flash memories assume that the program iteration and verification are a single atomic operation. This is required to avoid the charge loss [30] during the suspension triggering an additional program pulse that could shift the cells to an unwanted threshold voltage level. The resumption from the suspension starts then with a program pulse. The programming voltage before the suspension is stored in the on-chip control logic to avoid the repetition of the same pulse when exiting from the suspension phase [10].

The suspension paradigm dwells in a perfect understanding of the page-buffer role. This on-chip peripheral circuit is shared between the program and the read operation. In the former, it contains the data to be written, and, in the latter, it stores the memory retrieved content from a page before transferring it to the host via the channel controller. Since the controller is responsible for managing all the data transfers to the memory, it can be naively used for storing a copy of the data to write until the program operation is finished.

In the case of a suspension request to preempt a read, it will serve it and then re-send the program data upon program resumption. However, this method performs poorly from the timing standpoint since it overloads the communication resources of the channel controller. In high performance Flash devices, it is embodied a *shadow buffer* that contains an on-chip replica of the page buffer and that loads itself in the page buffer upon a program request [10].

3.2. Characterization and Simulation Tools

The program suspend operation has been characterized on an off-the-shelf sub-100 layer TLC 3D NAND Flash chip including this feature in its command set. The experimental setup described in [31,32] was exploited in this work. Such an electrical characterization tool interfaces with the memory at a data rate of 400 MT/s and is capable of extracting timing information with 1 µs precision about the on-going program operation by monitoring a dedicated pin of the memory chip (the Ready/Busy pin as the one described in [16]).

The overall program time t_{PGM} is the main parameter that we investigated in all the characterizations. As shown in Figure 3a, 3D NAND Flash devices exhibit an intrinsic variability of this parameter. We also performed a measurement to retrieve the inherent variability of t_{PGM} by re-programming the same wordline multiple times. Tests demonstrated that, if the number of re-programming cycles was in a range that did not alter the programming characteristics of the wordline, the t_{PGM} stayed almost constant. This setup can also extract the number of Fail Bits Count (FBC) on 16 kB pages and, therefore, retrieve reliability information from the memory.

Finally, the characterization system exploits two current probes to infer the memory power consumption (at a given supply voltage) from the VDD_{core} (i.e., the power supply for the 3D NAND Flash peripheral circuitry) and from the VDD_Q (i.e., the power supply from the memory I/O pins) supplies. For confidentiality reasons, we can only state that VDD_{core} is between 2.7 V and 3.6 V, whereas VDD_Q is between 1.2 V and 1.8 V. Figure 3b shows an example (a zoom) of an extracted power trace from a 3D NAND Flash chip during a program suspend operation. In the figure, we can appreciate the single iterations of the program algorithm and the idle consumption during the suspension time since, in this example, no read operations were preempted. As we considered a TLC memory in this study, we collected different power traces for each page type and operation.

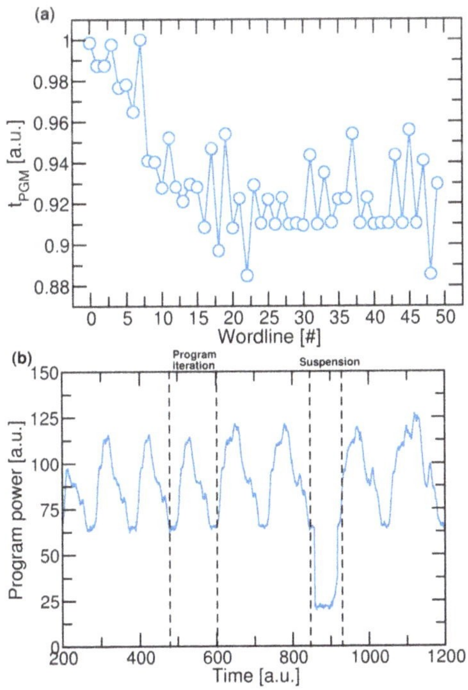

Figure 3. (**a**) t_{PGM} measurements performed on 50 randomly sampled wordlines in a 3D NAND Flash block. (**b**) Power consumption measured during a program suspend operation.

To explore the QoS, performance, and power consumption features of SSD architectures integrating 3D NAND Flash memories with program suspend capabilities, we exploited the SSDExplorer co-simulator [33]. The tool allows a fine-grained design space exploration of a drive by allowing modifications of its micro-architectural parameters, including the command queues, interaction mechanisms with the host system, and error recovery policies. The simulations performed considered the 3D NAND Flash timing characteristics as well as the power consumption during the serviced program and read operations through a back annotation process as shown in Figure 4.

The assessment of the program suspend role was performed through a microbenchmark that is common in SSDs latency and QoS evaluations [34], namely a 4-kB-aligned workload. A queue depth (QD) either equal to 16 or to 64 was considered with a different mixture of read and write transactions. All the SSD architectural parameters considered in this work are presented in Table 1 and mimic those of an enteprise-class SSD [35].

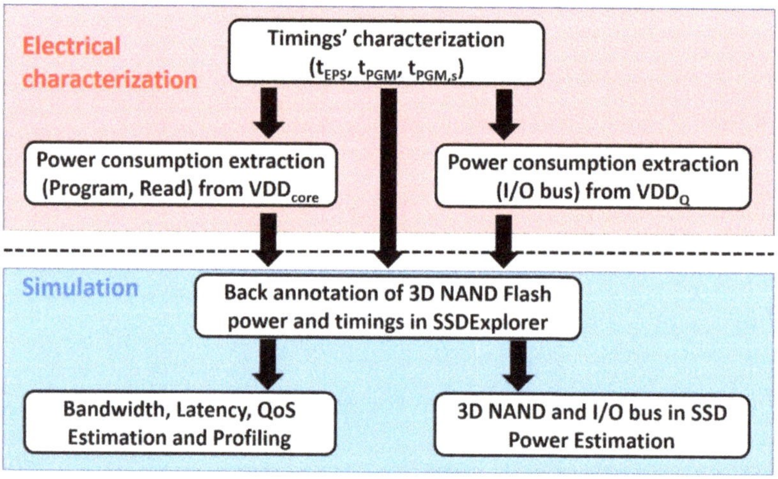

Figure 4. Electrical characterization of TLC 3D NAND Flash samples and SSD co-simulation flow adopted in this work for the program suspend study.

Table 1. 3D NAND Flash and SSD architectural parameters considered in the simulations.

Parameter	Value
Host interface	PCIe gen3 X4
Frame buffer size	1024
QD	16–32–64
Channels	16
Targets (number of Flash die per chip)	8
DRAM cache	64–512 MB
Error Correction Code (ECC)	Low Density Parity Check (LDPC)
Over-Provisioning	30%
Write Amplification Factor	2.4
GC threshold	10%
3D NAND Flash Storage Medium	Charge trapping
3D NAND Flash Storage paradigm	TLC
3D NAND Flash page	16 kB + parity
Layers	[48–100]
Pages per block	256
Blocks per die	2048

4. Exploring the Program Suspend in 3D NAND Flash

The assessment of the program suspend characteristics in a TLC 3D NAND Flash product used the test flow depicted in Figure 5. After the reception of the program confirm command, the memory starts the internal program algorithm (e.g., via ISPP) and applies the required program voltage bias to the cells to drive them to the desired threshold voltage level.

To avoid any topology-related effects that could alter the program characteristics [36], we programmed the TLC pages of the memory with a random pattern. Concurrently, we started a timer (i.e., *Delay1*) that represented the elapsed time before a read operation should preempt the program in-service. To allow fine-grained exploration, we considered several *Delay1* times ranging from 10% up to 100% using 10% steps of the t_{PGM} measured without suspension (for confidentiality reasons we cannot report the exact t_{PGM} value of the product).

When *Delay1* ended, we submitted a program suspend command that was processed by the memory with a time t_{EPS}, which represents the time to enter suspend mode. A read operation was then serviced. When all read data were available in the memory page buffer, we resumed the suspended program operation and concurrently started another timer (i.e., *Delay2*) that represents the elapsed time before servicing another read operation during the program.

Here, we also considered *Delay2* times ranging from 10% up to 100% of the t_{PGM}. At the end of the *Delay2*, we were free to service another preempting read operation. This last step involving the *Delay2* parameter, as evidenced in Figure 5, was repeated *N* times until the memory effectively reached the completion of the suspended program operation. This test flow was applied for all *Delay1* and *Delay2* combinations on 50 different wordlines and TLC pages.

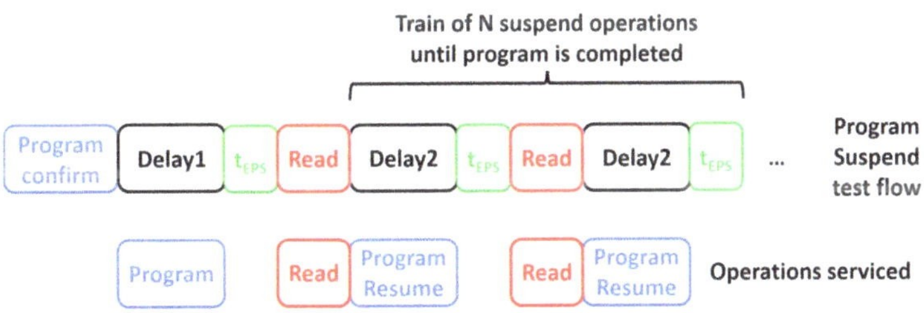

Figure 5. The program suspend test flow exploited in this work to explore theoperation characteristics.

4.1. Number of Suspend Operations

After executing the test flow described thus far, we extracted the actual number of program suspensions that the memory was servicing by exercising all the possible combinations of *Delay1* and *Delay2*. Figure 6 summarizes the results of this characterization. Lower delay times lead to a higher number of suspensions that can be serviced prior the completion of the program operation (up to 10 in our measurements).

Interestingly, due to the t_{PGM} variability experienced in all the wordlines of a 3D NAND Flash block, there was a slight variability in the number of sustainable program suspensions. If we pick two different wordlines and apply the same delay timing combination in the experiment, we are likely to observe a different number of issued suspend operations due to the different times required to reach the completion of the program phase.

We want to highlight that the results shown in Figure 6 are strongly affected by several factors: (i) the intrinsic variability of the t_{PGM} that makes possible for different wordlines to sustain a different number of suspends even when tested with the same *Delay1* and *Delay2* parameters; and (ii) the small statistics in the overall number of issuable suspends served by the memory. Both factors contribute to a non-monotone decreasing histogram of the number of issued suspend.

As a general rule, we found that, if both *Delay1* and *Delay2* assume a value higher than 60% of t_{PGM}, there is room for servicing only one program suspension.

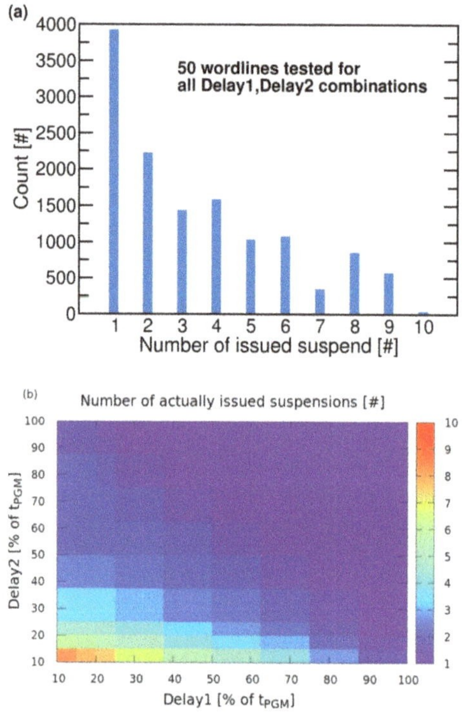

Figure 6. (**a**) Histogram of the actual number of issued suspensions considering all the delay combinations and for all 50 wordlines in the study. (**b**) Number of program suspensions as a function of *Delay1* and *Delay2*.

4.2. Time to Enter in Suspension

The t_{EPS} was characterized to assess the impact of the suspend operations on the overall programming time as t_{EPS} is the price to pay when a suspend operation is issued. This can be useful in assessing the timings required prior servicing a read operation. We performed two quantifications of the t_{EPS}: (i) considering the average value extracted among the 50 tested wordlines in a memory block; and (ii) considering the worst value that also represents the maximum stall time to expect before the read.

The characterization was performed for each combination of the *Delay1* and *Delay2* values exploited during the tests. Figure 7 demonstrates that, in general, low *Delay2* values provide a larger t_{EPS} time, although we found some specific combinations of the two delays in which both the average and worst t_{EPS} were larger than expected. The magnitude of the t_{EPS} was, in general, in a range of few tens of µs, which is 3–4% of the t_{PGM} measured without suspensions. That behavior could be ascribed to peculiar 3D NAND Flash internal algorithms in managing the suspension through multiple page buffers.

As speculated in [10], the memory completes the program pulse in the ISPP algorithm independently by the exact point where the user decides to suspend the operation. This is mandatory to avoid the presence of some memory cells in the page to program that receives incomplete pulses and, therefore, has poor control of their program dynamics. The goal is to provide, to all the memory cells, the same ISPP routine even if the iterations in the algorithm are interrupted by the suspension.

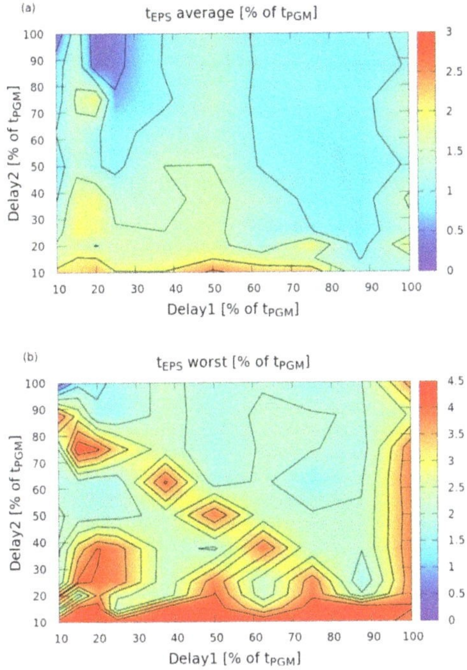

Figure 7. (**a**) The average of the t_{EPS} measurements performed on 50 randomly sampled wordlines in a 3D NAND Flash block. (**b**) The worst t_{EPS} in each measure.

4.3. Total Program Time

Suspending the program operation straightforwardly increases the overall program time depending on the number of performed suspensions. Naively, this relationship is assumed to be linear since the higher the suspension number is, the higher the time to complete the program. However, the results previously shown concerning the t_{PGM} variability in 3D NAND Flash wordlines and their impact on the number of suspend issued makes this relationship more complex.

Figure 8 shows the results of the characterization performed on the overall program time in the presence of suspensions ($t_{PGM,s}$) achieved through different *Delay1* and *Delay2* combinations. In general, the number of ISPP iterations for each cell to achieve a desired threshold voltage level is the same with or without suspension (the program dynamics of the cells is not altered by the suspension). Every time we suspend, we have to wait for the t_{EPS} time; therefore, the higher the number of suspensions issued is (occurring when the delay timings are low), the larger the overhead on $t_{PGM,s}$ imposed by that metric.

If both delay timings approach 100% of the t_{PGM}, the $t_{PGM,s}$ is more than twice as expected (t_{EPS} is also considered as part of the programming operation since it is time required to enter the suspension mode). However, this result was not achieved for all combinations. We observed that, if *Delay1* is relatively small (below 20% of t_{PGM}) and *Delay2* is higher than 70% of t_{PGM}, we suffer from an increased $t_{PGM,s}$ due to the early described t_{EPS} overhead and to an additional combination of factors.

We speculate that, if *Delay1* is small (in the range below 20% of t_{PGM}), the memory cells within the wordline we are programming are still in an unstable region of the ISPP algorithm where the relationship between the voltage applied for programming and the threshold voltage of the cells is not linear. For high *Delay2* values, an early retention loss of the cells may trigger additional ISPP iterations to reach a target threshold voltage. Both contributions can play a role in the overall $t_{PGM,s}$ increase.

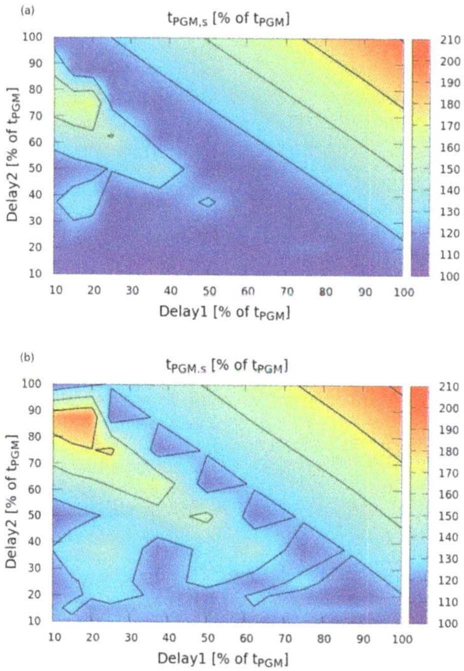

Figure 8. (**a**) The average of the overall programming time $t_{PGM,s}$. (**b**) The worst $t_{PGM,s}$ extracted from the 50 wordlines involved in the experiment. Both measurements are plotted as a function of different *Delay1* and *Delay2* combinations.

4.4. RBER in Suspended and Not Suspended Blocks

An additional characterization to perform on the program suspend test flow is related to the reliability of the wordlines belonging either to the memory block for which the program is suspended or to other blocks that are not involved by the suspend operation. When the program operation is performed on a specific memory block, all the other blocks in the memory undergo a retention period at a certain temperature (the higher the temperature is, the higher the impact on reliability). The effect of the program suspension is to increase the overall program time of a single page; therefore, we expect that the overall programming time of a block will increase accordingly as well as the retention time experienced by non-programmed blocks.

To better understand, we performed a test in which we programmed, exercising all the possible delay timings combinations, all the wordlines of a memory block. This test induced many program/erase cycles on the same block as, every time we chose a *Delay1* and *Delay2* combination, we reprogrammed the entire block. We then compared the RBER metric extracted before and after the experiment. At the same time, we extracted the RBER from a randomly chosen block in which no program operations were performed to check whether the program suspension could induce a reliability degradation in other, non-accessed portions of a 3D NAND Flash memory.

As shown in Figure 9a, the program suspension does not alter the reliability characteristics of the wordlines in the block where the operation took place, since the RBER stays the same before and after the experiment. Concerning the wordlines in the memory block (see Figure 9b) for which no programs are executed, we actually see a slight increase of the RBER from the beginning of the experiment; however, this is likely due to the retention loss that we are experiencing at room temperature and is well below the correction limit of the many Error Correction Code engines employed in SSDs [35].

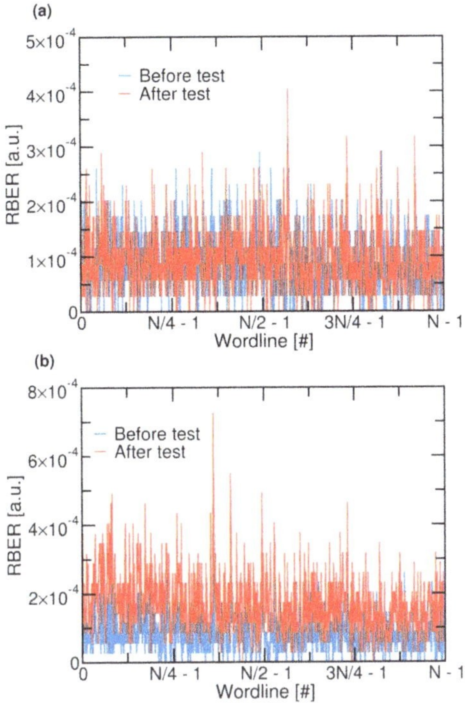

Figure 9. (**a**) The RBER measured in all the wordlines of a block where the program suspension took place. (**b**) The same measurements performed in a block where no program operation is issued.

Based on our experiments, we reached the conclusion that the program suspension did not induce specific reliability issues on the memory with the test conditions adopted. Possible solutions to reduce the RBER in non-suspended blocks could rely on state-of-the-art techniques ranging from advanced Error Correction Code policies (e.g., read retry and soft decoding) or by implementing periodic data refresh strategies to avoid retention-induced charge loss.

5. SSD Simulations to Understand the Role of Program Suspend

We performed a set of SSD simulations using the SSDExplorer platform to understand the impact at the system level of the program suspend operation in 3D NAND Flash memories by back annotating all the electrical characterization metrics presented so far into the simulation environment. As we cannot disclose the full details of the operation timings featured by the tested 3D NAND Flash product, we assume the generic values for the program, read, and erase operations as defined in Table 2, which were extracted from literature referring to a similar TLC 3D NAND Flash technology [9].

Table 2. The 3D NAND Flash operation timings and voltage intervals considered in the simulations.

Parameter	Value
t_{PGM}	2 ms
t_{READ}	80 µs
t_{ERS}	5 ms
VDD_{core}	[2.7–3.6] V
VDD_Q	[1.2–1.8] V

The specific values of t_{EPS} and $t_{PGM,s}$ shown during the characterization results are taken into account since they are expressed as in terms of relative deviation with respect to t_{PGM}, therefore, being applicable in this study case. The workload used as the input stimulus for the SSD to extract all the drive's performance metrics is either a 75% write −25% read or a 25% write −75% read according to the evaluation guidelines for suspension operations provided in [9].

In all the simulations performed, we compared three different use cases for the SSD: (i) the case in which the 3D NAND Flash memories integrated do not implement the program suspend functionality; (ii) the case where the program suspend is enabled and allows for a maximum of five suspensions to preempt an incoming read operation; and (iii) the same as in the previous case, but allowing up to 10 suspensions.

5.1. SSD Bandwidth and Latency

First, we extracted the achieved SSD's bandwidth measured in Input/Output operations per second (IOPS). This is a required measurement unit since we are dealing with a random workload. The results of Figure 10a show that the introduction of the program suspend operation increased the sustainable bandwidth irrelevant from the amount of write operations submitted to the drive (i.e., independent from the write/read ratio of the workload).

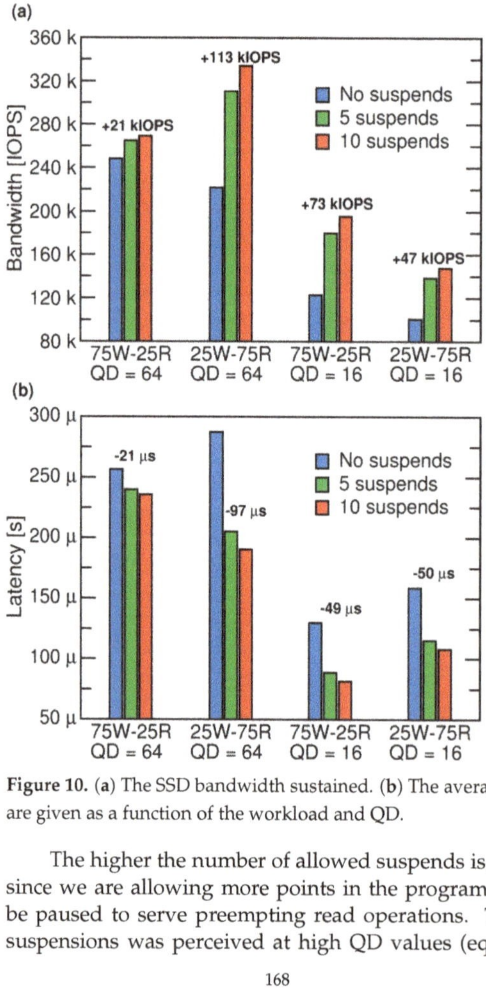

Figure 10. (**a**) The SSD bandwidth sustained. (**b**) The average total latency of the drive. Both metrics are given as a function of the workload and QD.

The higher the number of allowed suspends is, the higher the achieved bandwidth since we are allowing more points in the program operation where the operation can be paused to serve preempting read operations. The biggest advantage in using the suspensions was perceived at high QD values (equal to 64 in our study) and with a workload mostly dominated by the read operations. The achieved bandwidth increase was 113 kIOPS. Such behavior is ascribed to the fact that, in read-intensive workloads, there is a larger pool of preemptible operations and, therefore, enabling program suspension straightforwardly to allocate more servicing time.

The program suspension also provides, as expected from the theoretical standpoint, benefits in terms of the latency experienced by the drive. In Figure 10b, we report the SSD average latency calculated among all the read and write transactions involving the drive. Similarly, as observed earlier, the larger the amount of the allowable suspensions is, the more the experienced latency decreases as well as the performance benefit. For the QD = 64 case, in which the workload is read intensive, there is a latency decrease of 97 µs when up to 10 suspensions are allowed.

5.2. SSD Power Consumption

To understand the role of the program suspension on the SSD's power consumption, we investigated two different contributions in the drive: (i) the sole 3D NAND Flash memory modules constituting the storage media sub-system; (ii) the internal SSD I/O bus on which both the write and the read flows to and from the memories. As we considered a TLC memory in this study, we collected different power traces for each page type (i.e., lower, central, and upper pages) using our experimental setup applied on a TLC 3D NAND Flash product.

The assumption behind the calculation of the overall memory power consumption in drive is the following: when multiple memory chips are accessed in parallel on an SSD, the Kirchoff's current law (KCL) holds on the power supply of the drive so that the memory sub-system power consumption is the sum of the single power contributions [8]. The results in Figure 11a show one of the negative sides of the program suspension. When the 3D NAND Flash memories in the SSD allows for preempting read operations, it means that the memory chips are available to sustain more operations per second (as shown by the results in the previous section concerning the bandwidth), hence, being in a sort of "active" state drawing current from the power supply for a longer period.

The larger the number of operations served is, the higher is the power consumption (up to 400 mW increase in the QD = 64 25W-75R condition). A similar result can be appreciated in Figure 11b. Indeed, more suspensions in the program operation means that more read data are to be requested from the memory chips requiring the activation of the internal I/O bus of the drive for a longer period (up to 17 mW in the QD = 16 75W-25R condition).

The only counter-intuitive result is related to the power consumption extracted from simulations with a write intensive workload and QD = 64, in which the former metric displays an inverted trend with respect to all the other cases. We interpret this result as a combination of several factors. The case of QD = 64 represents a situation in which all the SSD resources (i.e., channels) are allocated to serve write/read transactions.

Write (program) operations in 3D NAND Flash devices consume more power than the read one, and, since the former operation is longer, there is a large probability, especially in the high QD scenario, that multiple 3D NAND Flash chips concurrently serve a write function, and therefore the total power consumption is the sum of each operation. By including program suspensions, we are deliberately reducing that probability and allowing a power consumption reduction of almost 1.5 W.

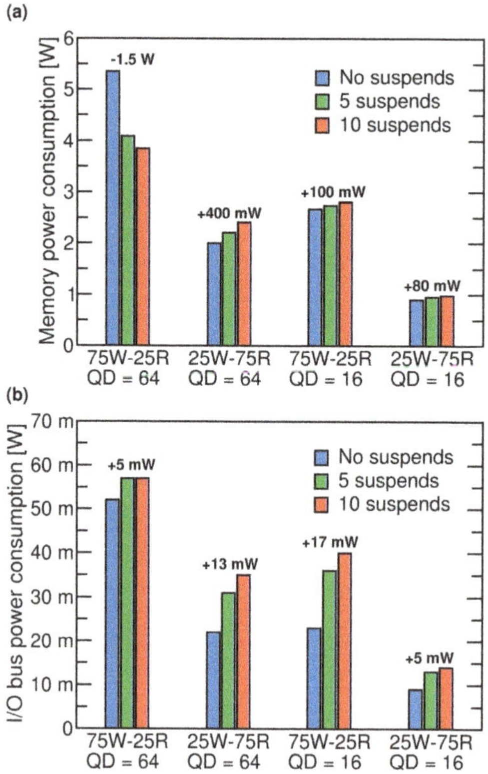

Figure 11. (**a**) The power consumption of the memory sub-system in the SSD. (**b**) The power consumption of the internal I/O bus in the drive.

5.3. *The Impact on QoS*

The QoS study can be performed in multiple ways since it depends on the definition of the QoS level and on the latency that is being profiled for reaching that. In this study, we considered, without a lack of generality, a QoS calculated at the 99.99% percentile of the latency Cumulative Distribution Function (CDF) [32] either considering the total transactions sustained by the drive (i.e., total QoS) or only considering the read transactions (i.e., read QoS), which are those benefiting from the program suspension. Figure 12 shows a profile of a latency CDF in which the read, write, and total contributions of the QoS are appreciable.

Figure 13a shows that the total QoS performed 630 µs better when the program suspension was exploited in workloads largely dominated by read operations that were prioritized with respect to write. Workloads performed on lower QD values still displayed an advantage even if it became marginal (i.e., a few hundred µs). Increasing the number of allowable suspensions from 5 to 10 slightly penalized the QoS, but in the high QD scenario there was still a large advantage compared with the no supensions case. Similar considerations can be extrapolated from the results of the read QoS shown in Figure 13b; however, here, the QoS gain increased up to 1.03 ms, proving that the program suspension is one of the most suitable ways to reduce the latency in read intensive scenarios.

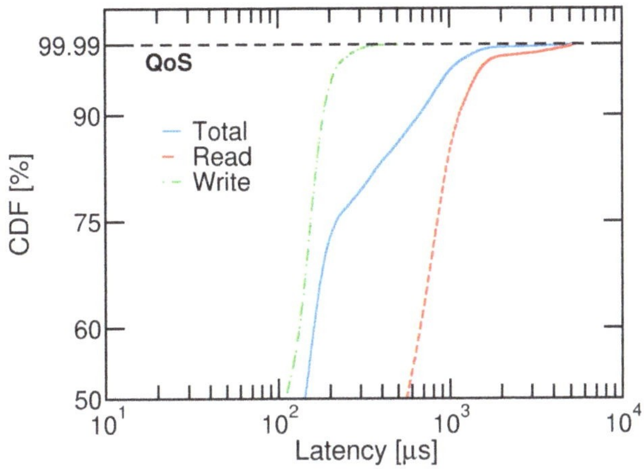

Figure 12. The latency CDF profile of a 75W-25R workload with QD = 64 simulated with the SSDExplorer framework.

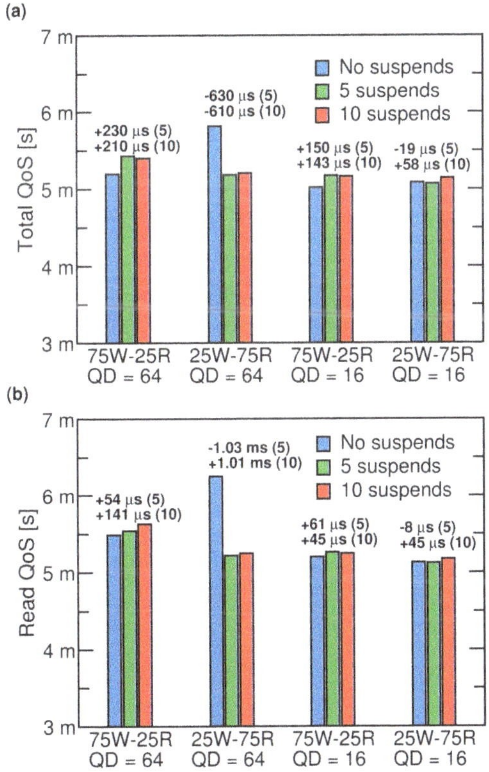

Figure 13. (**a**) The total QoS. (**b**) The read QoS. Both metrics were extracted at the 99.99-percentile of the latency CDF.

6. Conclusions

In this work, we assessed the role of the program suspend operation in TLC 3D NAND Flash memories envisioned as the storage medium of SSDs platforms. The electrical characterization performed on an off-the-shelf product demonstrated that the overall program time, including the suspension and the time required to enter in the suspension phase, was not a linear function of the wait time before the arrival of a preempting read operation. We also observed that the impact on the memory RBER in suspended and non-suspended blocks was negligible, therefore, proving the safe applicability of the suspend paradigm.

By back annotating the timing and the power consumption features of the product in the SSDExplorer co-simulation environment, we were able to run a design space exploration using different workloads and micro-architectural parameters , such as the QD. The simulation results demonstrated that the program suspend led to a 113 kIOPS increase in terms of the achieved drive's bandwidth and an average latency reduction of 97 μs.

Finally, we exposed the drawbacks of the program suspend operation in terms of the SSD power consumption, and we observed that the real advantage on the QoS (1.03 ms) was appreciable mainly in read intensive scenarios at a high QD range.

Author Contributions: The individual contributions to this paper are the following: conceptualization, C.Z. and R.M.; methodology, C.Z. and R.M.; software, L.Z.; validation, C.Z., L.Z. and R.M.; electrical measurements and resources, A.A. and S.S.; investigation, C.Z., L.Z., A.A. and R.M.; data curation, C.Z. and R.M.; writing—original draft preparation, C.Z. and R.M.; writing—review and editing, R.M. and P.O.; visualization, C.Z.; supervision, R.M. and P.O. All authors have read and agreed to the published version of the manuscript.

Funding: This research received no external funding.

Conflicts of Interest: The authors declare no conflict of interest.

Abbreviations

The following abbreviations are used in this manuscript:

GC	Garbage Collection
HDD	Hard Disk Drive
HPC	High Performance Computing
IOPS	Input/Output Operations Per Second
MLC	Multi Level Cell
QD	Queue Depth
QoS	Quality of Service
RBER	Raw Bit Error Rate
SLC	Single Level Cell
SSD	Solid State Drive
TLC	Triple Level Cell

References

1. Zuolo, L.; Zambelli, C.; Micheloni, R.; Olivo, P. Solid-State Drives: Memory Driven Design Methodologies for Optimal Performance. *Proc. IEEE* **2017**, *105*, 1589–1608. [CrossRef]
2. Schroeder, B.; Merchant, A.; Lagisetty, R. Reliability of nand-Based SSDs: What Field Studies Tell Us. *Proc. IEEE* **2017**, *105*, 1751–1769. [CrossRef]
3. Zuolo, L.; Zambelli, C.; Micheloni, R.; Bertozzi, D.; Olivo, P. Analysis of reliability/performance trade-off in Solid State Drives. In Proceedings of the IEEE International Reliability Physics Symposium (IRPS), Waikoloa, HI, USA, 1–5 June 2014; pp. 4B.3.1–4B.3.5. [CrossRef]
4. SNIA—Solid State Storage Initiative. An Introduction to Solid State Drive Performance, Evaluation and Test. 2013. Available online: https://www.snia.org/sites/default/files/SNIASSSI.SSDPerformance-APrimer2013.pdf (accessed on 21 April 2021).
5. Grossi, A.; Zuolo, L.; Restuccia, F.; Zambelli, C.; Olivo, P. Quality-of-Service Implications of Enhanced Program Algorithms for Charge-Trapping NAND in Future Solid-State Drives. *IEEE Trans. Device Mater. Reliab.* **2015**, *15*, 363–369. [CrossRef]
6. Gugnani, S.; Lu, X.; Panda, D.K. Analyzing, Modeling, and Provisioning QoS for NVMe SSDs. In Proceedings of the IEEE/ACM 11th International Conference on Utility and Cloud Computing (UCC), Zurich, Switzerland, 17–20 December 2018; pp. 247–256. [CrossRef]

7. Sun, C.; Le Moal, D.; Wang, Q.; Mateescu, R.; Blagojevic, F.; Lueker-Boden, M.; Guyot, C.; Bandic, Z.; Vucinic, D. Latency Tails of Byte-Addressable Non-Volatile Memories in Systems. In Proceedings of the IEEE International Memory Workshop (IMW), Monterey, CA, USA, 14–17 May 2017; pp. 1–4. [CrossRef]
8. Zambelli, C.; Zuolo, L.; Crippa, L.; Micheloni, R.; Olivo, P. Mitigating Self-Heating in Solid State Drives for Industrial Internet-of-Things Edge Gateways. *Electronics* **2020**, *9*, 1179. [CrossRef]
9. Pletka, R.; Papandreou, N.; Stoica, R.; Pozidis, H.; Ioannou, N.; Fisher, T.; Fry, A.; Ingram, K.; Walls, A. Achieving Latency and Reliability Targets with QLC in Enterprise Controllers. In Proceedings of the Flash Memory Summit, Santa Clara, CA, USA, 10–12 November 2020.
10. Wu, G.; He, X. Reducing SSD Read Latency via NAND Flash Program and Erase Suspension. In Proceedings of the USENIX Conference on File and Storage Technologies, San Jose, CA, USA, 14 February 2012; pp. 1–10. [CrossRef]
11. Micheloni, R.; Aritome, S.; Crippa, L. Array Architectures for 3-D NAND Flash Memories. *Proc. IEEE* **2017**, *105*, 1634–1649. [CrossRef]
12. Yamada, T.; Sun, C.; Takeuchi, K. A high-performance solid-state drive by garbage collection overhead suppression. In Proceedings of the Non-Volatile Memory Technology Symposium (NVMTS), Jeju, Korea, 27–29 October 2014; pp. 1–2. [CrossRef]
13. Shin, W.; Kim, M.; Kim, K.; Yeom, H.Y. Providing QoS through host controlled flash SSD garbage collection and multiple SSDs. In Proceedings of the International Conference on Big Data and Smart Computing (BIGCOMP), Jeju, Korea, 9–11 February 2015; pp. 111–117. [CrossRef]
14. Kim, S.; Bae, J.; Jang, H.; Jin, W.; Gong, J.; Lee, S.; Ham, T.J.; Lee, J.W. Practical Erase Suspension for Modern Low-Latency SSDs. In Proceedings of the USENIX Conference on Usenix Annual Technical Conference, Renton, WA, USA, 10 July 2019; pp. 813–820. [CrossRef]
15. Li, Y. 3 Bit Per Cell NAND Flash Memory on 19nm Technology. In Proceedings of the Flash Memory Summit, San Jose, CA, USA, 22–24 August 2012.
16. Maejima, H.; Kanda, K.; Fujimura, S.; Takagiwa, T.; Ozawa, S.; Sato, J.; Shindo, Y.; Sato, M.; Kanagawa, N.; Musha, J.; et al. A 512Gb 3b/Cell 3D flash memory on a 96-word-line-layer technology. In Proceedings of the IEEE International Solid-State Circuits Conference (ISSCC), San Francisco, CA, USA, 11–15 February 2018; pp. 336–338. [CrossRef]
17. Mielke, N.R.; Frickey, R.E.; Kalastirsky, I.; Quan, M.; Ustinov, D.; Vasudevan, V.J. Reliability of Solid-State Drives Based on NAND Flash Memory. *Proc. IEEE* **2017**, *105*, 1725–1750. [CrossRef]
18. Elyasi, N.; Arjomand, M.; Sivasubramaniam, A.; Kandemir, M.T.; Das, C.R.; Jung, M. Exploiting Intra-Request Slack to Improve SSD Performance. *SIGPLAN Not.* **2017**, *52*, 375–388, [CrossRef]
19. Park, J.; Lee, J.; Kim, M.; Chun, M.; Kim, J. Reducing read latency fluctuations of flash storage systems using preemptible programs and erases. In Proceedings of the Conference on File and Storage Technologies, Work-in-Progress Reports (WiPs), FAST'18. USENIX Association, Oakland, CA, USA, 12–15 February 2018.
20. Dimitrijevic, Z.; Rangaswami, R. Design and Implementation of Semi-preemptible IO. In Proceedings of the Conference on File and Storage Technologies—FAST, San Francisco, CA, USA, 31 March–2 April 2003.
21. Qureshi, M.K.; Franceschini, M.M.; Lastras-Montaño, L.A. Improving read performance of Phase Change Memories via Write Cancellation and Write Pausing. In Proceedings of the International Symposium on High-Performance Computer Architecture (HPCA), Bangalore, India, 9–14 January 2010; pp. 1–11. [CrossRef]
22. Zambelli, C.; Navarro, G.; Sousa, V.; Prejbeanu, I.L.; Perniola, L. Phase Change and Magnetic Memories for Solid-State Drive Applications. *Proc. IEEE* **2017**, *105*, 1790–1811. [CrossRef]
23. Liu, C.Y.; Kotra, J.B.; Jung, M.; Kandemir, M.T.; Das, C.R. SOML Read: Rethinking the Read Operation Granularity of 3D NAND SSDs. In Proceedings of the International Conference on Architectural Support for Programming Languages and Operating Systems (ASPLOS), Providence, RI, USA, 13–17 April 2019; pp. 955–969. [CrossRef]
24. Hyun, J.W.; Brinicombe, M.; Sun, H.; Zhong, H.; Strasser, J.; Wood, R. Program Suspend/Resume for Memory. U.S. Patent 9021158B2, 2015. Available online: https://patents.google.com/patent/US9021158B2/en (accessed on 2 May 2021).
25. Micheloni, R.; Aldarese, A.; Scommegna, S. Method and Apparatus with Program Suspend Using Test Mode. U.S. Patent 9892794B2, 2018. Available online: https://patents.google.com/patent/US9892794B2/en (accessed on 2 May 2021).
26. Bjorling, M. Open-Channel Solid State Drives NVMe Specification (rev 1.2). 2018. Available online: http://lightnvm.io/docs/OCSSD-2_0-20180129.pdf (accessed on 23 April 2021).
27. Takeuchi, K.; Tanaka, T. A dual-page programming scheme for high-speed multigigabit-scale NAND flash memories. *IEEE J. Solid-State Circ.* **2001**, *36*, 744–751. [CrossRef]
28. Suh, K.D.; Suh, B.H.; Um, Y.H.; Kim, J.K.; Choi, Y.J.; Koh, Y.N.; Lee, S.S.; Kwon, S.C.; Choi, B.S.; Yum, J.S.; et al. A 3.3 V 32 Mb NAND flash memory with incremental step pulse programming scheme. In Proceedings of the International Solid-State Circuits Conference, San Francisco, CA, USA, 15–17 February 1995; pp. 128–129. [CrossRef]
29. Zambelli, C.; Micheloni, R.; Scommegna, S.; Olivo, P. First Evidence of Temporary Read Errors in TLC 3D-NAND Flash Memories Exiting From an Idle State. *IEEE J. Electron Devices Soc.* **2020**, *8*, 99–104. [CrossRef]
30. Chen, C.P.; Lue, H.T.; Hsieh, C.C.; Chang, K.P.; Hsieh, K.Y.; Lu, C.Y. Study of fast initial charge loss and it's impact on the programmed states Vt distribution of charge-trapping NAND Flash. In Proceedings of the International Electron Devices Meeting (IEDM), San Francisco, CA, USA, 6–8 December 2010; pp. 5.6.1–5.6.4. [CrossRef]

31. Zambelli, C.; King, P.; Olivo, P.; Crippa, L.; Micheloni, R. Power-supply impact on the reliability of mid-1X TLC NAND flash memories. In Proceedings of the IEEE International Reliability Physics Symposium (IRPS), Pasadena, CA, USA, 17–21 April 2016; pp. 2B-3-1–2B-3-6. [CrossRef]
32. Zambelli, C.; Micheloni, R.; Crippa, L.; Zuolo, L.; Olivo, P. Impact of the NAND Flash Power Supply on Solid State Drives Reliability and Performance. *IEEE Trans. Device Mater. Reliab.* **2018**, *18*, 247–255. [CrossRef]
33. Zuolo, L.; Zambelli, C.; Micheloni, R.; Indaco, M.; Carlo, S.D.; Prinetto, P.; Bertozzi, D.; Olivo, P. SSDExplorer: A Virtual Platform for Performance/Reliability-Oriented Fine-Grained Design Space Exploration of Solid State Drives. *IEEE Trans. Comput. Aided Des. Integr. Circ. Syst.* **2015**, *34*, 1627–1638. [CrossRef]
34. Hady, F.T.; Foong, A.; Veal, B.; Williams, D. Platform Storage Performance With 3D XPoint Technology. *Proc. IEEE* **2017**, *105*, 1822–1833. [CrossRef]
35. Microsemi PM8609 NVMe2032 Flashtec NVMe Controller. 2019. Available online: https://www.microsemi.com/product-directory/storage-ics/3687-flashtec-nvme-controllers (accessed on 2 May 2021).
36. Monzio Compagnoni, C.; Ghetti, A.; Ghidotti, M.; Spinelli, A.S.; Visconti, A. Data Retention and Program/Erase Sensitivity to the Array Background Pattern in Deca-nanometer nand Flash Memories. *IEEE Trans. Electron Devices* **2010**, *57*, 321–327. [CrossRef]

electronics

Article

A Systematic Assessment of W-Doped CoFeB Single Free Layers for Low Power STT-MRAM Applications

Siddharth Rao *, Sebastien Couet, Simon Van Beek, Shreya Kundu, Shamin Houshmand Sharifi, Nico Jossart and Gouri Sankar Kar

IMEC, Kapeldreef 75, 3001 Leuven, Belgium; sebastien.couet@imec.be (S.C.); simon.vanbeek@imec.be (S.V.B.); shreya.kundu@imec.be (S.K.); shamin.houshmandsharifi@imec.be (S.H.S.); nico.jossart@imec.be (N.J.); gouri.kar@imec.be (G.S.K.)

* Correspondence: siddharth.rao@imec.be

Abstract: Spin-transfer torque magnetoresistive random access memory (STT-MRAM) technology is considered to be the most promising nonvolatile memory (NVM) solution for high-speed and low power applications. Dual MgO-based composite free layers (FL) have driven the development of STT-MRAMs over the past decade, achieving data retention of 10 years at the cost of higher write power consumption. In addition, the need for tunnel magnetoresistance (TMR)-based read schemes limits the flexibility in materials beyond the typical CoFeB/MgO interfaces. In this study, we propose a novel spacerless FL stack comprised of CoFeB alloyed with heavy metals such as tungsten (W) which allows effective modulation of the magnet properties (M_s, H_k) while retaining compatibility with MgO layers. The addition of W results favours a delayed crystallization process, in turn enabling higher thermal budgets up to 180 min at 400 °C. The presence of tungsten reduces the total FL magnetization (M_s) but simultaneously increasing its temperature dependence, thus, enabling a dynamic write current reduction of ~15% at 2 ns pulse widths. Reliable operation is demonstrated with a WER of 1 ppm and endurance >10^{10} cycles. These results pave the way for alternative designs of STT-MRAMs for low power electronics.

Keywords: STT-MRAM; spintronics; CoFeB; composite free layer; low power electronics

Citation: Rao, S.; Couet, S.; Van Beek, S.; Kundu, S.; Sharifi, S.H.; Jossart, N.; Kar, G.S. A Systematic Assessment of W-Doped CoFeB Single Free Layers for Low Power STT-MRAM Applications. *Electronics* **2021**, *10*, 2384. https://doi.org/10.3390/electronics10192384

Academic Editor: Dongseok Suh

Received: 9 August 2021
Accepted: 18 September 2021
Published: 29 September 2021

1. Introduction

Recent advancements in the computing community such as cloud computing and the Internet-of-Things (IoT) have increasingly created a heavy demand for an on-board memory solution in terms of speed and reduced power consumption. To tackle this growing problem, researchers have investigated several nonvolatile memory (NVM) technologies. STT-MRAM technology is considered to be the most viable solution owing to its attractive properties such as CMOS process compatibility, high operation speeds, superior endurance, and negligible leakage, thus, making it an ideal solution for low power, embedded electronics [1–3]. Significant resources have been invested in the past decade at major foundries and tool suppliers to optimize this technology for last-level cache (LLC) memory, microcontroller units (MCU), eFLASH, and automotive applications [2,4–12]. Despite these excellent advancements, some challenges remain. Further reduction of the write power consumption is required to ensure product lifetimes and enable further scaling at advanced logic nodes. This outstanding challenge persists due to an apparent tradeoff between the required write current for an achievable data retention (or thermal stability) of the free (or storage) layer in STT-MRAM devices.

The switching dynamics of the free layer (FL) have been extensively explored and are well described by the Landau–Lifshitz–Gilbert–Slonczeweski (LLGS) equation. For perpendicular magnetic tunnel junctions (pMTJs), reasonable predictions for the write current can be made using the coherent reversal model [13,14]. Then, the write current (I_c) and the switching time (t) in the precessional switching regime can be expressed as:

$$\frac{I_c}{I_{co}} - 1 = \frac{\ln(\pi/2\theta_o)}{t/t_o} \quad (1)$$

$$I_{co} = \frac{4}{\eta}\frac{2\alpha e}{\hbar}\frac{M_s t}{\pi d^2} H_k \quad (2)$$

where I_{co} refers to the critical write current at zero temperature and is purely dependent on material properties, θ_o is the initial angle between the magnetic moment and the easy axis of the magnetic anisotropy, and t_o is the characteristic relaxation time of the FL magnetization. M_s, H_k, α, and η correspond to the saturation magnetization, effective magnetic anisotropy field, damping, and spin polarization, respectively. The FL properties are described by the critical dimension (CD, d) and thickness (t). To improve write performance, optimization of one or more material parameters has been the conventional route. Given $\Delta = M_s H_k t d / 2k_B T$, Equation (1) can be rewritten as:

$$I_{c0} = \frac{4ek_B T}{\hbar}\frac{\alpha}{\eta}\Delta \quad (3)$$

Thus, the tradeoff between write current and thermal stability (Δ) becomes the key limiter. In addition, tunnel magnetoresistance (TMR)-based read schemes require a high TMR for fast reading, and thus are limited to the interfacial CoFeB/MgO system, which in turn results in a limited scope for materials exploration. As a result, most efforts in write current reduction have focused on optimizing the conventional dual MgO-based composite FL stack with spacer and cap layer engineering [6,15–19].

Here, we propose an alternative route for write current reduction through alloying conventional STT-MRAM materials such as CoFeB with nonmagnetic metals such as tungsten (W). The key advantage of this method is that it enables a precise and continuous variation of the FL properties (M_s, H_k) and the overall MTJ crystallization temperature. In this study, we demonstrate improved STT-MRAM write performances with a novel FL design based on a single W-CoFeB alloyed layer. The addition of W enables precise control of the FL magnetization (M_s) and its temperature dependence, aiding a current reduction during the write process. As I_{c0} and Δ are largely dependent on static magnetic properties of the FL (M_s, H_k), we find the coherent model of reversal, as described in Equations (1)–(3), a convenient and reasonably accurate model to elucidate the general performance trends for W-based alloyed FL designs as compared with conventional FLs. The introduction of metallic impurities or dopants in or near the FL has been explored recently, but this has been limited to enabling higher thermal budgets for compatibility of STT-MRAMs with Si-based backend-of-the-line (BEOL) processing [17,20]. Some device-level assessments of Δ have been carried out within the framework of spacer material optimization in conventional FLs; however, the lack of absolute metrics limits a fair evaluation of the true benefits of alloying CoFeB FLs with nonmagnetic dopants. Here, we systematically discuss the impact of heavy metal (such as W) doping from hypothesis and stack development to a full device-level assessment including write performance as low as a 1 ppm level, thermal stability, breakdown, and cycling. We demonstrate that this tuneable FL design opens up avenues for production-level development of STT-MRAMs, while maintaining application-relevant figures of merit.

2. Materials and Methods

Bottom-pinned pMTJ stacks were deposited on thermally oxidized Si (001) wafers by magnetron sputtering in a Canon-ANELVA EC7800 cluster tool at IMEC's state-of-the-art 300 mm assembly line. Two pMTJ stacks were deposited. The two stacks shared the same hard layer (HL) and reference layer (RL) composition, while the free layer (FL) composition was varied. One stack comprised the conventional composite FL CoFeB/Ta/CoFeB

(t = 2.5 nm with a 3 Å-thick Ta spacer layer) and served as the reference stack for this study. The second stack comprised the proposed new FL W_xCoFeB_{1-x} (t = ~3 nm). The compositional variation of the new FL was achieved by varying the deposition power of the W and composite $Co_{46.7}Fe_{23.3}B_{30}$ targets. The hard layer (HL) was a $[Co/Pt]_x$-based multilayer, and the reference layer was a [Co/spacer/FeB]-based trilayer stack. The HL and RL were coupled through a metallic spacer to form a synthetic antiferromagnet (SAF). A 1 nm thick MgO tunnel barrier was retained for both stacks. The reference stack was annealed at 400 °C for 30 min, while the W-based stack was annealed at 400 °C for 180 min. Patterned MTJ devices of critical dimension CD (or d) = 50 nm were fabricated by 193 nm immersion lithography, and a modified ion beam etch process [11] that suppressed etch-induced MTJ damage. All samples were subjected to a 2 T magnetic field at the end-of-line (EOL) to orient the SAF magnetization direction.

Vibrating sample magnetometry (VSM) was used at the film level to measure the out-of-plane magnetization (M_z) as a function of magnetic field and temperature. For device-level characterization, we used a Hprobe MRAM prober equipped with a Keysight M8190A pulse generator (t_{rise} = 70 ps, $\tau_{PW,\ min}$ = 200 ps), a Keithley 2450 source and measurement unit (SMU), and digital multimeter (DMM). The interval time between the read and write pulses was optimized to 700 µs, while the read pulse duration was fixed at 200 µs. A write-verify scheme was adopted for both the write performance and WER studies, while for the endurance measurement, MTJ resistance reads were limited to five read sequences per decade of cycling events. Positive voltage bias ($V_{MTJ} > 0$) resulted in AP-to-P switching, while negative bias ($V_{MTJ} < 0$) resulted in P-to-AP switching.

The breakdown statistics were collected by electrically stressing the MTJ devices with a ramped step function voltage waveform, as described previously [21,22].

3. Results and Discussion

In this section, we evaluate the device-level performance of the $W_xCoFeB_{(1-x)}$ alloy as a free layer for STT-MRAM devices in the context of low power applications. The results and discussion are presented in a three-fold manner: First, the fundamental MTJ properties are assessed and the role of W in enabling extended BEOL compatibility is discussed; in the second subsection, we investigate the writing performance by quasi-static switching experiments as low as 2 ns current pulses, and the impact of W doping on anomalous behaviour such as backhopping; finally, the device reliability is assessed by means of breakdown trends, data retention, write error rates, and lifetime extrapolations at operating conditions.

3.1. Stack Development and Fundamental MTJ Properties

Backend-of-line (BEOL) compatible bottom-pinned pMTJ stacks with two different FLs were prepared—the first, with the conventional composite FL, and the second, with the proposed FL W_xCoFeB_{1-x} (W concentration range = 5–10%). Figure 1a illustrates the schematic representation of the full pMTJ stack, and the two free layer stack combinations investigated in this study. For the composite FL stack, a Ta spacer-based design was chosen for this study instead of a W spacer-based design, owing to a ~2x reduction in damping with the former, as shown in our previous studies [23]. Thus, a Ta spacer-based composite FL is a more appropriate reference for switching current evaluation in alloyed FL designs. Prior to device fabrication, the saturation magnetization (M_s) of the film stacks is estimated by vibrating sample magnetometry (VSM) measurements for temperatures up to 200 °C. As the W concentration increases, a monotonically decreasing trend in M_s and a significant reduction in the Curie temperature (T_c) is observed, as seen in Figure 1b, in line with recent reports [20]. The STT-driven writing in scaled pMTJ devices (CD < 100 nm) is expected to result in significant self-heating in the vicinity of the MgO barrier [24]. Thus, the temperature induced M_s loss can be exploited to dynamically reduce the required switching current during the write operation while retaining its thermal stability after removing the write stress.

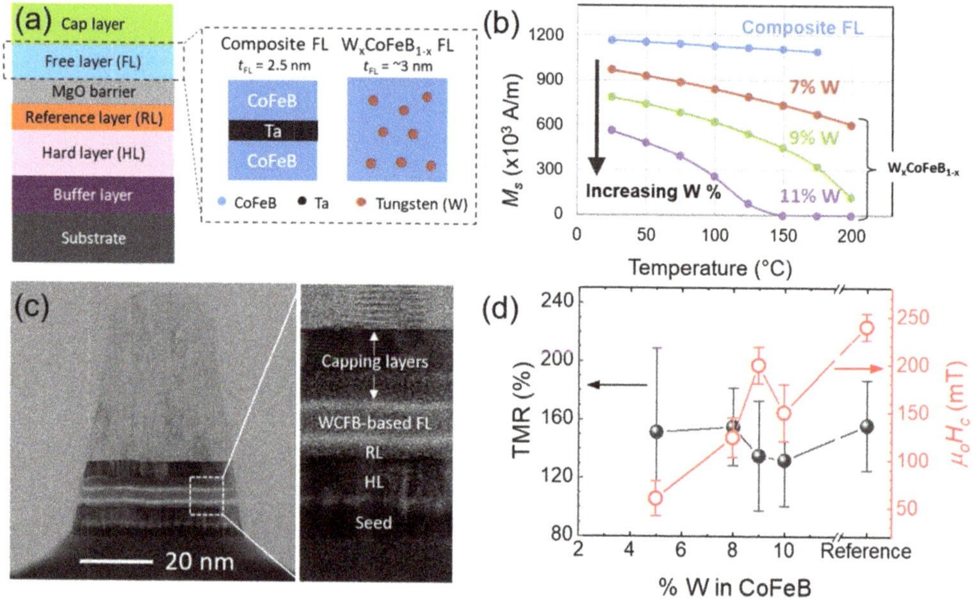

Figure 1. (**a**) Schematic of the pMTJ stack and the two free layer (FL) stack options. The hard layer (HL) is a $[Co/Pt]_x$-based multilayer, and the reference layer is a [Co/spacer/FeB]-based trilayer stack. The HL and RL are coupled through a metallic spacer to form a synthetic antiferromagnet (SAF). The composite FL serves as the reference for this study; (**b**) temperature dependence of the saturation magnetization (M_s) measured for the composite FL and W_xCoFeB_{1-x} FL with varying W concentrations; (**c**) cross-sectional TEM image of a fully fabricated 50 nm CD device with FL = $W_{0.09}(CoFeB)_{0.91}$. The zoomed-in image shows good polycrystalline order at the RL/MgO/FL/MgO interfaces; (**d**) tunnel magnetoresistance (TMR) and magnetic coercivity (μ_oH_c) measured on 50 nm CD devices for varying W concentrations and the reference stack.

To evaluate this hypothesis, cylindrical devices of 50 nm CD are fabricated (see Materials and Methods). Cross-sectional TEM images, as shown in Figure 1c, at a W concentration of 9% depict a broad polycrystalline nature of the FL/MgO interface and minimal diffusion across the different layers of the pMTJ stack. To corroborate these morphological assessments, tunnel magnetoresistance (TMR) and FL coercivity (μ_oH_c) were extracted from standard magnetoresistance measurements under the influence of an external out-of-plane (μ_oH_{ext}) magnetic field, as shown in Figure 1d. At a low W concentration of 5%, the measured coercivity (μ_oH_c ~110 mT) drops steeply as compared with the composite FL reference (μ_oH_c ~210 mT). This drop in μ_oH_c is attributed to the degradation of the CoFeB/MgO interfacial anisotropy for longer annealing times (3 h at 400 °C). As the concentration of W increases to 10%, we observe an increase in the μ_oH_c up to 170 mT, while TMR drops by 20% as compared with low W concentrations. This counter-intuitive observation can be understood by the impact of W doping on the crystallization of the FL from the CoFeB/MgO interface. It has been previously reported that an increase in boron (B) concentration leads to a delayed crystallization of the CoFeB/MgO interface [25,26]. The co-sputtering of W alongside CoFeB increases intermixing and limits B diffusion out of the FL. As both TMR and μ_oH_c are dependent on the crystalline nature of the FL/MgO interface, we attribute the reduction in both metrics at W content >8% to a partially crystallized interface. By carefully tuning the W content to target the application-specific thermal budget, high TMR (>130%) and μ_oH_c (>150 mT) metrics can be achieved.

3.2. STT Write Performance

STT-MRAM devices are typically expected to operate in the 20–200 MHz frequency range for the embedded memory market. Figure 2a,b illustrates the electrical scheme realized to enable both *dc* and pulsed switching measurements. We investigate the write performance for pulse durations from 2–100 ns, wherein one can expect a transition from pure precessional switching regime (τ_{PW} < 10 ns) to an intermediate region comprising both switching regimes (τ_{PW} ~10–100 ns). The switching probability (P_{sw}) is extracted from 2000 switching events for each stack and pulse duration.

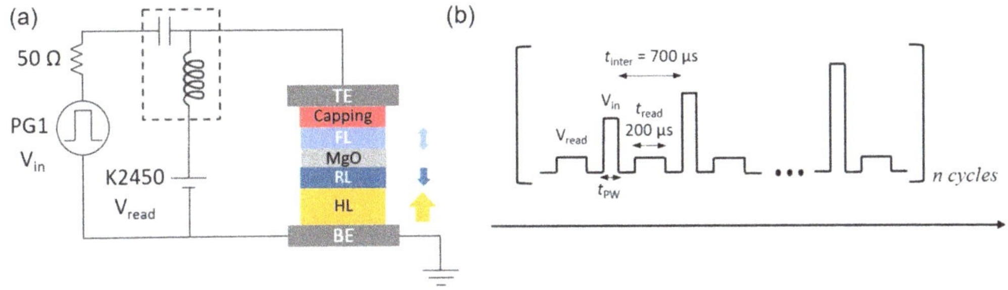

Figure 2. (**a**) Electrical measurement scheme for *dc* and pulsed measurements. A Keithley K2450 and Keysight M8190A pulse generator are connected through a bias tee (up to 6 GHz) to enable fast switching and DC readout of the resistance states; (**b**) waveform sequence for read and write pulses. The interval time between the pulses (t_{inter}) and integration time of the read pulse (t_{read}) have been optimized to ensure accurate readout.

Figure 3a shows the average switching currents over the operating frequency range ($1/\tau_{PW}$) for the different W-based FLs as compared with the composite FL as a reference. For the 5% W-based FL, we observe a reduction in switching currents for all frequencies with respect to the reference, but this can be attributed to the significantly reduced coercivity ($\mu_0 H_c$) reported for this FL (25% of the reference FL). As the doping concentration increases to 9% W, we observe a crossover in the switching current trends as the operating frequency increases. At τ_{PW} = 5–10 ns, the switching currents for higher W concentrations (9% and 10%) are in line with the reference and can achieve a further 25% reduction in switching energy as τ_{PW} reduces to 2 ns. These results can be understood in a two-fold manner, i.e., the increase in the W concentration gradually increases the effective anisotropy ($H_{k,\,eff}$) of the FL leading to a faster switching process in the precessional regime, while the increased thermal sensitivity of the W-based FL at higher concentrations also reduces the required switching current during the pulse application. To further corroborate these findings, we can extract a linear dependence of the form $1/\tau = A(I_c - I_{c0})$ between the pulse duration and the average switching currents [27]. The parameter 'A' is a figure of merit for the efficiency of the spin angular momentum transfer, and 'I_{c0}' is the zero-temperature critical switching current. These results are depicted in Figure 3b as a function of the W concentration in the FL. As compared with the composite FL, the efficiency for W concentration up to 8% is similar but experiences a four-fold increase for 9% W in the FL. A further increase in the W concentration results in a minor drop in the efficiency by ~10%. It must also be noted that I_{c0} increases along with the efficiency reaching a local maximum at 9% W which suggests an increase in the effective FL anisotropy ($H_{k,eff}$) in line with previous reports [13]. The measured I_c trends scale in a linear fashion with FL M_s reduction for increasing W%. Damping in such FL stacks can also influence the switching metrics, although in our proposed alloyed FL design, the M_s variation under the influence of local self-heating (during write current pulse) is believed to be the dominant factor during switching [24]. Understanding damping in such FL designs would require a more extensive design of experiments in terms of W concentrations and annealing conditions, which we believe falls outside the stated scope of this study. Deterministic switching is achieved across the tested pulse duration range for the 10% W-doped FL, as shown in Figure 3c, with switching voltages (V_{MTJ}) <0.8 V to as low as 5 ns and <1 V at 2 ns, thus reinforcing the suitability of the W-doped FL for high-speed and low power applications.

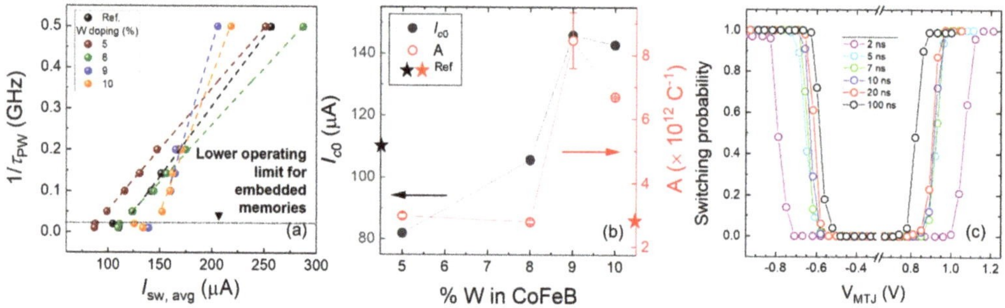

Figure 3. (**a**) Comparison of average write currents at different frequencies ($1/\tau_{PW}$) for the different FL stack options estimated from 2000 switching events/data point; (**b**) extracted critical switching currents (I_{c0}) and spin transfer efficiency (A) as a function of W doping concentration; (**c**) switching probability curves on showing 100% deterministic switching across the entire pulse duration testing range (2 – 100 ns) for FL = $W_{0.1}CoFeB_{0.9}$.

We also evaluate the operating margin between deterministic switching and the onset of 'backhopping', where the device is observed to switch back to its initial state in apparent opposition to the majority spin torques during the writing pulse. The observation of 'backhopping' is typically attributed to an unintentional STT-induced reversal of the reference layer magnetization at higher bias voltages [28,29]. Backhopping is detrimental to the market adoption of STT-MRAM devices, owing to the limitation in achievable write error rates (WER). This anomalous behaviour can be addressed either by increasing the RL/SAF pinning fields, and/or by reducing the switching current requirements of the FL to enlarge the error-free switching window. The operating margin for error-free switching is evaluated for both switching directions as a function of the pulse duration and W doping concentration, as shown in Figure 4a,b. In the thermally assisted switching regime ($\tau_{PW} \geq 50$ ns), the W-doped wafers achieve a wider operating margin with increasing W concentration as compared with the composite FL reference. In a purely precessional regime ($\tau_{PW} \leq 10$ ns), the 10% W-doped FL performs comparably to the reference FL but has a larger operating margin as the pulse duration decreases. Comparing the trends for both switching directions (AP-to-P and P-to-AP), we observe a monotonic dependence on the W concentration, though in opposite directions. This is attributed to the effective stray magnetic field seen at the FL of the devices due to contributions of the RL and SAF layers, which we believe can be resolved by careful control of the stack deposition and lithography processes. Interestingly, the average operating margin considering both switching directions is ~300 mV at τ_{PW} = 2 ns regardless of the amount of W doping. These results confirm that W doping has a negligible impact on the stability of the RL and allows independent engineering of the MTJ stack to enable both low switching currents and robustness to backhopping.

3.3. *Reliability*

3.3.1. Breakdown Characteristics

To further investigate the impact of W doping, we evaluate the breakdown characteristics of the ~1 nm thin MgO barrier by electrically stressing the devices. The distribution of the breakdown times for a group of pMTJ devices are typically well-described by a Weibull distribution as:

$$F(t_{bd}) = 1 - \exp\left(\frac{t_{bd}}{t_{63}}\right)^{\beta} \quad (4)$$

where t_{bd} is the time to breakdown, β is the Weibull slope indicative of the dispersion of the breakdown times, and t_{63} is the mean time to failure of approximately 63.2% of all devices under test. Figure 5a depicts the Weibull distributions of the breakdown voltages estimated for different W doping concentrations under a ramped voltage stress scheme (see Materials and Methods). Each distribution is extracted from 36 randomly selected devices, allowing for an accurate estimation of the dielectric failure characteristics. As compared with the composite FL, the W-doped stacks demonstrate a 100 mV reduction in the 63% breakdown voltage (V_{bd}) except for the 9% W-doped wafer, where the higher V_{bd} is found to arise due a higher device-level R.A product (not shown). This anomalous increase in the RA is attributed to non-uniformities during the fabrication process, which is highlighted by the wider dispersion observed for the 9% W-doped wafer. The addition of W can impact the defectivity in and around the MgO dielectric barrier, which may explain the lower breakdown voltage performance. The observed bimodality in some of the distributions indicates that a careful tuning of the W content and post-patterning treatments can achieve breakdown characteristics similar to those of the composite FL alongside a write current reduction.

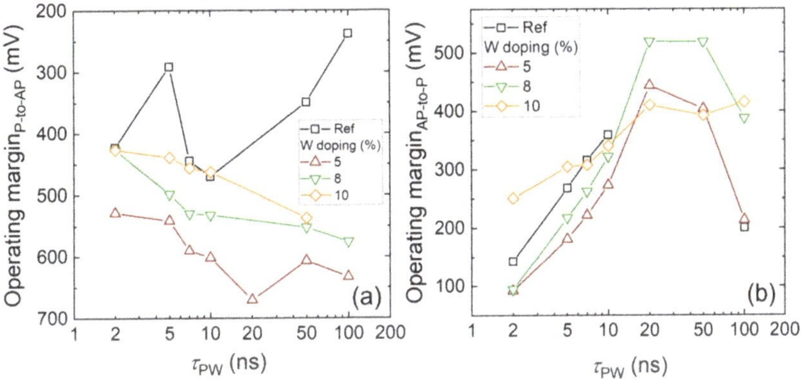

Figure 4. Absolute operating margins ($V_{backhopping} - V_{write}$) estimated for the different FL options: (**a**) AP-to-P switching; (**b**) P-to-AP switching across the tested pulse duration range (2–100 ns). The margins are estimated from the median values of switching voltages and backhopping voltages. Negligible backhopping was observed for the 9% W-doped CoFeB FL, likely due to a stronger RL/MgO interface arising from the higher-than-expected RA product.

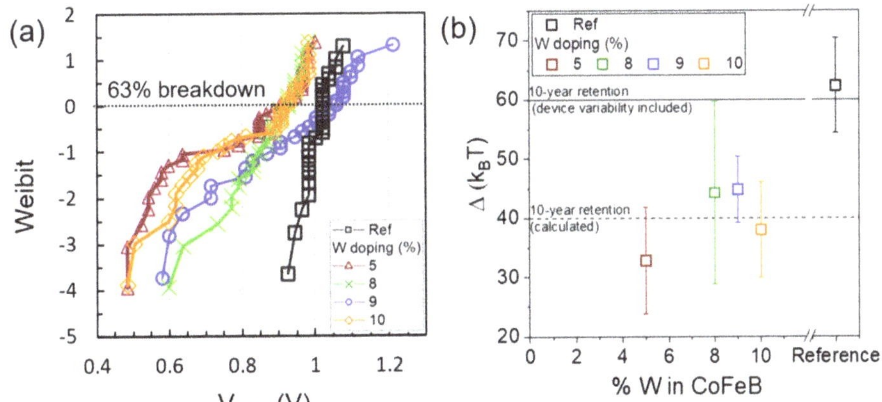

Figure 5. (**a**) Weibull distributions of DC breakdown performance for the different FL stack options. The dotted line indicates the point where 63% of the tested devices will fail. (**b**) Thermal stability (in k_BT) estimated from fits of room temperature (RT) switching field distributions for the different FL stack options.

3.3.2. Thermal Stability and Retention

In addition to the oxide reliability, data retention is another key reliability figure of merit for STT-MRAMs to replace conventional embedded memories. The required retention time is estimated based on the hierarchical separation between the CPU and the memory storage device. Memories for the last level cache (LLC) or embedded flash (eFLASH) applications typically require ~10 years of retention time per bit of information, while those closer to the CPU require less retention times. The retention time can be estimated from a standard Arrhenius rate equation described as $t_{ret} = t_0 exp\left(\frac{E_b}{k_B T}\right) = t_0 exp(\Delta)$, where t_0 is the attempt time (~1 ns), E_b is the energy barrier between the two stable states of the MTJ, k_B is the Boltzmann's constant, and T is the temperature in Kelvin. The thermal stability (Δ), expressed in the units of (k_BT), is a convenient figure of merit and, hereafter, is used to describe the retention of the pMTJ devices in this paper. Figure 5b shows the thermal stability estimated from 50–100 devices for every FL stack from switching field distributions [13]. We observe a 30% reduction in Δ for the doped W-CoFeB FLs with respect to the conventional composite FL across the doping range. The reduction in Δ is attributed to the weakening of the interfacial CoFeB/MgO PMA with increasing W content, as is also seen in the coercivity trend in Figure 1d. While an estimation of the absolute Δ can also be done in the framework of a domain wall-mediated reversal model for these CDs [16], we do not expect the general trends seen in Figure 5b to change. To accurately quantify and elucidate the impact of W doping on Δ, further in-depth investigation of the temperature-dependent FL properties must be carried out. A careful optimization of the W doping content between 8 and 10% can achieve a healthy margin over the required retention specification. For memory applications closer to the CPU, this target can be further relaxed, thus falling within the scope of the W-doped FLs.

3.3.3. Write Error Rates (WER) and Endurance

To fully realize STT-MRAMs for embedded memories, it is critical to demonstrate the robustness in terms of bit error rates and endurance (or cycling). LLC applications require a minimum write error rate (WER) of one failure in 1 million write operations (or 1 ppm) and 10^{10}–10^{12} cycling events. We postulated in Section 3.2 that the operating margin to backhopping is skewed due to the stray magnetic fields experienced by the FL due to the RL and SAF layers within the pMTJ device. To validate this hypothesis, we performed one million write operations with and without an external magnetic field ($\mu_o H_{ext}$) to compensate the effects of the internal stray field ($\mu_o H_{stray}$ ~−20 mT). As it is not feasible to demonstrate the required specifications over several stacks and devices in a realistic testing timeframe, we choose to focus on the most promising stack, which is the FL with 10% W doping concentration. Figure 6a shows the WER achieved on a single bit on the above chosen stack with and without field compensation. In the case of $\mu_o H_{ext}$ = 0, 1 ppm error-free write is demonstrated for P-to-AP switching, while severe backhopping for the AP-to-P switch limits the achievable WER = 2×10^{-3} only. By roughly compensating the stray field contribution in this device ($\mu_o H_{ext}$ = +20 mT), we observe a shift in the WER curves for both switching directions, i.e.,1 ppm error-free write is now achieved for AP-to-P switching, while P-to-AP switching achieves WER = 2×10^{-5}. These results show that in addition to backhopping suppression, good control of the magnetic stray fields is necessary to realize 1 ppm WER with sufficient margin.

Figure 6b showcases the robustness of the proposed stack at typical write conditions (τ_{PW} = 5 ns); no reduction of the TMR window is observed over the course of 2×10^{10} write operations (for AP-to-P and P-to-AP independently). The 1 ppm failure rate in our devices can be extrapolated with reasonable accuracy from accelerated testing at overdrive voltages, as has been shown in the past, by fitting to a power law model [22]. Figure 6c shows the 63% failure rates extracted from cycling tests for different overdrive conditions. For the applied write bias ($V_{write, average}$ = 0.79 V with 20% variability), it can be seen that most devices survived a write operation of at least 10^{10} cycles, thus, meeting all of the basic requirements for embedded memory applications.

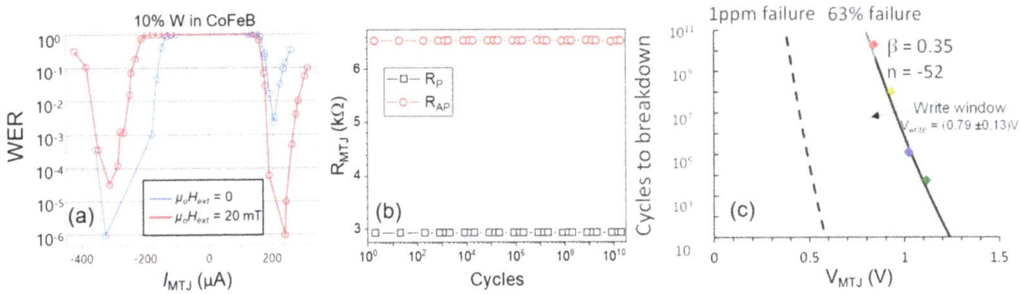

Figure 6. Reliability assessment of devices with $W_{0.1}CoFeB_{0.9}$ FL at τ_{PW} = 5 ns: (**a**) Experimentally measured write error rates as a function of external magnetic field ($\mu_o H_{ext}$). 1 ppm (WER = 1 × 10^{-6}) is demonstrated for both switching directions through compensation of the stray field ($\mu_o H_{stray}$); (**b**) evolution of device resistances (R_P, R_{AP}) during 2 × 10^{10} cycling events, no TMR degradation is observed; (**c**) lifetime extrapolations of the tested devices for 1 ppm failure based on four independent stress tests on fresh devices. Each colour indicates the biasing voltage at which 63% of the tested devices fail.

4. Conclusions

In conclusion, we clearly demonstrate the benefits of alloying conventional FL materials such as CoFeB with nonmagnetic metals (such as W) as an alternative approach to achieve write performance improvement in STT-MRAM devices. The W_xCoFeB_{1-x} FL design is shown to achieve precise control of the FL magnetization (M_s) and its temperature dependence, in addition to enhanced thermal budgets up to 180 min at 400 °C. These effects result in a 15% reduction of the write current at short pulse widths, as low as 2 ns. While backhopping is observed in these devices, we believe the enhancement of the RL pinning fields through stronger RL/SAF coupling should address these concerns satisfactorily. A successful 1 ppm write error rate and an endurance >10^{10} cycles is also demonstrated at the bit level. We also propose a step-by-step optimization strategy to address the observed reduction in data retention and breakdown voltage metrics. As commercial STT-MRAM products must operate in a wide temperature range, it is of significant interest to further this study by exploring the temperature dependence of the W_xCoFeB_{1-x} FL properties to evaluate trends in thermal stability and switching behaviour. We believe that this new alloyed FL concept is of great interest to the STT-MRAM community as it presents a viable alternative route towards low power STT-MRAM coupled with a high degree of controllability in terms of production, while retaining most of its inherent advantages in terms of performance.

Author Contributions: S.C. conceptualized the stack; S.R. and S.C. designed the study; S.C., S.K., S.H.S. and N.J. supervised and/or fabricated the devices for the study; S.R., S.V.B. and S.C. performed the measurements, associated data analyses, and interpreted the findings; G.S.K. was the project leader and enabled the funding and project administration for this study; S.R. wrote the manuscript. All authors have read and agreed to the published version of the manuscript.

Funding: This work is supported by IMEC's Industrial Affiliation Program (IIAP) on STT-MRAM devices.

Data Availability Statement: The data presented in this study are available from the corresponding author on reasonable request.

Acknowledgments: This work is supported by IMEC's Industrial Affiliation Program (IIAP) on STT-MRAM devices. The authors would also like to acknowledge the support of IMEC's pilot line support teams in realizing the devices for this study, and the MCA team for providing the TEM images.

Conflicts of Interest: The authors declare no conflict of interest.

References

1. Ikeda, S.; Miura, K.; Yamamoto, H.; Mizunuma, K.; Gan, H.D.; Endo, M.; Kanai, S.; Hayakawa, J.; Matsukura, F.; Ohno, H. A Perpendicular-Anisotropy CoFeB–MgO Magnetic Tunnel Junction. *Nat. Mater.* **2010**, *9*, 721–724. [CrossRef] [PubMed]
2. Kang, S.H. Embedded STT-MRAM for Energy-Efficient and Cost-Effective Mobile Systems. In Proceedings of the 2014 Symposium on VLSI Technology (VLSI-Technology): Digest of Technical Papers, Honolulu, HI, USA, 9–12 June 2014; pp. 1–2.
3. Slaughter, J.M.; Rizzo, N.D.; Janesky, J.; Whig, R.; Mancoff, F.B.; Houssameddine, D.; Sun, J.J.; Aggarwal, S.; Nagel, K.; Deshpande, S.; et al. High Density ST-MRAM Technology (Invited). In Proceedings of the 2012 International Electron Devices Meeting, San Francisco, CA, USA, 10–13 December 2012; pp. 29.3.1–29.3.4.
4. Lee, T.Y.; Yamane, K.; Otani, Y.; Zeng, D.; Kwon, J.; Lim, J.H.; Naik, V.B.; Hau, L.Y.; Chao, R.; Chung, N.L.; et al. Advanced MTJ Stack Engineering of STT-MRAM to Realize High Speed Applications. In Proceedings of the 2020 IEEE International Electron Devices Meeting (IEDM), San Francisco, CA, USA, 12–18 December 2020; pp. 11.6.1–11.6.4.
5. Lee, Y.K.; Song, Y.; Kim, J.; Oh, S.; Bae, B.-J.; Lee, S.; Lee, J.; Pi, U.; Seo, B.; Jung, H.; et al. Embedded STT-MRAM in 28-Nm FDSOI Logic Process for Industrial MCU/IoT Application. In Proceedings of the 2018 IEEE Symposium on VLSI Technology, Honolulu, HI, USA, 18–22 June 2018; pp. 181–182.
6. Jan, G.; Thomas, L.; Le, S.; Lee, Y.-J.; Liu, H.; Zhu, J.; Iwata-Harms, J.; Patel, S.; Tong, R.-Y.; Sundar, V.; et al. Demonstration of Ultra-Low Voltage and Ultra Low Power STT-MRAM Designed for Compatibility with 0x Node Embedded LLC Applications. In Proceedings of the 2018 IEEE Symposium on VLSI Technology, Honolulu, HI, USA, 18–12 June 2018; pp. 65–66.
7. Naik, V.B.; Yamane, K.; Lee, T.Y.; Kwon, J.; Chao, R.; Lim, J.H.; Chung, N.L.; Behin-Aein, B.; Hau, L.Y.; Zeng, D.; et al. JEDEC-Qualified Highly Reliable 22nm FD-SOI Embedded MRAM For Low-Power Industrial-Grade, and Extended Performance Towards Automotive-Grade-1 Applications. In Proceedings of the 2020 IEEE International Electron Devices Meeting (IEDM), San Francisco, CA, USA, 12–18 December 2020; pp. 11.3.1–11.3.4.
8. Dixit, H.; Naik, V.B.; Yamane, K.; Lee, T.; Kwon, J.-H.; Behin-Aein, B.; Soss, S.; Taylor, W.J. TCAD Device Technology Co-Optimization Workflow for Manufacturable MRAM Technology. In Proceedings of the 2020 IEEE International Electron Devices Meeting (IEDM), San Francisco, CA, USA, 12–18 December 2020; pp. 13.5.1–13.5.4.
9. Golonzka, O.; Alzate, J.-G.; Arslan, U.; Bohr, M.; Bai, P.; Brockman, J.; Buford, B.; Connor, C.; Das, N.; Doyle, B.; et al. MRAM as Embedded Non-Volatile Memory Solution for 22FFL FinFET Technology. In Proceedings of the 2018 IEEE International Electron Devices Meeting (IEDM), San Francisco, CA, USA, 1–5 December 2018; pp. 18.1.1–18.1.4.
10. Kan, J.J.; Park, C.; Ching, C.; Ahn, J.; Xue, L.; Wang, R.; Kontos, A.; Liang, S.; Bangar, M.; Chen, H.; et al. Systematic Validation of 2x Nm Diameter Perpendicular MTJ Arrays and MgO Barrier for Sub-10 Nm Embedded STT-MRAM with Practically Unlimited Endurance. In Proceedings of the 2016 IEEE International Electron Devices Meeting (IEDM), San Francisco, CA, USA, 3–7 December 2016; pp. 27.4.1–27.4.4.
11. Rao, S.; Kim, W.; van Beek, S.; Kundu, S.; Perumkunnil, M.; Cosemans, S.; Yasin, F.; Couet, S.; Carpenter, R.; O'Sullivan, B.J.; et al. STT-MRAM Array Performance Improvement through Optimization of Ion Beam Etch and MTJ for Last-Level Cache Application. In Proceedings of the 2021 IEEE International Memory Workshop (IMW), Dresden, Germany, 16–19 May 2021; pp. 1–4.
12. Xue, L.; Ching, C.; Kontos, A.; Ahn, J.; Wang, X.; Whig, R.; Tseng, H.; Howarth, J.; Hassan, S.; Chen, H.; et al. Process Optimization of Perpendicular Magnetic Tunnel Junction Arrays for Last-Level Cache beyond 7 Nm Node. In Proceedings of the 2018 IEEE Symposium on VLSI Technology, Honolulu, HI, USA, 18–12 June 2018; pp. 117–118.
13. Khvalkovskiy, A.V.; Apalkov, D.; Watts, S.; Chepulskii, R.; Beach, R.S.; Ong, A.; Tang, X.; Driskill-Smith, A.; Butler, W.H.; Visscher, P.B.; et al. Basic Principles of STT-MRAM Cell Operation in Memory Arrays. *J. Phys. D Appl. Phys.* **2013**, *46*, 074001. [CrossRef]
14. Liu, H.; Bedau, D.; Sun, J.Z.; Mangin, S.; Fullerton, E.E.; Katine, J.A.; Kent, A.D. Dynamics of Spin Torque Switching in All-Perpendicular Spin Valve Nanopillars. *J. Mag. Mag. Mater.* **2014**, *358–359*, 233–258. [CrossRef]
15. Iwata-Harms, J.M.; Jan, G.; Serrano-Guisan, S.; Thomas, L.; Liu, H.; Zhu, J.; Lee, Y.-J.; Le, S.; Tong, R.-Y.; Patel, S.; et al. Ultrathin Perpendicular Magnetic Anisotropy CoFeB Free Layers for Highly Efficient, High Speed Writing in Spin-Transfer-Torque Magnetic Random Access Memory. *Sci. Rep.* **2019**, *9*, 19407. [CrossRef] [PubMed]
16. Santos, T.S.; Mihajlović, G.; Smith, N.; Li, J.-L.; Carey, M.; Katine, J.A.; Terris, B.D. Ultrathin Perpendicular Free Layers for Lowering the Switching Current in STT-MRAM. *J. Appl. Phys.* **2020**, *128*, 113904. [CrossRef]
17. Chatterjee, J.; Sousa, R.C.; Perrissin, N.; Auffret, S.; Ducruet, C.; Dieny, B. Enhanced Annealing Stability and Perpendicular Magnetic Anisotropy in Perpendicular Magnetic Tunnel Junctions Using W Layer. *Appl. Phys. Lett.* **2017**, *110*, 202401. [CrossRef]
18. Hu, G.; Gottwald, M.G.; He, Q.; Park, J.H.; Lauer, G.; Nowak, J.J.; Brown, S.L.; Doris, B.; Edelstein, D.; Evarts, E.R.; et al. Key Parameters Affecting STT-MRAM Switching Efficiency and Improved Device Performance of 400 °C-Compatible p-MTJs. In Proceedings of the 2017 IEEE International Electron Devices Meeting (IEDM), San Francisco, CA, USA, 2–6 December 2017; pp. 38.3.1–38.3.4.
19. Kan, J.J.; Gottwald, M.; Park, C.; Zhu, X.; Kang, S.H. Thermally Robust Perpendicular STT-MRAM Free Layer Films Through Capping Layer Engineering. *IEEE Trans. Magn.* **2015**, *51*, 1–5. [CrossRef]
20. Iwata-Harms, J.M.; Jan, G.; Liu, H.; Serrano-Guisan, S.; Zhu, J.; Thomas, L.; Tong, R.-Y.; Sundar, V.; Wang, P.-K. High-Temperature Thermal Stability Driven by Magnetization Dilution in CoFeB Free Layers for Spin-Transfer-Torque Magnetic Random Access Memory. *Sci. Rep.* **2018**, *8*, 14409. [CrossRef] [PubMed]

21. O'Sullivan, B.J.; Van Beek, S.; Roussel, P.J.; Rao, S.; Kim, W.; Couet, S.; Swerts, J.; Yasin, F.; Crotti, D.; Linten, D.; et al. Extended RVS Characterisation of STT-MRAM Devices: Enabling Detection of AP/P Switching and Breakdown. In Proceedings of the 2018 IEEE International Reliability Physics Symposium (IRPS), Burlingame, CA, USA, 11–15 March 2018; pp. P-MY.5-1–P-MY.5-6.
22. Van Beek, S.; Rousse, P.; O'Sullivan, B.; Degraeve, R.; Cosemans, S.; Linten, D.; Kar, G.S. Study of Breakdown in STT-MRAM Using Ramped Voltage Stress and All-in-One Maximum Likelihood Fit. In Proceedings of the 2018 48th European Solid-State Device Research Conference (ESSDERC), Dresden, Germany, 3–6 September 2018; pp. 146–149.
23. Couet, S.; Devolder, T.; Swerts, J.; Lin, T.; Liu, E.; Van Elshoct, S.; Kar, G.S. Impact of Ta and W-based spacers in double MgO STT-MRAM free layers on perpendicular anisotropy and damping. *Appl. Phys. Lett.* **2017**, *107*, 152406. [CrossRef]
24. Van Beek, S.; O'Sullivan, B.J.; Roussel, P.J.; Degraeve, R.; Bury, E.; Swerts, J.; Couet, S.; Souriau, L.; Kundu, S.; Rao, S.; et al. Impact of Self-Heating on Reliability Predictions in STT-MRAM. In Proceedings of the 2018 IEEE International Electron Devices Meeting (IEDM), San Francisco, CA, USA, 1–5 December 2018; pp. 25.2.1–25.2.4.
25. Endoh, T.; Honjo, H. A Recent Progress of Spintronics Devices for Integrated Circuit Applications. *J. Low Power Electron. Appl.* **2018**, *8*, 44. [CrossRef]
26. Pellegren, J.P.; Furuta, M.; Sundar, V.; Liu, Y.; Zhu, J.-G.; Sokalski, V. Increased Boron Content for Wider Process Tolerance in Perpendicular MTJs. *AIP Adv.* **2017**, *7*, 055901. [CrossRef]
27. Bedau, D.; Liu, H.; Sun, J.Z.; Katine, J.A.; Fullerton, E.E.; Mangin, S.; Kent, A.D. Spin-Transfer Pulse Switching: From the Dynamic to the Thermally Activated Regime. *Appl. Phys. Lett.* **2010**, *97*, 262502. [CrossRef]
28. Hu, G.; Nowak, J.J.; Gottwald, M.G.; Sun, J.Z.; Houssameddine, D.; Bak, J.; Brown, S.L.; Hashemi, P.; He, Q.; Kim, J.; et al. Reliable Five-Nanosecond Writing of Spin-Transfer Torque Magnetic Random-Access Memory. *IEEE Magn. Lett.* **2019**, *10*, 1–4. [CrossRef]
29. Kim, W.; Couet, S.; Swerts, J.; Lin, T.; Tomczak, Y.; Souriau, L.; Tsvetanova, D.; Sankaran, K.; Donadio, G.L.; Crotti, D.; et al. Experimental Observation of Back-Hopping With Reference Layer Flipping by High-Voltage Pulse in Perpendicular Magnetic Tunnel Junctions. *IEEE Trans. Magn.* **2016**, *52*, 1–4. [CrossRef]

electronics

MDPI

Article

A Noise-Resilient Neuromorphic Digit Classifier Based on NOR Flash Memories with Pulse–Width Modulation Scheme

Gerardo Malavena *, Alessandro Sottocornola Spinelli and Christian Monzio Compagnoni

Dipartimento di Elettronica, Informazione e Bioingegneria, Politecnico di Milano, 20133 Milan, Italy; alessandro.spinelli@polimi.it (A.S.S.); christian.monzio@polimi.it (C.M.C.)
* Correspondence: gerardo.malavena@polimi.it

Abstract: In this work, we investigate the implementation of a neuromorphic digit classifier based on NOR Flash memory arrays as artificial synaptic arrays and exploiting a pulse-width modulation (PWM) scheme. Its performance is compared in presence of various noise sources against what achieved when a classical pulse-amplitude modulation (PAM) scheme is employed. First, by modeling the cell threshold voltage (V_T) placement affected by program noise during a program-and-verify scheme based on incremental step pulse programming (ISPP), we show that the classifier truthfulness degradation due to the limited program accuracy achieved in the PWM case is considerably lower than that obtained with the PAM approach. Then, a similar analysis is carried out to investigate the classifier behavior after program in presence of cell V_T instabilities due to random telegraph noise (RTN) and to temperature variations, leading again to results in favor of the PWM approach. In light of these results, the present work suggests a viable solution to overcome some of the more serious reliability issues of NOR Flash-based artificial neural networks, paving the way to the implementation of highly-reliable, noise-resilient neuromorphic systems.

Keywords: artificial neural networks; neuromorphic computing; NOR Flash memory arrays; program noise; random telegraph noise; pulse-width modulation

Citation: Malavena, G.; Sottocornola Spinelli, A.; Monzio Compagnoni, C. A Noise-Resilient Neuromorphic Digit Classifier Based on NOR Flash Memories with Pulse–Width Modulation Scheme. *Electronics* **2021**, *10*, 2784. https://doi.org/10.3390/electronics10222784

Academic Editor: Changhwan Shin

Received: 27 September 2021
Accepted: 10 November 2021
Published: 13 November 2021

1. Introduction

Artificial neural networks (ANNs) are computing systems that take inspiration from biological neural networks to address many problems involving unstructured data, such as image recognition and classification [1,2]. What makes ANNs different from classical CMOS systems based on the Von Neuman architecture is that they do not feature distinct computing and memory units that communicate with each other through a bus; rather, they implement an in-situ computational paradigm, which is based on the matrix-by-vector multiplication (MVM) operation [1]. For that reason, ANNs represent a promising solution to achieve a performance and efficiency improvement in those systems designed to perform data-intensive tasks, for which the memory bottleneck, arising from the continuous data exchange between memory and CPU, represents a limiting factor.

A convenient way to implement ANNs consists in exploiting non-volatile memory (NVM) arrays as artificial synaptic arrays connecting adjacent layers of artificial neurons. To that purpose, different memory solutions have been investigated for their adoption in neuromorphic systems and presented in literature. They include works based on crossbar arrays of resistive elements, mainly resistive switching random access memories (RRAM) [3–5] and phase change memories (PCM) [6,7], or based on memory arrays of charge storage devices, such as NAND and NOR Flash memory arrays [8–15].

Among those different solutions, the adoption of Flash memory technologies sounds appealing for many reasons, such as their reduced power consumption, the virtually analog tuning of the synaptic weights stored in the memory array, and their mature and reliable CMOS-compatible manufacturing process. In particular, even though some ANNs implementations based on NAND Flash arrays have been presented [15,16], the parallel architecture of NOR Flash memory arrays makes them the most straightforward solution to implement the MVM at the basis of the operation of neuromorphic systems [1,17].

For this reason, different implementations of neuromorphic systems based on NOR Flash memory arrays have been analyzed, including both supervised and unsupervised networks, which usually rely on some modification to the cell design [18–20], to the array design [2,8–10] or to cells program/erase voltage schemes [11–13] to make the memory array operation compliant with the desired application. Notably, in [2] a fully integrated three-layer ANN (with dimensions 784 × 64 × 10) was implemented and tested for handwritten digits recognition via the gradient-descent method based on the backpropagation algorithm [21] reaching a 94.7% classification fidelity with a single-pattern classification time and energy equal to 1 μs and less than 20 nJ, respectively.

In this work, we take inspiration from [16], where a working scheme based on PWM is adopted for the implementation of a neuromorphic image classifier based on NAND Flash memory arrays, and show that a similar PWM-based approach can be employed to operate a NOR Flash-based neuromorphic digit classifier, replacing the typically adopted PAM scheme. In particular, by means of a simulation-based analysis, we demonstrate, thanks to that PWM scheme, the possibility to achieve a tremendous reduction in classifier sensitivity to noise sources such as PN, RTN, and temperature variations. The results of this analysis present a way to strongly relieve those issues, thus enabling the development of noise-resilient artificial neural networks based on scaled NOR Flash memory arrays.

After a brief review on related works dealing with various techniques to reduce noise sensitivity in ANNs (Section 2), PAM and PWM encoding schemes are introduced in Section 3 and Section 4, respectively. After that, in Section 5, we will present the architecture of the investigated neuromorphic digit classifier and the simulation results when no noise sources are accounted for. Then, in Section 6 we will discuss the impact of PN, RTN, and temperature variations on the classifier performance. Finally, conclusions will be drawn in Section 7.

2. Noise-Sensitivity Reduction Techniques in ANNs

Despite the advantages coming with ANNs with respect to Von Neunman architecture-based systems, their analog computing paradigm is inherently affected by noise, regardless the type of NVM arrays adopted. In fact the synaptic weights stored in the memory cells are inevitably impacted by several non-idealities, either occurring during the program phase and leading to a limited program accuracy, or following it, undermining the synaptic weights stability over time. The deviation of the stored synaptic weights from their ideal value will lead to a degradation of the network performance that may ultimately compromise its functionality.

For this reason, a considerable research effort is being devoted to the conception of various techniques to design neural networks with low noise sensitivity. For example, conductance variability in PCMs due to $1/f$ noise, drift noise [22], program noise (PN, [23]), and device variability have been recently addressed. In [24], additive noise is injected during the learning phase to train networks that are more robust to cells conductances noise; in [25], instead, the combination of multiplicative noise injection and drift regularization is presented to reduce network performance degradation due to PN and drift noise of about a factor 10.

Other techniques to relieve network reliability issues due to stochastic variations of RRAM devices resistance are presented in [26,27], resorting again on noise injection during training; in particular in [27] non-idealities arising from the IR drops along the resistance network are addressed. In [28], instead, the basic idea of biasing the learning phase to encourage the network to learn large synaptic weights is proposed; even though that results in a reduction in noise because large synaptic weights are less sensitive to conductance instabilities, it comes with the drawback of an increase in the network power consumption.

In the case of NOR Flash memory arrays, PN and RTN are two major issues of concern for network performance. Their impact on the classification accuracy of a NOR Flash-based neuromorphic digit classifier was studied in [29]. In that work, it was shown that stringent requirements on the memory array programming scheme and on the memory cells scaling are needed to limit the network performance degradation within an acceptable interval. The investigation of techniques that allow to relieve the impact of PN and RTN on ANNs based on NOR Flash memory arrays is therefore crucial to overcome those limitations in future implementations and to improve their performance.

3. Pulse Amplitude Modulation

When employed in neuromorphic applications, NOR Flash cells are typically operated in subthreshold regime [30,31], where the drain-to-source current I_{DS} displays an exponential dependence on the word-line (WL) voltage V_{WL} according to:

$$\underbrace{I_{DS}}_{\text{output}} = \underbrace{I_0 \cdot \exp\left[\frac{q\alpha_G\left(V_{WL} - V_T^{ref}\right)}{mkT}\right]}_{\text{input}} \cdot \underbrace{\exp\left[-q\frac{\alpha_G \Delta V_T}{mkT}\right]}_{\text{weight}} \tag{1}$$

In the previous equation, I_0 is the current prefactor, q is the elementary charge, α_G is the control-gate–to-floating-gate capacitive coupling ratio, m is the subthreshold slope ideality factor, kT is thermal energy, V_T is the cell threshold voltage, V_T^{ref} an arbitrary chosen reference cell V_T, and ΔV_T is the cell V_T shift from V_T^{ref}. According to this approach, the input weight synaptic multiplication is implemented by considering $w = \exp[-q\alpha_G \Delta V_T / mkT]$ as the weight of the artificial synapse, while the remaining factor, which is a function of V_{WL} but not of the cell V_T, plays the role of the input presynaptic signal. Since the input signal is modulated by the amplitude of V_{WL}, this approach can be referred to as pulse-amplitude modulation (PAM).

As shown in Figure 1a, by changing cell V_T it is possible to modulate I_{DS} at a given V_{WL} and, therefore, the cell synaptic weight, with positive and negative ΔV_T leading to $w < 1$ and $w > 1$, respectively. Even though the synaptic weights could in principle assume analog values, their tuning resolution is ultimately limited by charge granularity, becoming relevant especially in very scaled devices [32]. In addition to this, an upper limitation for them is needed to keep the cell from exiting the subthreshold regime. In fact, according to the PAM mode the input weight synaptic multiplication is implemented thanks to the exponential law of Equation (1), and the polynomial $I_{DS} - V_{WL}$ characteristic of cells operating in the on-state regime would not allow such a behavior. For the same reason, once a maximum synaptic weight w^{max} is chosen, also an upper bound for V_{WL} must be selected. It follows that only a small portion of the $I_{DS} - V_{WL}$ curve can be exploited, resulting in a limited acceptable ΔV_T range.

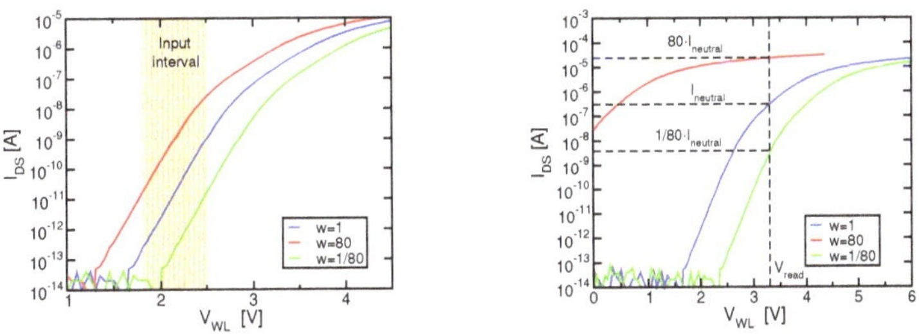

(a) (b)

Figure 1. (**a**) $I_D - V_{WL}$ curve measured on a 40 nm embedded technology [33] and used to simulate the implementation of a NOR Flash-based digit classifier adopting a PAM approach. $I_D - V_{WL}$ curves corresponding to different values of the synaptic weights and the corresponding limited input range are shown. (**b**) Same as (**a**) but when PWM is employed; no input limitation is required in the latter case.

In addition to that, another drawback peculiar of the PAM approach is the strong sensitivity of w to ΔV_T due to the exponential law that relates these two quantities. This represents an issue not to be overlooked, especially for scaled technologies, as several non-idealities could determine fluctuations in the value of ΔV_T, resulting in even stronger variations of w. Indeed, in [29] the impact of program accuracy and of V_T instabilities due to RTN on the performance of a NOR Flash-based neuromorphic digit classifier operated according to the PAM approach was investigated. The study was conducted by modeling the cell V_T placement during a program-and-verify (P&V) scheme based on ISPP [34,35], and by performing a parametric analysis of the main dependences that affect the program phase, i.e., the V_T discretization step (V_s) and the control-gate–to–floating-gate capacitance (C_{pp}). In addition to that, the role of cell V_T instabilities due to RTN [36] was addressed for different values of the single-trap fluctuation amplitude (λ). Results revealed that to keep the degradation of the network performance within an acceptable range, a stringent upper bound to V_s and λ and a lower one to C_{pp} are mandatory. This poses a severe limitation on the array programming time, mainly ruled by V_s [34,37], and on the possibility to employ deeply scaled NOR Flash technologies, that would lead to lower C_{pp} and larger λ values [38–40]. In addition, for similar reasons, also unwanted temperature variations are expected to be detrimental for the classifier truthfulness when the PAM scheme is employed.

4. Pulse Width Modulation

The PWM approach is based on the idea of exploiting a different encoding scheme for the presynaptic signals. According to it, these are not encoded in the amplitude of the V_{WL} pulse (such as the PAM approach), but rather in its duration with a constant amplitude. This means that V_{WL} is kept constant to a value V_{read}, and its time duration changes according to the input signal: large signals correspond to long pulses and the other way round. To keep the proportionality between input and output signals, the overall charge that flows through the cell (and not the current) during the input pulse is taken as the output signal, and therefore Equation (1) is modified in:

$$\underbrace{\text{Charge}}_{\text{output}} = \underbrace{\text{Time}}_{\text{input}} \times \underbrace{\text{Current}}_{\text{weight}}. \tag{2}$$

Since no limitation of the cell working regime is required in the PWM scheme, even though the cell weight is still related to the ΔV_T value, the link between them cannot be expressed using a single mathematical form anymore. As shown in Figure 1b, I_{DS} for $\Delta V_T = 0$ and $V_{WL} = V_{read}$ corresponds to $w = 1$, and is referred to as $I_{neutral}$; then, when the cell is programmed with $\Delta V_T \neq 0$, I_{DS} will be larger or lower than $I_{neutral}$, leading to an operative definition of the synaptic weight as $w = I_{DS}/I_{neutral}$.

What really makes such a PWM approach appealing is that one cell working point can span both the subthreshold and the on-state regime, allowing the ΔV_T range to be much wider than that of the PAM scheme. In addition to this, those memory cells that operate in the on-state regime present a polynomial relation between I_{DS} and V_T, with a reduced sensitivity of their w to ΔV_T. For those reasons, the PWM approach looks more promising when dealing with the previously mentioned noise sources and surely deserves some attention.

The obvious downside of PWM, on the other hand, is that the implementation of large cell weights requires memory cells to be biased with higher I_{DS} if compared to the PAM case, leading to a reduction in the classifier energy efficiency. Even though this, in principle, may be a limiting factor, from a practical standpoint the maximum current never exceeds 30 μA in our work (see the next section), which is comparable with typical operating current in PCM-based ANNs [25].

5. Neuromorphic Digit Classifier Based on PWM

In order to prove the benefits coming during adoption of the PWM scheme, we considered the NOR Flash-based neuromorphic digit classifier investigated in [29] and operated according to the PAM approach (see Figure 2a). It consists of a three-layer fully-connected feed-forward ANN trained to recognize the hand-written digits belonging to the MNIST database [41]. Since the MNIST digits are represented as a 28×28 greyscale images and range from "0" to "10", the input layer is made of 784 neurons, each one providing an analog input for a pixel, and the output layer features 10 neurons, equal to the number of digits to be identified; the number of neurons in the hidden layer, instead, was set to 40 to keep the classifier dimension as small as possible. According to such architecture, the NOR Flash memory arrays employed to connect each couple of adjacent layers of neurons must have dimensions equal to $(784 + 1) \times 40$ and $(40 + 1) \times 10$, respectively, where one more WL is needed to implement the bias b of each neuron. It is worth mentioning that, according to the presented design choice with a single memory cell storing a synaptic weight, only positive weights can be reproduced. Following [29], the impact of this limitation on the classifier truthfulness was strongly mitigated by employing a complementary encoding of network outputs, which consists in identifying the digit that is shown at the input of the classifier not by looking at the neuron in the output layer with the highest output signal, but at the one with the lowest. Note that, even though a differential implementation of cell synaptic weights (see [1,2]) would easily lead to a even larger recognition truthfulness due to the possibility to reproduce negative weights too (at the expense, however, of a double array area occupancy), we preferred to keep a single-cell synaptic weight implementation, to focus on the impact of various noise sources on the classifier recognition accuracy, rather than on the maximization of its absolute value.

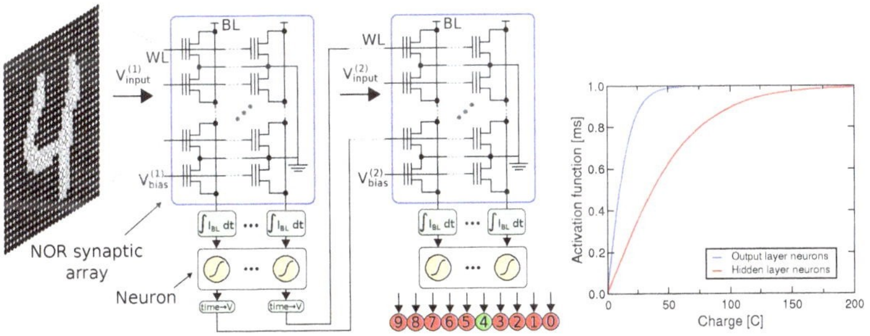

(a) (b)

Figure 2. (**a**) Schematic of the NOR Flash-based neuromorphic digit classifier investigated in this work when PWM is employed (adapted from [29], © 2019 IEEE). (**b**) Activation functions used for the neurons in the hidden and output layers; different scaling parameters were chosen to match the different dimensions of the respective NOR Flash arrays.

As explained in Section 4, for the classifier to be operated according to the PWM scheme, the input signals (x_i) must be encoded in the time duration of the voltage pulses applied to the WLs of NOR Flash arrays (t_i). Differently from the PAM case, in which the amplitude of input pulses must be confined in a well-defined interval to keep the memory cells in the subthreshold regime, no strict limitation affects this design choice. Therefore we picked $T = 0\,\text{ms}$ and 1 ms as the minimum and maximum pulse durations, respectively, leading to $t_i = x_i \cdot T^{max}$. Then, the working point of the memory cells was defined by considering the trans-characteristic shown in Figure 1b, measured on a NOR Flash cell developed with a 40 nm embedded technology [33] at a drain-to-source voltage $V_{DS} = 200\,\text{mV}$. In particular, the reference I–V curve was taken for $V_T^{ref} = 3.1\,\text{V}$, measured with a constant current criterion at $I_{DS} = 100\,\text{nA}$; V_{read} was chosen to be equal to 3.3 V, leading to $I_{neutral} = 300\,\text{nA}$. Finally, the tanh–neuron activation function shown in Figure 2b was adopted; it was properly chosen to be consistent with the PWM scheme, therefore receiving a charge signal at its input and delivering a time signal at its output in the range [0 ms, 1 ms].

Figure 3a shows the recognition truthfulness of the PWM-based classifier during its training with the standard stochastic gradient-descent method based on the backpropagation algorithm [21] (blue curve), compared to the final value resulting from the classically adopted PAM approach. In both cases, a cross-entropy error function, a mini-batch size equal to 10 and a learning rate equal to 0.01 were adopted [21]. When no noise source are taken into account, the PWM and PAM schemes are practically equivalent from the performance standpoint at the end of the training phase, confirming the validity of the former. The red curve in Figure 3a, on the other hand, refers to the simulation results achieved with the PWM approach when the synaptic weights are forced in the interval $[w^{min}, w^{max}] = [0.01, 100]$; this means that if $w < w^{min}$ ($w > w^{max}$) results from the network training, its value is clamped to w^{min} (w^{max}). Since no relevant variations in the classifier truthfulness are observed, this limitation will be safely applied to all the results presented in the following, allowing to limit I_{DS} for each cell in the interval [3 nA, 30 µA].

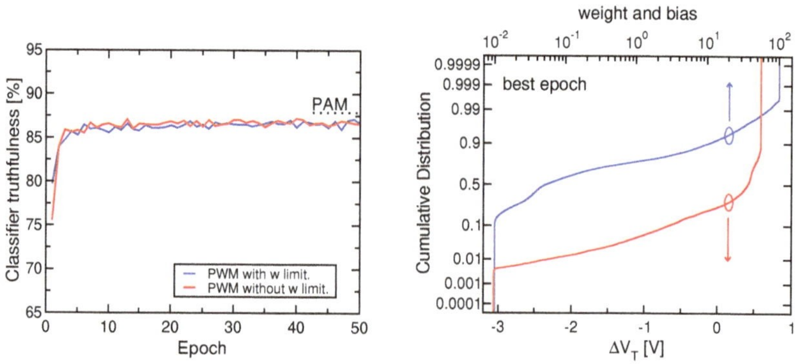

(**a**) (**b**)

Figure 3. (**a**) Simulation results of the PWM-based classifier truthfulness during training with the stochastic gradient-descent method; results with and without weights limitation are shown and compared to the maximum accuracy achieved in the PAM case. (**b**) Cumulative distribution of w and ΔV_T in NOR Flash memory arrays for the training epoch corresponding to the maximum classifier truthfulness; the steep increase in the curves at their extremes is due the weights limitation.

The cumulative distribution of w and ΔV_T for the epoch corresponding to the maximum classifier truthfulness is reported in Figure 3b. The accumulation of both quantities at the extremes occurs at quite low and high probabilities, confirming that only a small number of the NOR Flash cells are affected by the limitation of the synaptic weights. In addition to this, it is worth noting that the ΔV_T values resulting from training are distributed over quite a large range (>3.5 V); this represents an encouraging result suggesting the possibility to keep good classifier performance even in presence of different noise sources.

6. Noise-Sensitivity Analysis of the Classifier Performance

Even though the PAM and PWM schemes ideally lead to comparable values of classifier truthfulness, for the reasons explained in the previous sections significant differences are expected when a more realistic analysis of the network is carried out. In this section, the impact of PN, RTN, and temperature variations on the network performance is addressed.

6.1. Impact of Program Noise

The program operation in NOR Flash memory arrays is based on a P&V scheme relying on ISPP. According to it, each memory cell is set to an erase state first, (i.e., with a very low-V_T). Then, a program pulse with WL and BL voltages equal to $(V_{WL}^{P,1}, V_{BL}^{P})$ is applied, triggering channel hot-electron injection into cell floating-gate that makes cell V_T increase. Right after the program pulse, cell V_T is read (this is referred to as verify phase) and compared with a target level V_{PV}. If V_T is lower than V_{PV}, another program pulse is applied to the cell, with the WL voltage increased by a quantity V_s, that is $V_{WL}^{P,2} = V_{WL}^{P,1} + V_s$. This procedure is repeated until the condition $V_T > V_{PV}$ is verified, with $V_{WL}^{P,i} = V_{WL}^{P,i-1} + (i-1) \cdot V_s$ ($V_{WL}^{P,i}$ represents V_{WL} at the i-th step and $i > 1$), when the program phase stops. It can be shown that, after a sufficiently large number of pulses, the average V_T variation for each cell due to each programming pulse is exactly equal to V_s and the final cell V_T is expected to be in the $[V_{PV}, V_{PV} + V_s]$ range [34,35]. This is shown schematically in Figure 4a, where each vertical rectangle corresponds to a programming pulse with V_{WL} of increasing amplitude and the white squares represents the increasing cell V_T.

However, due to the stochastic nature of the physics ruling the injection of electrons into the cell FG, memory cells sharing the same V_{PV} are affected by PN (see Figure 4b), which manifests itself as a dispersion of the values of the final V_Ts of those cells [37]. For example, Figure 4b shows the simulated V_T evolution of four different memory cells with the same V_{PV}. Due to PN, each cell ends up with a different final V_T value at the end of the program phase.

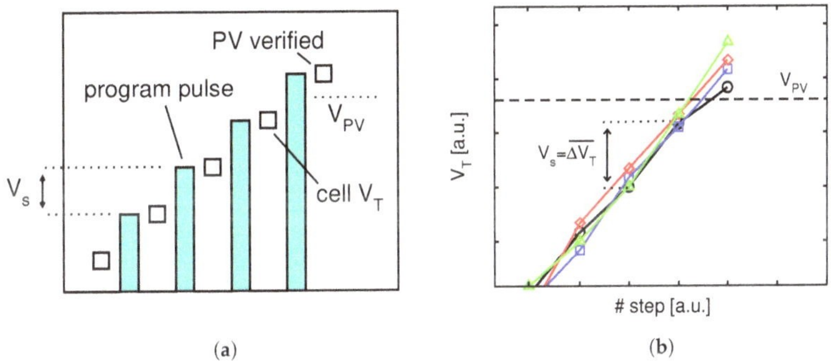

Figure 4. (**a**) Schematic of the ISPP phase and (**b**) simulated evolution during ISPP of the V_T of four different cells with the same V_{PV}; due to PN each cell has a different V_T at the end of program.

To evaluate the impact of PN on the performance of our NOR Flash-based digit classifier, the V_{PV} levels were discretized with step equal to V_s, and each memory cell was associated with the closest V_{PV} value lower than its target V_T. Then, the program operation was simulated, accounting for the randomness of the number of electron injected into cell FG during each ISPP pulse in a Monte Carlo fashion, as described in [32]. In particular, the analysis was repeated for different values of V_s and C_{pp}, as PN is shown to be stronger in presence of large V_s and small C_{pp} [32]; for each (V_s, C_{pp}) couple, the Monte Carlo simulation was repeated 50 consecutive times, testing the classifier truthfulness after each repetition.

Simulation results are reported in Figure 5a, showing the classifier truthfulness after program as a function of V_s and C_{pp} when the network is operated according to the PWM scheme. Results reveal that just a weak reduction in the average classifier performance occurs in all cases, being slightly larger than 1% only for V_s approaching the value of 1000 mV. In that case, even though cell weights sensitivity to noise sources affecting V_T is weak thanks to the polynomial relation between I_{DS} and V_T itself, still such large V_s values result in an extremely coarse P&V levels discretization and a strong program noise that lead, in turn, to a non-negligible reduction in the classifier truthfulness. In addition, also C_{pp} has a certain impact on the average results, mainly for the $V_s = 1000$ mV case, since lower values of C_{pp} tend to further enhance PN, thus leading to a noticeable performance degradation. Finally, even though a stronger statistical spread is found as V_s becomes larger, just a few data points out of the 50 experiment repetitions results in a performance reduction lower than 2%, confirming again the robustness of the PWM scheme with respect to PN. Results appear even more striking if they are compared with those calculated when the digit classifier is operated according to the PAM approach, as shown in Figure 5b. In fact, a much stronger performance degradation is displayed in the latter case, not only in terms of average accuracy but also in terms of the statistical dispersion of the data. It is important to stress that $V_s > 300\,\text{mV}$ should never be used in the PAM case, as cells ΔV_T discretization would be so coarse to make some of them exit the subthreshold working region, therefore severely compromising the network performance. In this sense, another advantage coming with the adoption of PWM is the possibility to speed up the program phase by exploiting larger V_s, without degrading the classifier accuracy too much.

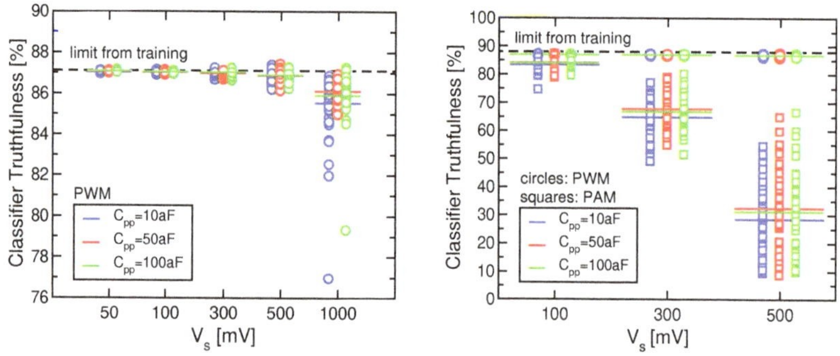

(**a**) (**b**)

Figure 5. (**a**) Simulated classifier truthfulness in presence of to PN for different values of V_s and C_{pp} when PWM is employed; for each condition, 50 consecutive simulations were performed (average values are indicated by horizontal lines). (**b**) Same as (**a**) but with the data points resulting from the adoption of the PAM scheme.

6.2. Impact of RTN

Even though ISPP allows a cell V_T tuning precise enough to keep network truthfulness degradation very limited, any source of V_T instabilities may still compromise the classifier performance at later times. In NOR Flash memory arrays a major source of V_T instabilities is represented by RTN, which arises from the capture and emission of electrons in tunnel-oxide traps. Both the amplitude and the timing of RTN-induced V_T fluctuations are non-deterministic, therefore RTN is usually addressed by looking at the statistical distribution of the V_T difference between two consecutive read operations (ΔV_T^{RTN}) [39,42,43]. As shown in Figure 6a, a clear signature of the ΔV_T^{RTN} cumulative distribution is its exponential tails; its slope λ, which is taken as the most representative RTN parameter, can be shown to approximate the single-trap fluctuation amplitude [38,42,44] and increases when single-cell dimensions are shrunk down. The remaining parameters describing RTN are the average number of traps per cell $\langle N_t \rangle$, which determines the height of the RTN tails, and the capture and emission trap time constants, which are typically uniformly distributed over the logarithmic time axis.

In order to evaluate the impact of RTN on the classifier performance, RTN-induced fluctuations were simulated following the Monte Carlo approach presented in [45], with N_t calibrated to reproduce the ΔV_T^{RTN} statistical distributions in Figure 6a and for different values of λ spanning from 20 mV/dec to 100 mV/dec. After that, a different RTN waveform was added to the V_T resulting from ISPP of each memory cell, and the network classification truthfulness was monitored periodically over time. The impact of such RTN-induced V_T oscillations on the statistical distributions of the memory cells weights is shown in Figure 6b, for the case in which the program phase was simulated with $C_{pp} = 50\,\text{aF}$ and

$V_s = 100\,\text{mV}$. Note that after 1000 s the enlargement in the weights distribution calculated in the PWM case is practically negligible, pointing towards a strong immunity to RTN for that scheme.

Simulation results reported in Figure 7a,b shows the RTN-induced instabilities in the classifier truthfulness for the cases in which the network is operated according to the PAM and PWM approaches, respectively. As already pointed out in [29], in the former case the classifier accuracy experiences strong instabilities, accompanied by an average degradation of the performance over time due to the increasing number of RTN traps coming into play as time goes by. This poses a limitation to the possibility to employ scaled NOR Flash memory arrays for neuromorphic applications, since the network performance degradation due to RTN becomes too severe as λ becomes too large. On the other hand, when PWM is employed, RTN results in oscillations of the classifier truthfulness of the order of 0.1% only, therefore not significantly contributing to degrade the classifier performance. It is worth stressing that the resilience of PWM to RTN demonstrated by our analysis allows to overcome the limitations of the PAM approach, paving the way to the development of high-density neuromorphic systems based on scaled NOR Flash memory array.

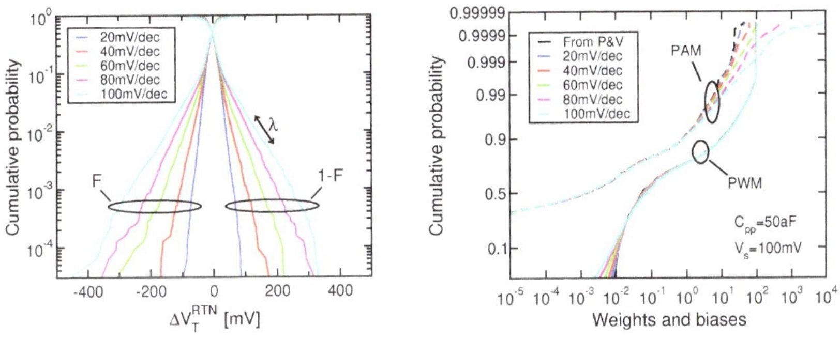

(a) (b)

Figure 6. (**a**) Simulated cumulative statistical distribution (F) of ΔV_T^{RTN} and its complementary (1-F) one for different values of λ. (**b**) Cumulative statistical distribution of the memory cells weights resulting from RTN Monte Carlo simulations after 1000 s from the end of the program phase for increasing values of λ; the distribution calculated at the end of the P&V phase is shown too.

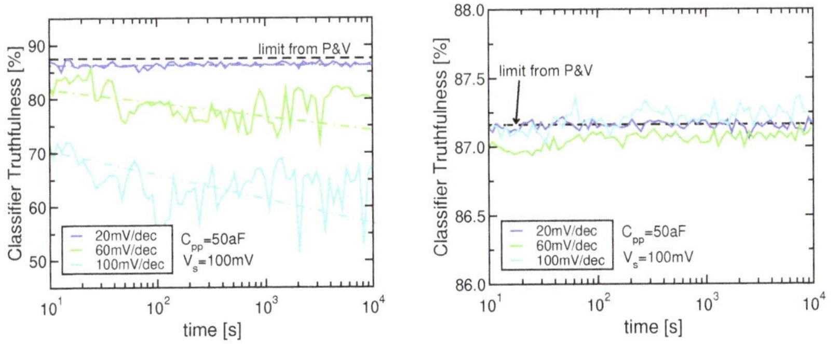

(a) (b)

Figure 7. Calculated classifier truthfulness in presence of RTN with increasing values of λ for (**a**) PAM (dashed-dotted lines highlight the average reduction in performance over time) and (**b**) PWM. $C_{pp} = 50\,\text{aF}$ and $V_s = 100\,\text{mV}$ were assumed for the program phase.

To further confirm our results, we performed a wafer-level experimental test involving a NOR Flash array test structure with 8 WLs and 1 BL (see Figure 8a). The 8 memory cells along the BL were programmed by ISPP with $V_s = 300\,\text{mV}$ to V_T values corresponding to weights uniformly distributed in the [1/80,80] range; each memory cell was then associated with an input signal randomly drawn in the [0,1] range. With all the input signals simultaneously applied to the WLs, the output signal was monitored for 300 s following the program phase. Results are reported in Figure 8b, for the two cases in which the experiment is carried out employing the PAM and PWM encoding schemes. As expected, PWM assures superior performance in terms of output signal stability, which is reflected in a very low RTN-sensitivity in those large-scale ANNs based on larger NOR Flash memory arrays.

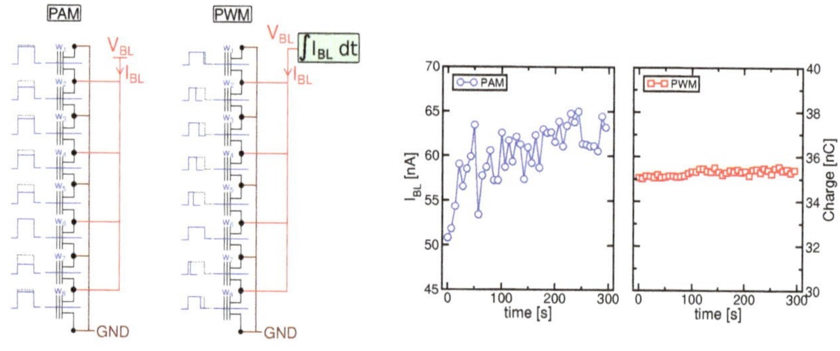

(**a**) (**b**)

Figure 8. (**a**) Experimental test devised to reproduce the main features of a NOR Flash-based ANN in a 8WLs NOR Flash string. Each memory cell is programmed to have a weight in the [1/80, 80] range and is associated with an input signal between 0 and 1. The experiment is repeated twice, implemented first according to the PAM (left) and then according to the PWM (right) scheme. (**b**) Evolution of the output signals measured during the experiment shown in (**a**). Much larger instabilities of the output signal are measured in the former case with respect to the latter, confirming experimentally the strong immunity of PWM to noise sources affecting cells V_Ts, such as RTN.

6.3. *Impact of Temperature Variations*

Another source of truthfulness instabilities is represented by temperature variations that may occur after the program phase, affecting the synaptic weights and, therefore, the classifier operation. To investigate this point, we measured the single-cell $I_{DS} - V_{WL}$ transcharacteristic already shown in Figure 1 not only at 300 K, but also at temperatures as high as 420 K, as shown in Figure 9a. Then, assuming the training phase in our network to take place at 300 K, the impact of temperature variations on the classifier behavior was accounted for by transforming the $I_{DS} - V_{WL}$ curve of each NOR Flash memory cell in agreement with Figure 9a, and testing the network accuracy at the remaining temperatures. Note that the impact of PN on each cell V_T resulting from program was considered negligible to focus just on the role played by temperature variations.

Figure 9b shows the results of the previous analysis, for the PAM and the PWM cases. As a consequence of the stronger temperature dependence of the curves of Figure 9a in their subthreshold region, temperature variations have a detrimental effect on the classification accuracy when the PAM scheme is employed. In the PWM case, instead, even though some memory cells operate in subthreshold regime, those with the largest weights are the ones in the on-state, displaying a much weaker temperature dependence. This results in a stronger immunity for PWM, allowing to keep the performance degradation within 10% for temperature variations close to 100 K.

A complete comparison between PAM and PWM approaches is finally reported in Table 1.

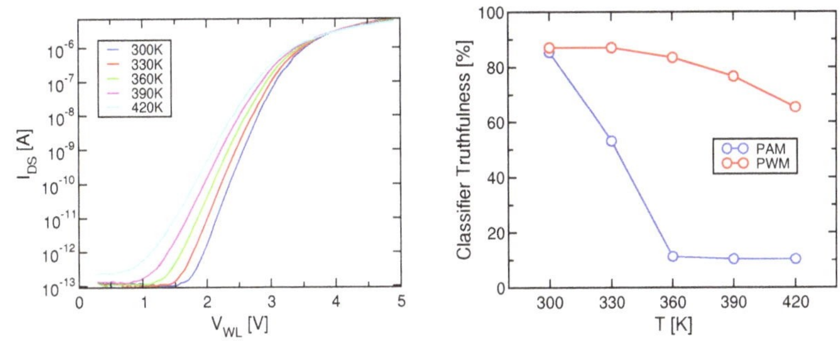

(a) (b)

Figure 9. (**a**) $I_{DS} - V_{WL}$ characteristics measured on a 40 nm embedded technology at increasing values of temperature. (**b**) Classifier truthfulness evaluated following an ideal program operation (i.e., without V_{PV} discretization and PN) in the PAM and PWM cases.

Table 1. Comparison between PAM and PWM approaches in terms of input–output encoding schemes, noise sensitivity, and energy efficiency.

	Input	Output	PN Sensitivity	RTN Sensitivity	Temperature Sensitivity	Energy Efficiency
PAM	V_{WL}	I_{DS}	strong	strong	strong	high
PWM	t_i	$\int_0^{t_i} I_{DS}dt$	weak	weak	weak	moderate

7. Conclusions

In this work, we have presented the implementation of a NOR Flash-based neuromorphic digit classifier based on PWM. By comparing our results with those previously reported for a similar classifier operated according to the classically adopted PAM scheme, we have shown that PWM and PAM are practically equivalent from the standpoint of the recognition accuracy when no noise sources are taken into account. Then, we have considered three distinct noise sources to affect the classifier performance, that is, PN, RTN and temperature variations, with the first limiting the cell V_T tuning precision and the remaining ones impacting the cell V_T stability over time after the program phase. When the impact of all those noise sources on the classifier performance is accounted for, PWM has been shown to lead to much a higher classification accuracy, representing a better choice with respect to PAM. In particular, we have shown that the superior noise immunity of PWM to PN and RTN enables the adoption of smaller V_s during ISPP, thus speeding up the program phase, and of more scaled NOR Flash memory cells, leading to an increase in the network integration density. Finally, the main conclusions resulting from our simulation activities were also confirmed experimentally considering a 8 WLs test NOR Flash memory string programmed according to the PAM scheme first, and then to the PWM one. In both cases I_{BL} was monitored at constant intervals over time, showing much more stable values in the latter case with respect to the former one. For all those reasons, results reported here represent an important step towards the development of large-scale high-density neuromorphic systems that employs NOR Flash memory arrays.

Author Contributions: Conceptualization, C.M.C.; software, G.M. and A.S.S.; investigation, C.M.C., G.M. and A.S.S.; original draft preparation, G.M.; review and editing G.M. All authors have read and agreed to the published version of the manuscript.

Funding: This research received no external funding.

Acknowledgments: The authors would like to thank Simone Gangemi and Riccardo Teri for discussions and support.

Conflicts of Interest: The authors declare no conflict of interest.

References

1. Burr, G.W.; Shelby, R.M.; Sebastian, A.; Kim, S.; Kim, S.; Sidler, S.; Virwani, K.; Ishii, M.; Narayanan, P.; Fumarola, A.; et al. Neuromorphic computing using non-volatile memory. *Adv. Phys. X* **2017**, *2*, 89–124. [CrossRef]
2. Merrikh-Bayat, F.; Guo, X.; Klachko, M.; Prezioso, M.; Likharev, K.K.; Strukov, D.B. High-performance mixed-signal neurocomputing with nanoscale floating-gate memory cell arrays. *IEEE Trans. Neural Netw. Learn. Syst.* **2017**, *29*, 4782–4790. [CrossRef] [PubMed]
3. Ielmini, D. Resistive switching memories based on metal oxides: Mechanisms, reliability and scaling. *Semicond. Sci. Technol.* **2016**, *31*, 063002. [CrossRef]
4. Prezioso, M.; Merrikh-Bayat, F.; Hoskins, B.; Adam, G.C.; Likharev, K.K.; Strukov, D.B. Training and operation of an integrated neuromorphic network based on metal-oxide memristors. *Nature* **2015**, *521*, 61–64. [CrossRef]
5. Yu, S.; Chen, P.Y.; Cao, Y.; Xia, L.; Wang, Y.; Wu, H. Scaling-up resistive synaptic arrays for neuro-inspired architecture: Challenges and prospect. In Proceedings of the 2015 IEEE International Electron Devices Meeting, Washington, DC, USA, 7–9 December 2015; pp. 451–454. [CrossRef]
6. Raoux, S.; Wełnic, W.; Ielmini, D. Phase change materials and their application to nonvolatile memories. *Chem. Rev.* **2010**, *110*, 240–267. [CrossRef] [PubMed]
7. Burr, G.W.; Shelby, R.M.; Sidler, S.; Di Nolfo, C.; Jang, J.; Boybat, I.; Shenoy, R.S.; Narayanan, P.; Virwani, K.; Giacometti, E.U.; et al. Experimental demonstration and tolerancing of a large-scale neural network (165,000 synapses) using phase-change memory as the synaptic weight element. *IEEE Trans. Electron Devices* **2015**, *62*, 3498–3507. [CrossRef]
8. Merrikh-Bayat, F.; Guo, X.; Om'Mani, H.; Do, N.; Likharev, K.K.; Strukov, D.B. Redesigning commercial floating-gate memory for analog computing applications. In Proceedings of the 2015 IEEE International Symposium on Circuits and Systems, Lisbon, Portugal, 24–27 May 2015; pp. 1921–1924. [CrossRef]
9. Guo, X.; Merrikh-Bayat, F.; Bavandpour, M.; Klachko, M.; Mahmoodi, M.; Prezioso, M.; Likharev, K.; Strukov, D. Fast, energy-efficient, robust, and reproducible mixed-signal neuromorphic classifier based on embedded NOR Flash memory technology. In Proceedings of the 2017 IEEE International Electron Devices Meeting, San Francisco, CA, USA, 2–6 December 2017; pp. 151–154. [CrossRef]
10. Guo, X.; Bayat, F.M.; Prezioso, M.; Chen, Y.; Nguyen, B.; Do, N.; Strukov, D.B. Temperature-insensitive analog vector-by-matrix multiplier based on 55 nm NOR Flash memory cells. In Proceedings of the 2017 IEEE Custom Integrated Circuits Conference, Austin, TX, USA, 30 April–3 May 2017; pp. 1–4. [CrossRef]
11. Malavena, G.; Sottocornola Spinelli, A.; Monzio Compagnoni, C. Implementing spike-timing-dependent plasticity and unsupervised learning in a mainstream NOR Flash memory array. In Proceedings of the 2018 IEEE International Electron Devices Meeting, San Francisco, CA, USA, 1–5 December 2018; pp. 35–38. [CrossRef]
12. Malavena, G.; Filippi, M.; Sottocornola Spinelli, A.; Monzio Compagnoni, C. Unsupervised learning by spike-timing-dependent plasticity in a mainstream NOR Flash memory array—Part I: Cell operation. *IEEE Trans. Electron Devices* **2019**, *66*, 4727–4732. [CrossRef]
13. Malavena, G.; Filippi, M.; Sottocornola Spinelli, A.; Monzio Compagnoni, C. Unsupervised learning by spike-timing-dependent plasticity in a mainstream NOR Flash memory array—Part II: Array learning. *IEEE Trans. Electron Devices* **2019**, *66*, 4733–4738. [CrossRef]
14. Lee, S.T.; Lim, S.; Choi, N.; Bae, J.H.; Kim, C.H.; Lee, S.; Lee, D.H.; Lee, T.; Chung, S.; Park, B.G.; et al. Neuromorphic technology based on charge storage memory devices. In Proceedings of the 2018 IEEE Symposium on VLSI Technology, Honolulu, HI, USA, 18–22 June 2018; pp. 169–170. [CrossRef]
15. Lee, S.T.; Lim, S.; Choi, N.Y.; Bae, J.H.; Kwon, D.; Park, B.G.; Lee, J.H. Operation scheme of multi-layer neural networks using NAND Flash memory as high-density synaptic devices. *IEEE J. Electron Devices Soc.* **2019**, *7*, 1085–1093. [CrossRef]
16. Lee, S.T.; Lee, J.H. Neuromorphic computing using NAND Flash memory architecture with pulse width modulation scheme. *Front. Neurosci.* **2020**, *14*, 945. [CrossRef]
17. Milo, V.; Malavena, G.; Monzio Compagnoni, C.; Ielmini, D. Memristive and CMOS devices for neuromorphic computing. *Materials* **2020**, *13*, 166. [CrossRef]
18. Kim, H.; Park, J.; Kwon, M.W.; Lee, J.H.; Park, B.G. Silicon-based floating-body synaptic transistor with frequency-dependent short-and long-term memories. *IEEE Electron Device Lett.* **2016**, *37*, 249–252. [CrossRef]
19. Kim, H.; Hwang, S.; Park, J.; Yun, S.; Lee, J.H.; Park, B.G. Spiking neural network using synaptic transistors and neuron circuits for pattern recognition with noisy images. *IEEE Electron Device Lett.* **2018**, *39*, 630–633. [CrossRef]
20. Kim, C.H.; Lee, S.; Woo, S.Y.; Kang, W.M.; Lim, S.; Bae, J.H.; Kim, J.; Lee, J.H. Demonstration of unsupervised learning with spike-timing-dependent plasticity using a TFT-type NOR Flash memory array. *IEEE Trans. Electron Devices* **2018**, *65*, 1774–1780. [CrossRef]
21. Nielsen, M.A. Neural Networks and Deep Learning. Determination Press. 2015. Available online: http://neuralnetworksanddeeplearning.com/ (accessed on 12 November 2021).
22. Ielmini, D.; Lacaita, A.L.; Mantegazza, D. Recovery and drift dynamics of resistance and threshold voltages in phase-change memories. *IEEE Trans. Electron Devices* **2007**, *54*, 308–315. [CrossRef]

23. Nandakumar, S.; Boybat, I.; Han, J.P.; Ambrogio, S.; Adusumilli, P.; Bruce, R.L.; BrightSky, M.; Rasch, M.; Le Gallo, M.; Sebastian, A. Precision of synaptic weights programmed in phase-change memory devices for deep learning inference. In Proceedings of the 2020 IEEE International Electron Devices Meeting, San Francisco, CA, USA, 12–18 December 2020; pp. 29.4.1–29.4.4. [CrossRef]
24. Joshi, V.; Le Gallo, M.; Haefeli, S.; Boybat, I.; Nandakumar, S.R.; Piveteau, C.; Dazzi, M.; Rajendran, B.; Sebastian, A.; Eleftheriou, E. Accurate deep neural network inference using computational phase-change memory. *Nat. Commun.* **2020**, *11*, 2473. [CrossRef]
25. Kariyappa, S.; Tsai, H.; Spoon, K.; Ambrogio, S.; Narayanan, P.; Mackin, C.; Chen, A.; Qureshi, M.; Burr, G.W. Noise-Resilient DNN: Tolerating Noise in PCM-Based AI Accelerators via Noise-Aware Training. *IEEE Trans. Electron Devices* **2021**, *68*, 4356–4362. [CrossRef]
26. Long, Y.; She, X.; Mukhopadhyay, S. Design of reliable DNN accelerator with un-reliable ReRAM. In Proceedings of the 2019 IEEE Design, Automation & Test in Europe Conference & Exhibition (DATE), Florence, Italy, 25–29 March 2019; pp. 1769–1774. [CrossRef]
27. He, Z.; Lin, J.; Ewetz, R.; Yuan, J.S.; Fan, D. Noise injection adaption: End-to-End ReRAM crossbar non-ideal effect adaption for neural network mapping. In Proceedings of the 56th Annual Design Automation Conference 2019, Las Vegas, NV, USA, 2–6 June 2019; pp. 1–6. [CrossRef]
28. Zheng, Q.; Kang, J.; Wang, Z.; Cai, Y.; Huang, R.; Li, B.; Chen, Y.; Li, H. Enhance the robustness to time dependent variability of ReRAM-based neuromorphic computing systems with regularization and 2R synapse. In Proceedings of the 2019 IEEE International Symposium on Circuits and Systems, Sapporo, Japan, 26–29 May 2019; pp. 1–5. [CrossRef]
29. Malavena, G.; Petrò, S.; Sottocornola Spinelli, A.; Monzio Compagnoni, C. Impact of program accuracy and random telegraph noise on the performance of a NOR Flash-based neuromorphic classifier. In Proceedings of the IEEE 2019 European Solid-State Device Research Conference, Cracow, Poland, 23–26 September 2019; pp. 122–125. [CrossRef]
30. Diorio, C.; Hasler, P.; Minch, A.; Mead, C.A. A single-transistor silicon synapse. *IEEE Trans. Electron Devices* **1996**, *43*, 1972–1980. [CrossRef]
31. Diorio, C.; Hasler, P.; Minch, B.A.; Mead, C.A. A floating-gate MOS learning array with locally computed weight updates. *IEEE Trans. Electron Devices* **1997**, *44*, 2281–2289. [CrossRef]
32. Monzio Compagnoni, C.; Sottocornola Spinelli, A.; Gusmeroli, R.; Beltrami, S.; Ghetti, A.; Visconti, A. Ultimate accuracy for the NAND Flash program algorithm due to the electron injection statistics. *IEEE Trans. Electron Devices* **2008**, *55*, 2695–2702. [CrossRef]
33. Boccaccio, C. Embedded 1T Flash NOR: Still alive at 40 nm. And beyond? In Proceedings of the Leti Memory Workshop, Grenoble, France, 25–28 June 2013.
34. Calligaro, C.; Manstretta, A.; Modelli, A.; Torelli, G. Technological and design constraints for multilevel Flash memories. In Proceedings of the IEEE International Conference on Electronics, Circuits, and Systems, Rhodes, Greece, 16 October 1996; Volume 2, pp. 1005–1008. [CrossRef]
35. Monzio Compagnoni, C.; Chiavarone, L.; Calabrese, M.; Ghidotti, M.; Lacaita, A.L.; Spinelli, A.S.; Visconti, A. Fundamental limitations to the width of the programmed V_T distribution of NOR Flash memories. *IEEE Trans. Electron Devices* **2010**, *57*, 1761–1767. [CrossRef]
36. Goda, A.; Miccoli, C.; Monzio Compagnoni, C. Time dependent threshold-voltage fluctuations in NAND Flash memories: From basic physics to impact on array operation. In Proceedings of the 2015 IEEE International Electron Devices Meeting, Washington, DC, USA, 7–9 December 2015; pp. 374–377. [CrossRef]
37. Monzio Compagnoni, C.; Gusmeroli, R.; Spinelli, A.S.; Visconti, A. Analytical model for the electron-injection statistics during programming of nanoscale NAND Flash memories. *IEEE Trans. Electron Devices* **2008**, *55*, 3192–3199. [CrossRef]
38. Ghetti, A.; Monzio Compagnoni, C.; Sottocornola Spinelli, A.; Visconti, A. Comprehensive analysis of random telegraph noise instability and its scaling in deca–nanometer Flash memories. *IEEE Trans. Electron Devices* **2009**, *56*, 1746–1752. [CrossRef]
39. Sottocornola Spinelli, A.; Monzio Compagnoni, C.; Gusmeroli, R.; Ghidotti, M.; Visconti, A. Investigation of the random telegraph noise instability in scaled Flash memory arrays. *Jpn. J. Appl. Phys.* **2008**, *47*, 2598. [CrossRef]
40. Adamu-Lema, F.; Monzio Compagnoni, C.; Amoroso, S.M.; Castellani, N.; Gerrer, L.; Markov, S.; Spinelli, A.S.; Lacaita, A.L.; Asenov, A. Accuracy and issues of the spectroscopic analysis of RTN traps in nanoscale MOSFETs. *IEEE Trans. Electron Devices* **2012**, *60*, 833–839. [CrossRef]
41. LeCun, Y.; Bottou, L.; Bengio, Y.; Haffner, P. Gradient-based learning applied to document recognition. *Proc. IEEE* **1998**, *86*, 2278–2324. [CrossRef]
42. Monzio Compagnoni, C.; Gusmeroli, R.; Spinelli, A.S.; Lacaita, A.L.; Bonanomi, M.; Visconti, A. Statistical model for random telegraph noise in Flash memories. *IEEE Trans. Electron Devices* **2007**, *55*, 388–395. [CrossRef]
43. Ghetti, A.; Amoroso, S.M.; Mauri, A.; Monzio Compagnoni, C. Impact of nonuniform doping on random telegraph noise in Flash memory devices. *IEEE Trans. Electron Devices* **2011**, *59*, 309–315. [CrossRef]
44. Amoroso, S.M.; Monzio Compagnoni, C.; Ghetti, A.; Gerrer, L.; Sottocornola Spinelli, A.S.; Lacaita, A.L.; Asenov, A. Investigation of the RTN distribution of nanoscale MOS devices from subthreshold to on-state. *IEEE Electron Device Lett.* **2013**, *34*, 683–685. [CrossRef]
45. Miccoli, C.; Paolucci, G.M.; Monzio Compagnoni, C.; Sottocornola Spinelli, A.S.; Goda, A. Cycling pattern and read/bake conditions dependence of random telegraph noise in decananometer NAND Flash arrays. In Proceedings of the 2015 IEEE International Reliability Physics Symposium, Monterey, CA, USA, 19–23 April 2015; pp. MY.9.1–MY.9.6. [CrossRef]

MDPI
St. Alban-Anlage 66
4052 Basel
Switzerland
Tel. +41 61 683 77 34
Fax +41 61 302 89 18
www.mdpi.com

Electronics Editorial Office
E-mail: electronics@mdpi.com
www.mdpi.com/journal/electronics

www.ingramcontent.com/pod-product-compliance
Lightning Source LLC
Chambersburg PA
CBHW061456210926
51850CB00008B/1738